Microbial Biotechnology

An Interdisciplinary Approach

Microbial Biotechnology

An Interdisciplinary Approach

Pratyoosh Shukla

CRC Press
Taylor & Francis Group
Boca Raton   London   New York

CRC Press is an imprint of the
Taylor & Francis Group, an **informa** business

CRC Press
Taylor & Francis Group
6000 Broken Sound Parkway NW, Suite 300
Boca Raton, FL 33487-2742

First issued in paperback 2020

© 2017 by Taylor & Francis Group, LLC
CRC Press is an imprint of Taylor & Francis Group, an Informa business

No claim to original U.S. Government works

Version Date: 20161019

ISBN 13: 978-0-367-57413-0 (pbk)
ISBN 13: 978-1-4987-5677-8 (hbk)

Visit the Taylor & Francis Web site at
http://www.taylorandfrancis.com

and the CRC Press Web site at
http://www.crcpress.com

Contents

Foreword

THE BOOK *Microbial Biotechnology: An Interdisciplinary Approach*, edited by Dr. Pratyoosh Shukla, covers some of the latest applications of microorganisms from a practical point of view. The field of microbial biotechnology, in the context of the so-called cell factories, is of great interest and the number of groups involved in such projects is growing exponentially.

Although the book chapters cover different aspects of microbial biotechnology, I want to highlight two of them. The first refers to the field of functional foods; several chapters deal with the production of probiotics and prebiotics, and their effect on gastrointestinal health. The second topic is microbial bioremediation, which is exemplified in this book by the use of microbes to clean up mining sites and by the optimization of wastewater treatments. Other issues having a significant impact are also addressed in the book: for example, the use of microbial enzymes in pulp and paper industries, the different applications of exopolysaccharides, or the latest developments in medical biotechnology, among others.

In summary, there is no doubt about the interest of the contents displayed in this book. I am sure that the book *Microbial Biotechnology: An Interdisciplinary Approach* will provide the scientific community with great benefits for the coming years.

Francisco Plou
Research Scientist at Spanish CSIC
Honorary Professor at Autonomous University of Madrid
Madrid, April 4, 2016

Preface

THE BOOK DESCRIBES THE INTERDISCIPLINARY SCOPE OF BIOTECHNOLOGY and discoveries thereof. This book briefs the reader on various novel and innovative ideas of emerging biotechnology. The key features are described below to highlight the important contents of the book:

1. The book envisages the recent ideas of novel findings in microbiology.

2. It also provides insights into various interdisciplinary research avenues.

3. There are very few books available covering the diversity of topics described in this book.

4. Some key areas of modern biotechnology are also covered in this book, which are not available in any such books in the market.

5. Enhanced and simplified descriptions are the key components of this book, which provide unique benefits to its readers.

This book will also act as an important means of information on researchers working in interdisciplinary areas of research. The chapters outlined in this book cater to the needs of researchers working in the areas of bacterial exopolysaccharides, microalgal proteomics, applications of microbial L-asparaginases, novel aspects of bioremediation, probiotics and their impact on society, microbial community analysis in wastewater treatment techniques, etc. The book focuses on describing the above-mentioned aspects and on diversifying the understanding of microbial biotechnology to an expanded level.

This book will be a valuable resource to senior undergraduate and graduate students, researchers, professionals, and other interested individuals or groups working in the areas mentioned in the book.

Pratyoosh Shukla, PhD
December 2016
Rohtak, India

Contributors

Folasade Adeyemo
Institute for Water and Wastewater
  Technology (IWWT)
Durban University of
  Technology
Durban, South Africa

Josiah Adeyemo
Department of Civil and
  Structural Engineering
Masinde Muliro University of
  Science and Technology
Kakamega, Kenya

Vinod Aggarwal
Dr. Rajendra Prasad Centre for
  Ophthalmic Sciences
A.I.I.M.S.
New Delhi, India

Renu Agrawal
CSIR-CFTRI and Rural
  Development Programme
Mysore, India

Nanthakumar Arumugam
Department of Biotechnology
  and Food Technology
Durban University of Technology
Durban, South Africa

Oluyemi Olatunji Awolusi
Institute for Water and Wastewater
  Technology
Durban University of Technology
Durban, South Africa

Rajib Bandopadhyay
Department of Botany
The University of Burdwan
Burdwan, West Bengal, India

Aparna Banerjee
Department of Botany
The University of Burdwan
Burdwan, West Bengal, India

Mayank Bansal
Dr. Rajendra Prasad Centre for
  Ophthalmic Sciences
A.I.I.M.S.
New Delhi, India

Nishi K. Bhardwaj
Avantha Centre for Industrial
  Research & Development
Yamuna Nagar, Haryana, India

Faizal Bux
Institute for Water and Wastewater
  Technology (IWWT)
Durban University of Technology
Durban, South Africa

Rohan Chawla
Dr. Rajendra Prasad Centre for
    Ophthalmic Sciences
A.I.I.M.S.
New Delhi, India

Nabin Kumar Dhal
Environment and Sustainability
    Department
CSIR-IMMT
Bhubaneswar, India

Kashyap Kumar Dubey
University Institute of
    Engineering and Technology
    (UIET)
Maharshi Dayanand University
Rohtak, Haryana, India

Abimbola Motunrayo Enitan
Institute for Water and Wastewater
    Technology (IWWT)
Durban University of Technology
Durban, South Africa

Varun Gogia
Dr. Rajendra Prasad Centre for
    Ophthalmic Sciences
A.I.I.M.S.
New Delhi, India

Jashan Gokal
Institute for Water and Wastewater
    Technology
Durban University of Technology
Durban, South Africa

Shika Gupta
Dr. Rajendra Prasad Centre for
    Ophthalmic Sciences
A.I.I.M.S.
New Delhi, India

Vijai Kumar Gupta
School of Natural Sciences
NUI Galway
Galway, Ireland

Prabhjot Kaur
Avantha Centre for
    Industrial Research &
    Development
Yamuna Nagar, Haryana, India

Dhirendra Kumar
Department of Biotechnology
University Institute of
    Engineering and Technology
    (UIET)
Maharshi Dayanand University
Rohtak, Haryana, India

Santhosh Kumar
Department of Biotechnology and
    Food Technology
Durban University of Technology
Durban, South Africa

Sheena Kumari
Institute for Water and Wastewater
    Technology (IWWT)
Durban University of Technology
Durban, South Africa

Swati Sucharita Panda
Environment and Sustainability
 Department
CSIR-IMMT
Bhubaneswar, India

Puneet Pathak
Avantha Centre for Industrial
 Research & Development
Yamuna Nagar, Haryana, India

Kugenthiren Permaul
Department of Biotechnology and
 Food Technology
Durban University of Technology
Durban, South Africa

Raju Poddar
Department of Bioengineering
Birla Institute of Technology
Ranchi, India

Vishal Prasad
Institute of Environment and
 Sustainable Development
Banaras Hindu University
Varanasi, India

Ajay Shankar
Institute of Environment and
 Sustainable Development
Banaras Hindu University
Varanasi, India

Pratyoosh Shukla
Department of Microbiology
Maharshi Dayanand University
Rohtak, Haryana, India

Anjali Singh
Institute of Environment and
 Sustainable Development
Banaras Hindu University
Varanasi, India

Gulshan Singh
Institute for Water and Wastewater
 Technology (IWWT)
Durban University of
 Technology
Durban, South Africa

Puneet Kumar Singh
Department of Microbiology
Maharshi Dayanand University
Rohtak, Haryana, India

Suren Singh
Department of Biotechnology and
 Food Technology
Durban University of
 Technology
Durban, South Africa

Pradeep Venkatesh
Dr. Rajendra Prasad Centre for
 Ophthalmic Sciences
A.I.I.M.S.
New Delhi, India

Ruby Yadav
Department of Microbiology
Maharshi Dayanand University
Rohtak, Haryana, India

CHAPTER 1

Bacterial Exopolysaccharides

Major Types and Future Prospects

Aparna Banerjee and Rajib Bandopadhyay

CONTENTS

ABSTRACT

Exopolysaccharide (EPS) is secreted by bacteria for their survival in harsh environmental conditions as a protective mechanism. Repeating sugar units, attached with proteins, lipids, organic and inorganic compounds, metal ions, and DNA are found in EPS. Bacterial EPSs have possible commercial applications in pharmaceutical industry, food processing, drug detoxification, bioremediation, and in many more. The most used and patented bacterial EPSs are xanthan, cellulose, gellan, alginate, etc. Varied applications of microbial EPSs are somewhat unexplored and their study is persistently enhancing toward isolation and characterization of novel EPSs as renewable capital. Downstream processing for purification and genetic engineering for increased EPS biosynthesis require more attention.

INTRODUCTION

Polysaccharides is an important content of microbial cell walls, either as storage capsular polysaccharides or as biofilm called as exopolysaccharides (EPSs) secreted by microbes in its surrounding. Presently, isolation and characterization (Figure 1.1) of new microbial EPS is of key scientific interest, because EPS has shown promising application as texture enhancers, gelling agents, emulsifiers, viscosifiers, and also as the newest nanovector for drug delivery, resulting in sustained release of drugs. Bacterial EPS own a varied range of property which is not found in traditional plant polymers. Although it competes for algal (alginates, carrageenans, and ulvan), crustacean (chitin) or plant polysaccharides, its production level is less due to green house effect, global warming, marine pollution, sea level increment, loss of key stone species, crop failure, and overall climate change impacts. Microorganisms provide a controlled production in bioreactors, without any variation due to the physiological states encountered for higher organisms (1). But downstream processing of bacterial polysaccharides has cost-intensive steps, as the expenses needed for substrates requirement for microbial growth and bioreactors are too high (2). Moreover, cultivation of microorganisms in a fermenter allows growth optimization and production yield either by physiological study or by genetic modification. For high-value pharmaceutical industry, bacterial polysaccharides can be produced at a feasible economic cost. Research in the field of bacterial EPS production is till now done on most available EPSs such as cellulose, xanthan gum, levan, glucan, cellulose, sphigan, hyaluronan, and succinoglycan, of which xanthan gum from *Xanthomonas* sp., gellan from *Sphingomonas*

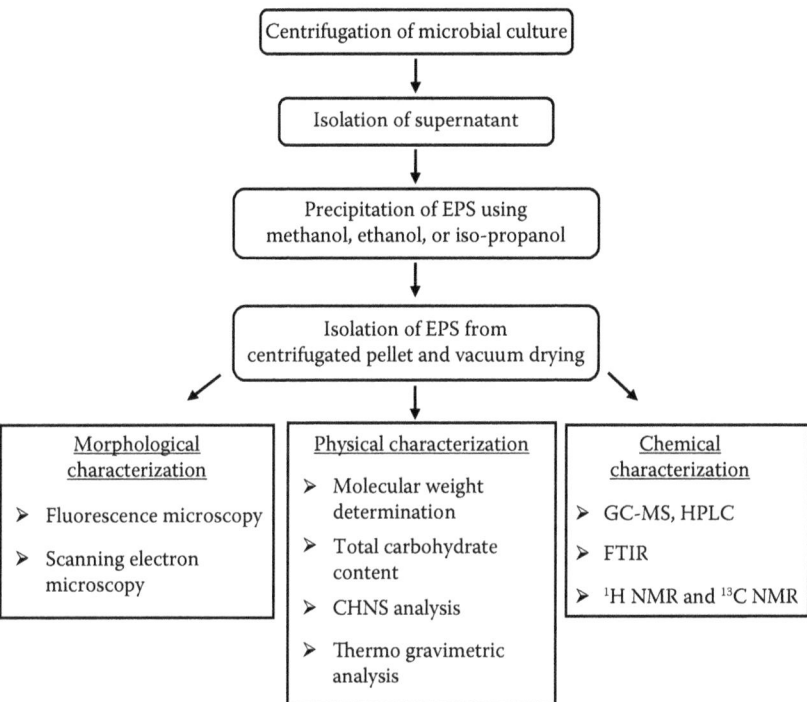

FIGURE 1.1 Flowchart elucidating the main steps of isolation and characterization of bacterial EPS.

elodea and *Sphingomonas paucimobilis,* cellulose from *Gluconacetobacter xylinus,* alginate from *Pseudomonas aeruginosa* and *Pseudomonas putida,* levan from *Zymomonas mobilis* and *Halomonas eurihalina,* succinoglycan from *Agrobacterium radiobacter, Rhizobium meliloti,* and *Agrobacterium tumefaciens,* Glucan from *Leuconostoc dextranicum* show strong potential commercial uses. All of these are very closely related structures and share high level of homology in many biosynthetic pathways.

The word EPS is first termed by Sutherland (3) for high molecular weight marine bacterial carbohydrate polymers. EPSs are produced in large amounts surrounding the microbial cells in extreme environments of Antarctic ecosystems, hypersaline lakes, hot water springs, or in deep-sea hydrothermal vents. Among all other adaptations for survival in extreme conditions, such as high temperatures, extreme salt levels, low pH, temperature variation, and high radiation zone, EPS biosynthesis is the most common protective mechanism by extremophilic microorganisms. EPS producing bacteria can be screened by the presence of glossy and slimy colony appearance, so that they can be further chosen for mass production (4).

EPS PRODUCING BACTERIA: MAJOR TYPES

Several groups of bacteria are capable of EPS production and they can be broadly categorized into five groups (Table 1.1): soil inhabitants, lactic acid bacteria, halophiles, thermophiles, and psychrophiles.

Soil Inhabitants

The most famous EPS producers are soil inhabiting rhizobia. It forms large amounts of polysaccharides when grown in pure cultures and also into the rhizosphere. *Rhizobium meliloti, Rhizobium leguminosarum,* and *Rhizobium tropici* are the three most studied EPS producing soil inhabiting bacteria (5,6). Regulation of motility related genes and presence of quorum sensing proteins in rhizobia resulted into a complex EPS biosynthesis pathway. EPS production by bacteria and biofilm formation enhances soil fertility and improved plant growth (7). *Pantoea* (formerly Enterobacter) *agglomerans* isolated from mangrove forest had very high ultraviolet radiation tolerance. The water-soluble EPS was extracted and was further tested for its ultraviolet radiation protection and free radical scavenging activities (8). *Micrococcus luteus* isolated from Egyptian soil, produced a maximum of 13 g/L EPS and the EPS showed high antioxidant activity (9).

Lactic Acid Bacteria

Lactic acid bacteria (LAB) improve the flavor, aroma, and texture of milk, meat, and vegetables and therefore they are used in food fermentations. LAB usually produce small amounts of EPS (100–200 mg/L) (10) but *Lactobacillus sakei* produced up to 4 g/L EPS (11). In fermented milk (12) and during cheese production (13), small amounts of EPS production change the texture and properties of the products. Several strains of *Streptococcus thermophilus* produce EPS in milk or in same type of media. The thermophilic LAB are also capable of producing EPS, such as *Streptococcus thermophilus* and *Lactobacillus delbrueckii* (12).

Halophiles

The diversity of halophilic bacteria so far isolated and characterized can be categorized into four different classes according to its salt requirement for their growth, which include slight halophiles, moderate halophiles, extreme

TABLE 1.1 Recent Studies on Bacterial Exopolysaccharides, Its Chemical Composition, and Probable Function

Bacterial Type	Bacterial Strain	Maximum EPS Production	Constituents of EPS	Function of EPS	References
Soil inhabitant	*Pseudomonas fluorescens*	4.5 g/L	Fructose:glucose:mannose = 4:1:0.6. NMR indicated EPS produced was levan with β-(2 → 6)-linked fructose units	Phosphate solubilizing activity	Taguett et al. (32)
	Rhizobium tropici	4.08 g/L	Mannose (0.86%, 1.49%, and 2.68%), rhamnose (2.58%, 2.49%, and 0.60%), glucuronic acid (8.6%, 5.97%, and 3.57%) and trace of galacturonic acid	Shear-resistant nature, soil stabilizing agent	Lemos et al. (5)
	Micrococcus luteus	13 g/L	Mannose:arabinose:glucose:glucuronic acid = 3,6:2.7:2.1:1.0 Main backbone consists of mannose units linked with (1 → 6) glycosidic bonds and arabinose units linked with (1 → 5) glycosidic bonds. There is a side chain consisting of mannose units linked with (1 → 6) glycosidic bonds at C3, when all glucose and most of glucuronic acid are found in the side chain.	In vitro DPPH radical-scavenging activity, with an EC_{50} value of 180 μg/mL	Asker et al. (33)
Lactic acid bacteria	*Lactobacillus sakei* and *Leuconostoc mesenteroides*	2 g/L and 1.4 g/L	IR spectra of both EPS were same as commercial dextran; specifically α-(1 → 6) glucan with approximately 6% substituted side chain glucose	Antiviral and immunomodulatory activity	Vázquez et al. (34)
	Lactobacillus plantarum	–	Mannose:fructose:galactose:glucose = 8.2:1:4.1:4.2	Antibacterial, antioxidant, and anticancerous activity	Wang et al. (35)
	Streptococcus phocae	12.14 g/L	–	Antibacterial, antioxidant, and flocculating activity	Kanmani et al.(36)

(Continued)

TABLE 1.1 (*Continued*) Recent Studies on Bacterial Exopolysaccharides, Its Chemical Composition and Probable Function

Bacterial Type	Bacterial Strain	Maximum EPS Production	Constituents of EPS	Function of EPS	References
Halophile	*Streptococcus thermophilus* and *Lactobacillus Delbrueckii*		Tetrasaccharide of glucose: galactose = 1:1 and heptasaccharide composed of galactose:glucose:rhamnose = 5:1:1	–	Marshall et al. (37)
	Halomonas stenophila	3.89 g/L	Monosaccharide (%): glucose (24 ± 1.73), glucuronic acid (7.5 ± 0.37), mannose (5.5 ± 0.17), fucose (4.5 ± 0.36), galactose (1.2 ± 0.17), and rhamnose (1 ± 0.05)	Flocculating and emulsifying activities	Amjres et al. (38)
	Halomonas almeriensis	1.7 g/L	Monosaccharide: 72% mannose, 27.5% glucose, and 0.5% rhamnose. Low-molecular-weight EPS: 1.1% protein, 70% mannose, and 30% glucose	Emulsifying hydrophobic substrates and pseudoplasticity	Llamas et al. (39)
Thermophile	*Brevibacillus thermoruber*	2.08 g/L	Glucose:galactose:mannose: galactosamine:mannosamine =57.7:16.3:9.2:14.2:2.4	Nonpathogenic promising EPS	Yildiz et al. (25)
	Geobacillus tepidamans	1.12 g/L	Carbohydrate content = 98% Protein content = 1.8% Uronic acids = 0.2%	Anti-cytotoxic activity against Avarol	Kambourova et al. (24)

(*Continued*)

TABLE 1.1 (*Continued*)   Recent Studies on Bacterial Exopolysaccharides, Its Chemical Composition and Probable Function

Bacterial Type	Bacterial Strain	Maximum EPS Production	Constituents of EPS	Function of EPS	References
Psychrophile	*Pseudoalteromonas* sp. S-5	–	–	Inhibitory activity against human leukemia K562 cells	Chen et al. (40)
	Pseudomonas sp. ID1	–	Carbohydrate content = 33.81% ± 2.59, composed of glucose (17.04% ± 0.32), galactose (8.57% ± 1.15), and fucose (8.21% ± 1.12). Total uronic acid content = 2.40% ± 0.33%.	Emulsifier and cryoprotector	Mercade et al. (41)
	Pseudoalteromonas sp. strain SM20310	8.61 g/L	–	Cryoprotector	Liu et al. (42)
	Zunongwangia profunda			Highly complex α-mannan polymer of 2-α-, 6-α-mannosyl residues, where 6-α-mannosyl residues are branched at 2nd position with 1–2 t-mannosyl residues	Cryoprotector	Liu et al. (28)

halophiles, and borderline halophiles. The halotolerant bacteria in contrast show tolerance to a wide range of salinity stress, from high salt concentration to zero salt requirement (14). Many halophilic Archaea such as *Haloferax, Haloarcula, Halococcus, Natronococcus,* and *Halobacterium* are also good EPS producers (15–17). *H. maura, H. eurihalina, H. ventosae,* and *H. Anticariensis* of *Halomonas* genus are the most dominant halophilic EPS producing group. The characteristic features of EPS synthesized by *Halomonas* strain are unusually high sulfate content and uronic acids in large amounts which result in good jellifying properties (18). Most halophiles show high levels of heavy metal resistance as it is found in coastal areas.

Thermophiles

Thermophiles are currently classified as: moderate thermophiles (50–70°C) and extremothermophiles (>70°C) based on its optimal growth temperatures; extremothermophiles grow optimally above 80°C and are also termed as "hyperthermophiles." They inhabit a wide range of habitats from geothermal springs and solfataric (sulfur) fields, shallow submarine hydrothermal systems, geothermally heated oil reservoirs to abyssal hot-vent environments or hot coal-refuse piles (19).

Different hyperthermophilic microorganisms such as *Thermotoga maritima, Archaeoglobus fulgidus,* and *Thermococcus litoralis* produce EPS significantly (20–22). Works has also been done on EPS producing *Bacillus licheniformis* from marine hot springs (23), glucan producing *Geobacillus tepidamans* from Bulgarian hot springs (24), *Brevibacillus thermoruber* from geothermal springs of Turkey and Bulgaria (25).

Psychrophiles

A large portion of reduced carbon reserve of the ocean is EPS and it enhances the survival rate of psychrophilic marine bacteria by modifying the physicochemical environment around the bacterial cell. Antarctic marine environments are rich in bacterial EPSs which help the microbial communities to survive under extreme cold temperature and salinity with least nutrient availability. EPS produced by a new genus of *Pseudoalteromonas,* isolated from Antarctic sea ice at −2°C and 10°C showed higher uronic acid content than EPS produced at 20°C (26).

EPS from Pathogenic Bacteria
Surface attaching bacteria come together in a hydrated polymeric matrix for biofilm formation. EPS biosynthesis by these virulent communities

results in resistance to antimicrobial drugs, which is the root to many chronic and persistent bacterial infections. Bacterial EPS is also the basis of growth of pathogenic microorganism in biofilms as it provides the substratum for microbial growth in mats; for example, dental caries by acidogenic Gram-positive cocci, cystic fibrosis pneumonia by *Pseudomonas aeruginosa* and *Burkholderia cepacia*, nosocomial infection by different Gram-negative bacteria are all biofilm-mediated infection. Nitric-oxide-mediated signaling results in EPS production in *Shewanella oneidensis*. LasI-dependent QS provides maturation signal and results in the formation of a thick and differentiated biofilm of *Pseudomonas aeruginosa*. Presently, diguanylate cyclase and its product cyclic di-guanosine monophosphate (c-di-GMP) are key biomedical targets for the inhibition of biofilm development.

REGULATION OF EPS PRODUCTION

A complex process involving the transport of organic and inorganic molecules to the outer surface, its adsorption to that surface and finally the formation of an unalterable attachment result in the growth of EPS as biofilm. The most studied regulatory mechanism controlling EPS production is QS. QS maintains bacterial intercellular communication and regulate gene-specific expression to increase cell density.

Two QS processes described in bacteria are AI-1 and AI-2. Autoinducer-1 (AI-1) regulates intraspecies communication and AI-2 regulates interspecies communication. Gram-negative bacteria secrete AI molecule: *N*-acyl homoserine lactone (AHL) that control cell density. The detection of accumulated AHL signal by bacteria switches on transcriptional effectors to activate silent genes above a certain threshold concentration. This results into cell density dependent gene expression and behavioral change. Gram-positive bacteria communicate with modified oligopeptides and membrane-bound histidine kinase sensor as receptors. Signaling is controlled by multiple phosphorylations which modify the activity of the regulator (27). The omnipresent bacterial c-di-GMP is an important messenger in controlling bacterial biofilm formation. Cyclic nucleotides are synthesized by external stimuli on various signaling domains within *N*-terminal region of dimeric diguanylate cyclase. It initiates condensation of two molecules of guanosine triphosphate opposite to each other within the C-terminal region of the enzyme. Additionally, the formation of a long-term stable aerobic granule, a superior biofilm for biological wastewater treatment, can be controlled by stimulating c-di-GMP.

Substrate (a) is first catabolized by glycolytic pathways into pyruvate while entering the cell. Under aerobic conditions, it is converted to acetyl-CoA and enters TCA cycle (b). UDP-Glc, UDP-Gal, and GDP-Man are interconverted by epimerization, oxidation, decarboxylation, reduction, and rearrangement (c) reactions to form energy-rich monosaccharide triphosphates. EPS biosynthesis and polymerization occurs through any of the following mechanisms (d).

Wzx–Wzy system (left) synthesize EPS by the sequentially transferring monosaccharides from NDP-sugars to a polyprenylphosphate lipid carrier. For polymerization by polymerase enzyme Wzy, modified repeating units are transported across the inner membrane by a flippase enzyme Wzx to the periplasmic place. Polysaccharide copolymerase (PCP) determines the polymer length, while an outer membrane polysaccharide export protein OPX forms the intermembrane channel.

The ABC-transporter system (right) results in EPS formation at the cytoplasmic side of the inner membrane by adding sugar residues to the nonreducing end of the polymer and export it across the inner membrane, followed by its translocation across the outer membrane by PCP and OPX (Figure 1.2).

PRESENT STUDIES ON BACTERIAL EPS

Microbial EPSs are usually linear molecules with different side chains. Association of high molecular weight polymer chains results in complex entanglement. These high molecular weight polymers show a tendency to form double-stranded helices (kappa carrageenans, xanthan, succinoglycan, and gellan) and sometimes triple stranded (curdlan, schizophyllan) (28,29) also. Commonly found monosaccharides in bacterial EPSs are D-glucose, D-galactose, and D-mannose; L-fucose and L-rhamnose; and N-acetylhexosamines, N-acetyl-D-glucosamine, and N-acetyl-D-galactosamine. Uronic acids such as D-glucuronic and D-galacturonic acid are also present in few microbial EPSs. Common and commercially most exploited types of bacterial EPSs (Figure 1.2) are xanthan, gellan, cellulose, sphingan, hyaluronan, alginate, etc.

Heteropolysaccharides

1. Xanthan gum: It is the first industrially produced biopolymer, which is extensively studied and widely accepted commercially. *Xanthomonas* genus of bacteria secretes this heteropolysaccharide

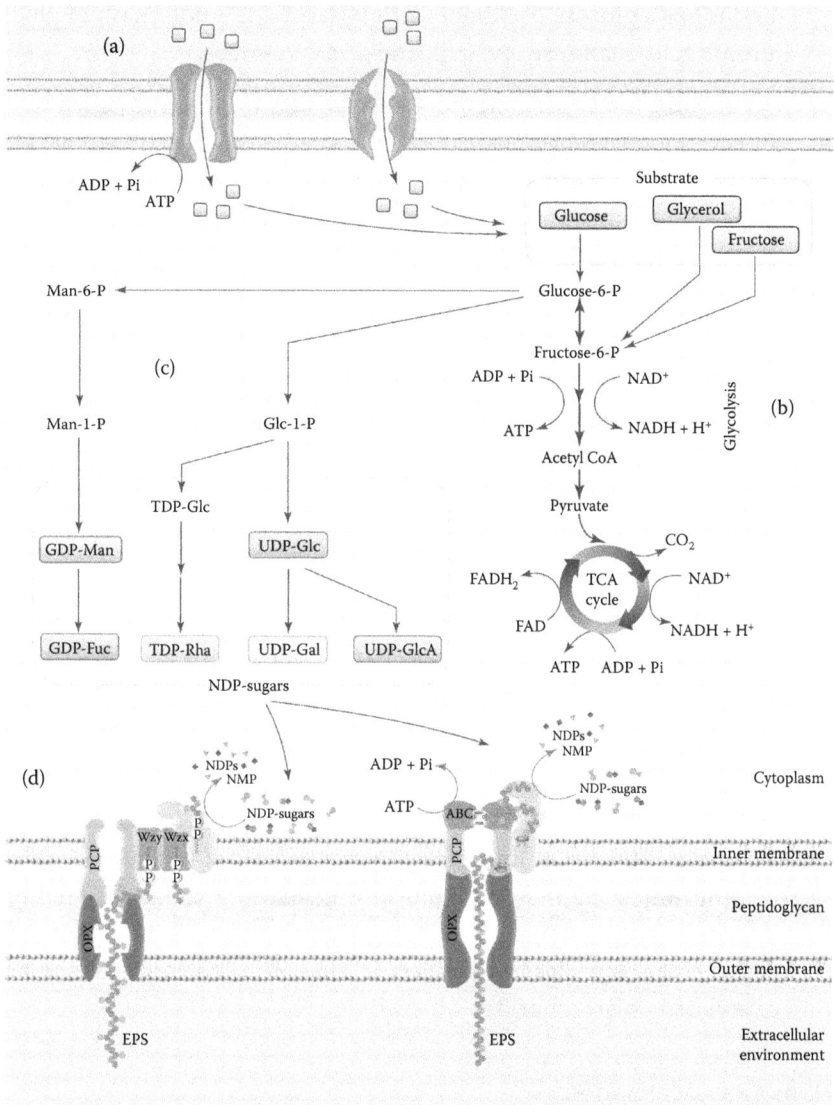

FIGURE 1.2 Schematic diagram showing biosynthetic pathways of EPS synthesis by Gram-negative bacteria. (a) Substrate entry to the cell, (b) entry of phosphorylated sugars in TCA cycle, (c) interconversion of sugars, and (d) EPS polymerization. Fuc, fucose; Gal, galactose; Glc, glucose; GlcA, glucuronic acid; Man, mannose; Rha, rhamnose; GDP, guanosine diphosphate; TDP, tyrosine diphosphate; NMP, nucleoside monophosphate. (Adapted from Freitas, F., Alves, D. V., and Reis, M. A. M. 2011. *Trends in Biotechnology* 29:388–398.)

and it has a glucose backbone with trisaccharide side chain of glucuronic acid, mannose, pyruvil, and acetyl residues.

2. Sphingans: These are heteropolysaccharides produced by members of the genus *Sphingomonas* having a characteristic tetrasaccharide backbone of rhamnose, glucose, and glucuronic acid. Different variations of sphingans are gellan, welan, rhamsan, and diutan. They have difference in composition and linkage of the side chains, for example, gellan contains acetyl and glyceryl residues in side chain, whereas welan has a rhamnose or mannose containing a side chain branch.

3. Hyaluronan: Repeating disaccharide units of glucuronic acid and *N*-acetylglucosamine form a linear structured polymer named hyaluronan. Bacterial strains, for example, *Pseudomonas aeruginosa* and group A and C *Streptococci* attenuated strains are observed to produce this high market value EPS.

4. Succinoglycan: This is a branched bacterial EPS with glucose and galactose backbone and tetrasaccharide side chains of glucose residues. Succinate, pyruvate, and acetate are also commonly found in it. Many soil inhabiting bacteria such as *Rhizobium* sp., *Alcaligenes* sp., *Pseudomonas* sp., and *Agrobacterium* sp. are good producers of succinoglycan.

5. Alginate: This is a linear copolymer of block-structured poly-mannuronic acid and poly-guluronic acids. Alginate is secreted by *Pseudomonas aeruginosa* and *Azotobacter vinelandii*. The main difference between algal and bacterial alginate is the algal one is an acetylated polysaccharide.

Homopolysaccharides

1. Glucans: Glucans are glucose homopolysaccharides differing in glycosidic bond, degree and type of branching, chain length, molecular mass, and polymer conformations. Glucans are of two types—α-glucans (reuteran, dextran, mutan, and alternan) and β-glucans (e.g., cellulose and curdlan). Bacterial genera, such as *Gluconacetobacter, Agrobacterium, Aerobacter, Achromobacter, Azotobacter, Rhizobium, Sarcina,* and *Salmonella* are able to produce cellulose. Extracellular enzyme dextransucrase that form α-glucans are produced from sucrose in several bacterial genera

of *Lactobacillus Leuconostoc* and *Streptococcus.* Substrate synthesis for cellulose production starts from the glycolytic intermediate glucose-6-phosphate. Curdlan is glucans homopolysaccharide of β-(1 → 3)-linked glucose residues produced by bacteria such as *Agrobacterium biobar* and *A. tumefaciens,* and it forms characteristic elastic gels after heating in aqueous suspension. Curdlan production by *Alcaligenes faecalis* is developed for commercial use in gel production.

2. Levan: It is synthesized from sucrose by the extracellular enzyme levansucrase (EC 2.4.1.10), also known as sucrose 6-fructosyltransferase by several bacterial genera of *Bacillus, Rahnella, Aerobacter, Erwinia, Streptococcus, Pseudomonas,* and *Zymomonas* (30). Levan is highly branched water-soluble fructose homopolysaccharide. Due to the presence of β-(2 → 6) linkage, levan is soluble in oil.

FUTURE PROSPECTS OF BACTERIAL EPSs

The future of bacterial EPS production will be glorious, as immense potential applications of EPS are already established *in vitro.* The most accepted bacterial EPS as drug delivery systems are xanthan and cellulose. There are many others and the most promising are the ones mentioned below.

Food Industry

First industrially marketed EPS dextran is produced by LAB and is used in confectionary to improve moisture retention, maintenance of viscosity, and to inhibit sugar crystallization. It acts as gelling agents in jelly and gum. It inhibits water crystal formation in ice cream and also gives the desired body and mouth feel in pudding. Due to the growing demand in natural and minimally processed foods, the use of antimicrobial compounds produced by LAB is of huge scientific and commercial interest as a safe and natural food preservative. Nisin produced by *Lactococcus lactis* and Reuterin produced by *Lactococcus reuteri* are widely used as natural antibacterial food preservatives.

Pharmaceutical Industry

Recently there is a huge growing demand for LAB as probiotics. The characteristic features of LAB strains as probiotics are its acid and bile tolerance, producing antimicrobial compounds against pathogens and adherence, and colonization in human intestinal mucosa.

Bacterial EPS xanthan gum synthesized by *Xanthomonas campestris* is a sustained drug delivery system and increase drug effectivity. Bacterial cellulose nanocrystals produced by *Gluconacetobacter xylinus* showed superior properties than plant-derived cellulose and are recently patented by FDA for using as transdermal drug delivery system, tablet excipient, and aerogel–hydrogels.

Biomedical Application

Polysaccharides of marine *Vibrio, Pseudomonas,* and *Bacillus licheniformis* showed to have antitumor, antiviral, and immune stimulant activity. *Alteromonas infernus* isolated from deep-sea hydrothermal vent, produced a low-molecular weight heparin-like EPS with good anticoagulant property. An L-fucose containing polysaccharide Clavan showed promising roles in tumor cell colonization prevention in lung, regulation of white blood-cell formation, rheumatoid arthritis treatment, antigen synthesis for antibody production, and in cosmeceuticals as a skin moisturizing agent. Water-soluble EPS from *Pantoea agglomerans* showed protective activity against UV radiation by its free radical scavenging activity (8).

Bioremediation and Wastewater Treatment

Bacterial EPS can effectively perform bioremediation, as bacteria growing within the biofilm have higher adaptation to different extreme environments and increased survival rate. Biofilm reactors are mainly used to treat municipal and industrial wastewater. As EPSs have good flocculation activity and are able to bind metal ions in solutions, it is used in the removal of heavy metals from the environment. Sulfate reducing bacteria, *Enterobacter* and *Pseudomonas* species are the major group of bacteria commonly found in metal contaminated wastewaters and are highly efficient in anaerobic degradation of organic pollutants and heavy metal precipitation from wastewater. Recently *Acidithiobacillus thiooxidans* and *Acidithiobacillus ferroxidans* are proved to be potent accumulators of Fe^{3+} (4).

PATENTING IN THE FIELD OF BACTERIAL EPS

Bioactive microbial EPSs have shown enormous growth in patenting from last 30 years as the first patent in "recombinant DNA plasmid for xanthan gum synthesis" was published in 1987 by Merck. Till then

TABLE 1.2   Important Patents in the Field of Bacterial Exopolysaccharide in the
Last Two Years

Patent	Filing Date	Applicants	Title
20150150959	06.04.2015	Children's Medical Center Corporation	Bacterial biofilm matrix as a platform for protein delivery
20150079137	19.03.2015	Polymaris Biotechnology	Exopolysaccharide for the treatment and/or care of the skin, mucous membranes, and/or nails
20140348878	27.11.2014	Ai et al.	Strain of exopolysaccharide-secreting *Lactobacillus brevis* and application thereof
20140322273	30.10.2014	Ai et al.	Strain of exopolysaccharide secreting *Lactobacillus plantarum* and application thereof
20140171386	19.06.2014	Cheuk et al.	Method and composition to reduce diarrhea in a companion animal
20140057018	27.02.2014	Ashraf Hassan	Processed cheese with cultured dairy components and manufacturing
20140037687	06.02.2014	Apariin et al.	Exopolysaccharide of *Shigella sonnei* bacteria, method for producing same, vaccine and pharmaceutical composition containing same
20140037597	06.02.2014	Senni et al.	Sulfated depolymerized derivatives of exopolysaccharides (eps) from mesophilic marine bacteria, preparing same, and uses thereof in tissue regeneration

Source: http://tgs.freshpatents.com/Exopolysaccharide-bx1.php

bacterial xanthan gum from *Xanthomonas campestris* have different patents with varying applications. Some of the recent patenting trends in last 2 years have excelled with lactic acid bacteria, soil inhabitants, and halophiles (Table 1.2). More research is needed in the field of EPS producing thermophiles and psychrophiles, as EPS is produced in large quantities to survive in extreme harsh environment. Hyaluronan because of its highly hydrophilic nature is currently used in pharmaceutical industry and has the highest market value of 1 billion US$ among all other bacterial EPSs, which indicates a strong market demand for it. But the main hindrance in using hyaluronan commercially is its high cost, which is around 100,000 US$. Other important bacterial exopolysaccharides like xanthan, gellan,

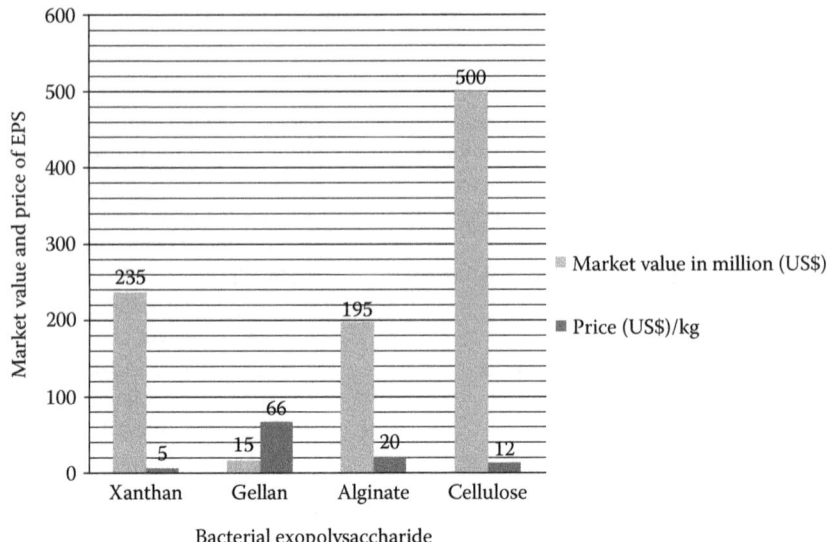

FIGURE 1.3   Commercially important bacterial EPS with their present market value and price in US$.

alginate, and cellulose are also illustrated with their market value and price in Figure 1.3.

CONCLUSION

Although diverse bacterial EPSs are studied in the last two decades, very less information is available for its increased production rate. Till date only dextran is the most utilized and commercialized bacterial exopolysaccharide used in different biomedical applications (31). To improve fermentation processes for bacterial EPS production, searching the model nutritional conditions will provide greater yield. Besides, finding the genes involved in sugar metabolism for EPS biosynthesis and transformation of bacterial cell with a strong promoter will have huge impact on maximum optimized EPS production. Better stability and biocompatibility of EPS can be obtained by either cross-linking it with salt or protein by physical processes or by cross-linking agents via chemical methods. As chemical cross-linking agents are highly toxic, genipin isolated from the fruit gardenia is a new horizon in the development of chemically safe microbial EPSs. In the present scenario, the bacterial EPS research is mainly confined into isolation, characterization, and native applications. More patenting of novel EPSs can enhance its utilization and sustained use.

ACKNOWLEDGMENT

The authors are thankful to the UGC Center of Advanced Study, Department of Botany, The University of Burdwan, for the research facilities and environment. Aparna Banerjee is also thankful to UGC for the financial assistance of JRF (State Funded) [Fc (Sc.)/RS/SF/BOT./2014-15/103 (3)].

REFERENCES

1. Bertagnolli, C., Espindola, A. P. D. M., Kleinübing, S. J., Tasic, L., and Silva, M. G. C. D. 2014. *Sargassum filipendula* alginate from Brazil: Seasonal influence and characteristics. *Carbohydrate Polymers* 111:619–623.
2. Kreyenschulte, D., Krull, R., and Margaritis, A. 2014. Recent advances in microbial biopolymer production and purification. *Critical Review in Biotechnology* 34:1–15.
3. Sutherland, I. W. 1972. Bacterial exopolysaccharides. *Advanced Microbial Physiology* 8:143–213.
4. Singha, K. T. 2012. Microbial extracellular polymeric substances: Production, isolation and applications. *IOSR Journal of Pharmacy* 2:276–281.
5. Lemos, E., Lemos, M., and Castellane, T. 2014. Evaluation of the biotechnological potential of *Rhizobium tropici* strains for exopolysaccharide production. *Carbohydrate Polymers* 111:191–197.
6. Marczac, M., Dzwierzynska, M., and Skorupska, A. 2013. Homo- and heterotypic interactions between Pss proteins involved in the exopolysaccharide transport system in *Rhizobium leguminosarum* bv.*trifolii*. *Biological Chemistry* 394:541–559.
7. Quarashi, W. A., and Sabri, N. A. 2012. Bacterial exopolysaccharide and biofilm formation stimulate chickpea growth and soil aggregation under salt stress. *Brazilian Journal of Microbiology* 43:1183–1191.
8. Wang, H., Jiang, X., Mu, H., Liang, X., and Guan, H. 2007. Structure and protective effect of exopolysaccharide from *P. Agglomerans* strain KFS-9 against UV radiation. *Microbiological Research* 162:124–129.
9. Ramadan, F. M., Mahmoud, G. M., Sayed, H. O., and Asker, S. M. M. 2014. Chemical structure and antioxidant activity of a new exopolysaccharide produced from *Micrococcus luteus*. *Journal of Genetic Engineering and Biotechnology* 12:121–126.
10. Cerning, J. 1995. Production of exopolysaccharides by lactic acid bacteria and dairy propionibacteria. Le *Lait* 75:463–472.
11. Robijn, G. W., Van den Berg, D. J. C., Haas, H., Kamerling, J. P., and Vliegenthart, J. F. G. 1995. Determination of the structure of the exopolysaccharide produced by *Lactobacillus sakei* 0-1. *Carbohydrate Research* 276:117–136.

12. Rawson, L. H., and Marshall, M. V. 1997. Effect of "ropy" strains of *Lactobacillus delbrueckii* ssp. *bulgaricus* and *Streptococcus thermophiles* on rheology of stirred yoghurt. *International Journal of Food Science and Technology* 32:213–220.

13. Low, D., Ahlgren, A. J., Horne, D., McMahon, J. D., Oberg, J. C., and Broadbent, R. J. 1998. Role of *Streptococcus thermophilus* MR-1C capsular exopolysaccharide in cheese moisture retention. *Applied and Environmental Microbiology* 64:2147–2151.

14. Valera, R. F., Berraquero, R. F., and Cormenzana, R. A. 1981. Characteristics of the heterotrophic bacterial populations in hypersaline environments of different salt concentrations. *Microbial Ecology* 7:235–243.

15. Ant'on, J. I., Meseguer, I, and Valera, R. F. 1988. Production of an extra-cellular polysaccharide by *Haloferax mediterranei*. *Applied Environmental Microbiology* 54:2381–2386.

16. Nicolaus, B., Lama, L., and Esposito, E. 1999. *Haloarcula* spp. able to bio-synthesize exo- and endopolymers. *Journal of Industrial Microbiology and Biotechnology* 23:489–496.

17. Paramonov, N. A., Parolis, L. A. S., Parolis, H., Bo, I. F., Ant'on, J., and Valera, R. F. 1998. The structure of the exocellular polysaccharide produced by the archaeon *Haloferax gibbonsii* (ATCC 33959). *Carbohydrate Research* 309:89–94.

18. Parolis, H., Parolis, L. A. S., and Bo, I. F. 1996. The structure of the exopoly-saccharide produced by the halophilic archaeon *Haloferax mediterranei* strain R4 (ATCC 33500). *Carbohydrate Research* 295:147–156.

19. Nicolaus, B., Kambourova, M., and Oner, E. T. 2010. Exopolysaccharides from extremophiles: From fundamentals to biotechnology. *Environmental Technology* 31:1145–1158.

20. Johnson, R. M., Conners, B. S., Montero, I. C., Chou, J. C., Shockley, R. K., and Kelly, M. R. 2006. The *Thermotoga maritima* phenotype is impacted by syntrophic interaction with *Methanococcus jannaschii* in hyperthermophilic coculture. *Applied and Environmental Microbiology* 72:811–818.

21. Lapaglia, C., and Hartzell, L. P. 1997. Stress-induced production of biofilm in the hyperthermophile *Archaeoglobus fulgidus*. *Applied and Environmental Microbiology* 63:3158–3163.

22. Rinker, D. K., and Kelly, M. R. 1996. Growth physiology of the hyper-thermophilic archaeon *Thermococcus litoralis*: development of a sulfur-free defined medium, characterization of an Exopolysaccharide, and evidence of biofilm formation. *Applied and Environmental Microbiology* 62:4478–4485.

23. Spano, A., Gugliandolo, C., Lentini, V., Maugeri, L. T., Anzelmo, G., Poli, A., and Nicolaus, B. 2013. A novel EPS-producing strain of *Bacillus licheni-formis* isolated from a shallow vent off Panarea Island (Italy). *Current Microbiology* 67:21–29.

24. Kambourova, M., Mandeva, R., Dimova, D., Poli, A., Nicolaus, B., and Tommonaro, G. 2009. Production and characterization of a microbial glucan, synthesized by *Geobacillus tepidamans* V264 isolated from Bulgarian hot spring. *Carbohydrate Polymers* 77:338–343.

25. Yildiz, Y. S., Anzelmo, G., Ozer, T., Radchenkova, N., Genc, S., Donato, D. P., Nicolaus, B., Oner, T. E., and Kambourova, M. 2013. *Brevibacillus themoruber*: A promising microbial cell factory for exopolysaccharide production. *Journal of Applied Microbiology* 116:314–324.

26. Liu, S. B., Chen, X. L., He, H. L., Zhang, X. Y., Xie, B. B., Yu, Y., Chen, B., Zhou, B. C., and Zhang, Y. Z. 2013. Structure and ecological roles of a novel exopolysaccharide from the arctic sea ice bacterium *Pseudoalteromonas* sp. strain SM20310. *Applied and Environmental Microbiology* 79:224–230.

27. Vu, B., Chen, M., Crawford, J. R., and Ivanova, P. E. 2009. Bacterial extracellular polysaccharides involved in biofilm formation. *Molecules* 14:2535–2554.

28. Mukherjee, S., Ghosh, S., Sadhu, S., Ghosh, P., and Maiti, T. K. 2011. Extracellular polysaccharide production by a *Rhizobium* sp. isolated from legume herb *Crotalaria saltiana* Andr. *Indian Journal of Biotechnology* 10:340–345.

29. Ghosh, S., Ghosh, P., Saha, P., and Maiti, T. K. 2011. The extracellular polysaccharide produced by *Rhizobium* sp. isolated from the root nodules of *Phaseolus mungo*. *Symbiosis* 53:75–81.

30. Freitas, F., Alves, D. V., and Reis, M. A. M. 2011. Advances in bacterial exopolysaccharides: From production to biotechnological applications. *Trends in Biotechnology* 29:388–398.

31. Banerjee, A. and Bandopadhyay, R. 2016. Use of dextran nanoparticle: A paradigm shift in bacterial exopolysaccharide based biomedical application. *International Journal of Biological Macromolecules* 87:295–301.

32. Taguett, F., Boisset, C., Heyraud, A., Boun, L., and Kaci, Y. 2015. Characterization and structure of the polysaccharide produced by *Pseudomonas fluorescens* strain TF7 isolated from an arid region of Algeria. *Comptes Rendus Biologies* 338:335–342.

33. Asker, A. S. M., Sayed, E. H. O., Mahmoud, G. M., and Ramadan, F. M. 2014. Chemical structure and antioxidant activity of a new exopolysaccharide produced from *Micrococcus luteus*. *Journal of Genetic Engineering and Biotechnology* 12:121–126.

34. Vázquez, N. M., Ballesterosa, N., Canalesa, A., Saint-Jeana, R. S., Pérez-Prietoa, I. S., Prietoa, A., Aznar, R., and López, P. 2015. Dextrans produced by lactic acid bacteria exhibit antiviral and immunomodulatory activity against salmonid viruses. *Carbohydrate Polymers* 124:292–301.

35. Wang, J., Zhao, X., Yang, Y., Zhao, A., and Yang, Z. 2014. Characterization and bioactivities of an exopolysaccharide produced by *Lactobacillus plantarum* YW32. *International Journal of Biological Macromolecules* 74:119–126.

36. Kanmani, P., Kumar, S. R., Yuvaraj, N., Paari, A. K., Pattukumar, V., and Arul, V. 2011. Production and purification of a novel exopolysaccharide from lactic acid bacterium *Streptococcus phocae* PI80 and its functional characteristics activity *in vitro*. *Bioresource Technology* 102:4827–4833.
37. Marshall, M. V., and Laws, P. A. 2001. The relevance of exopolysaccharides to the rheological properties in milk fermented with ropy strains of lactic acid bacteria. *International Dairy Journal* 11:709–721.
38. Amjres, H., Bejar, V., Quesada, E., Carranza, D., Abrini, J., Sinquine, C., Ratiskol, J., Joualt, C. S., and Llamas, I. 2015. Characterization of halo-glycan, an exopolysaccharide produced by *Halomonas stenophila* HK30. *International Journal of Biological Macromolecules* 72:117–124.
39. Llamas, I., Amjres, H., Mata, A. J., Quesada, E., and Bejar, V. 2012. The potential biotechnological applications of the exopolysaccharide produced by the halophilic bacterium *Halomonas almeriensis*. *Molecules* 17:7103–7120.
40. Chen, G., Quin, W., Li, J., Xu, Y., and Chen, K. 2014. Exopolysaccharide of Antarctic bacterium *Pseudoaltermonas* sp. S-5 induces apoptosis in K562 cells. *Carbohydrate Polymers* 121:107–114.
41. Mercade, E., Delgado, L., and Carrion, O. 2015. New emulsifying and cryoprotective exopolysaccharide from Antarctic *Pseudomonas* sp. ID1. *Carbohydrate Polymers* 117:1028–1034.
42. Liu, S. B., Qiao, L. P., He, H. L., Zhang, Q., Chen, X. L., Zhou, W. Z., Zhou, B. C., and Zhang, Y. Z. 2011. Optimization of fermentation conditions and rheological properties of exopolysaccharide produced by deep-sea bacterium *Zunongwangia profunda* SM-A87. *PLoS One* 6: e26825.

CHAPTER 2

Bioremediation

A Novel Green Technology to Clean Up the Highly Contaminated Chromites Mining Sites of Odisha

Swati Sucharita Panda and Nabin Kumar Dhal

CONTENTS

ABSTRACT

Heavy metal contamination is potentially a significant environmental issue in many parts of the world primarily due to rapid industrialization and urbanization. Bioremediation appears to be the consensus method of choice as it uses living organisms such as plants and microorganisms to degrade the environmental contaminants into less toxic forms. This chapter describes the chromium mining status and its impact on the environment, and also the possible remediation techniques and future research directions in the field with a relevant case study.

CHROMITES MINING STATUS

Mineral reserves are the economic and industrial backbone of a country. South Africa is the largest producer of chromite in the world contributing about 44%, followed by India (18%). In India, 98% of the total chromite reserves are found in Sukinda. Chromite is the principal ore of the element chromium and it has a wide range of uses in metallurgical, chemical, and refractory industries. The properties of chromium that make it most versatile are its resistance to corrosion, oxidation, and hardenability. In metallurgical industry, it is used for manufacturing low- and high-carbon, ferrochrome and charge chrome.

The chromite ore belt at Sukinda is spread over an area of approximately 200 sq. km. in Jajpur district having approximately 183 million tons of deposits (Das and Mishra, 2010). According to Indian Bureau of Mines, 2010, the South Kaliapani chromite mining area alone contributes about 97% of the total chromite reserve of the state. There is no doubt that the mining significantly contributes toward economic development but, on the other hand, it has also contributed to the deterioration of the natural environment (Tiwary et al., 2005). Sukinda is listed as the fourth most polluted place in the world (Blacksmith Institute report, 2007). The generation of millions of tons of waste rocks due to opencast mining contaminates the soil and water bodies in the vicinity of the mining area.

MINE WASTE

Different types of mine waste materials that were released during the mining activities vary in their physical and chemical composition. The types of mine waste include the following:

Overburden: It includes the soil and rock that is removed to gain access to the ore deposits at open pit mines. It had piled on the surface at mine sites and had a low potential for environmental contamination. It is usually used at mine sites for landscape contouring and revegetation during mine closure.

Waste rock: It contains minerals in concentrations considered too low to be extracted and stored in heaps or dumps on the mine site. The dumps are generally covered with soil and revegetated following mine closure.

Tailings: These are finely ground rock and mineral waste products of mineral processing operations. It contains leftover processing chemicals and deposited in the form of water-based slurry into tailings ponds.

Slags: These are nonmetallic by-products from metal smelting and were considered to be the waste. These are widely environmentally benign and increasingly used as aggregates in concrete and road construction.

Mine water: It is produced in many ways at mine sites and has potential for environmental contamination.

Water treatment sludge: It is produced by the active water in a treatment plant that consists of the solids that had been removed from water to improve the efficiency of the process. The majority of sludge has little economic value and is handled as waste.

Gaseous waste: It includes particulate matter (dust) and sulfur oxides. Emissions are mainly produced during high-temperature chemical processing such as smelting causing environmental pollution.

ENVIRONMENTAL IMPACT OF CHROMITE MINING

Chromium is a member of the transition metals and the most abundant element in the Earth's crust. It exhibits a broad range of oxidation states that is, +2, +3, and +6, with +3 being the most stable. Trivalent chromium does not induce any harmful effect whereas Cr + 6 beyond a certain concentration is toxic which has many adverse impacts on the environment. According to Indian standards the permissible limit of hexavalent chromium for portable water is 0.05 ppm and for industrial discharge water is 0.1 ppm (Mohanty and Patra, 2011). The locked up chromium in chromites carry the significant portion of the lithogenic Cr, and the refractoriness of the mineral is an environmental blessing. The waste rock materials are dumped in the open ground without considering the environmental aspects, resulting in the oxidation of Cr (III) to Cr (VI). The result has been detrimental to the topography and leaching of chromium into the groundwater as well as into surface water bodies (Tiwary et al., 2005).

TOXICITY OF Cr (IV)

Hexavalent chromium exposure in humans can induce allergies, irritations, eczema, ulceration, nasal and skin irritations, perforation of the eardrum, respiratory tract disorders, and lung carcinoma. Cr (VI) pollution in the environment alters the structure of soil microbial communities, reducing microbial growth and related enzymatic activities, with a consequent persistence of organic matter in soils and accumulation of Cr (VI). The toxic action of Cr (VI) is due to its capability to penetrate easily cellular membranes and damages the cell due to oxidative stress. It is transported into cells via the sulfate transport mechanisms, due to the similarity of sulfate and chromate in their structure and charge. It is a very dangerous chemical form on biological systems as it can induce mutagenic, carcinogenic, and teratogenic effects. Cr (III) is considered for the maintenance of glucose, lipid, and protein metabolism. In metabolism studies, injected and ingested chromium was found mainly in the liver, kidneys, and blood. Chromium (VI) can act as an oxidant directly on the skin surface or it can be absorbed through the skin and damage the cell.

REMEDIAL MEASURES

The conventional treatment for soils and groundwater contaminated with hexavalent chromium includes various methods such as excavation (or) pumping of the contaminated material, addition of chemical reductant, precipitation followed by sedimentation, or ion-exchange and/or adsorption. Bioremediation is one of the promising technologies that is expected to play a significant role in waste site cleanup. According to the

Environmental Protection Agency, bioremediation is a treatment that uses naturally occurring organisms to break down hazardous substances into less toxic or nontoxic substances. Technologies can be generally classified as *in situ* or *ex situ*.

IN-SITU BIOREMEDIATION

It involves the treatment of contaminants where they are located. It is the application of biological treatment to the cleanup of hazardous chemicals present in the subsurface. It involves supplying oxygen and nutrients by circulating aqueous solutions through contaminated soils to stimulate naturally occurring microbes to degrade contaminants. It can be used for soil and groundwater. This technique includes conditions such as the infiltration of water containing nutrients and oxygen or other electron acceptors for groundwater treatment.

Biosparging

It involves the injection of air under pressure below the water table to increase groundwater oxygen concentrations and enhance the rate of biological degradation of contaminants by naturally occurring microbes. It increases the mixing in the saturated zone and increases the contact between soil and groundwater. The ease and low cost of installing small-diameter air injection points allows considerable flexibility in the design and construction of the system.

Bioventing

It is a promising new technology that stimulates the natural *in-situ* bio-degradation of any aerobically degradable compounds within the soil by providing oxygen to the existing soil microorganisms. It uses low air-flow rates to provide only enough oxygen to sustain the microbial activity. Oxygen is most commonly supplied through direct air injection into residual contamination in soil with the help of wells. Both adsorbed fuel residuals and volatile compounds are biodegraded as vapors move slowly through the biologically active soil.

Bioaugmentation

It is the introduction of a group of natural microbial strains or a genetically engineered variant to treat the contaminated soil or water. It is commonly used in municipal wastewater treatment to restart activated sludge bioreactors. Monitoring of this system is difficult. It is a slow process and hence not the best approach when immediate site cleanup is desired.

EX-SITU BIOREMEDIATION

It is a different approach that utilizes a specially constructed treatment facility and is more expensive than *in-situ* bioremediation (Brar et al., 2006). It involves the excavation or removal of contaminated soil from the ground. It is classified as solid phase system (including land treatment and soil piles) and slurry phase systems (including solid–liquid suspensions in bioreactors.

PHYTOREMEDIATION

It is the use of higher plants to bioremediate contamination in soil, water, or sediments. Phytoremediation is classified into various types as shown in Figure 2.1 (Asha et al., 2013).

Phytodegradation

Phytodegradation is also called phytotransformation which involves subsequent breakdown, mineralization, or metabolization by the plant itself through various internal enzymatic reactions and metabolic processes.

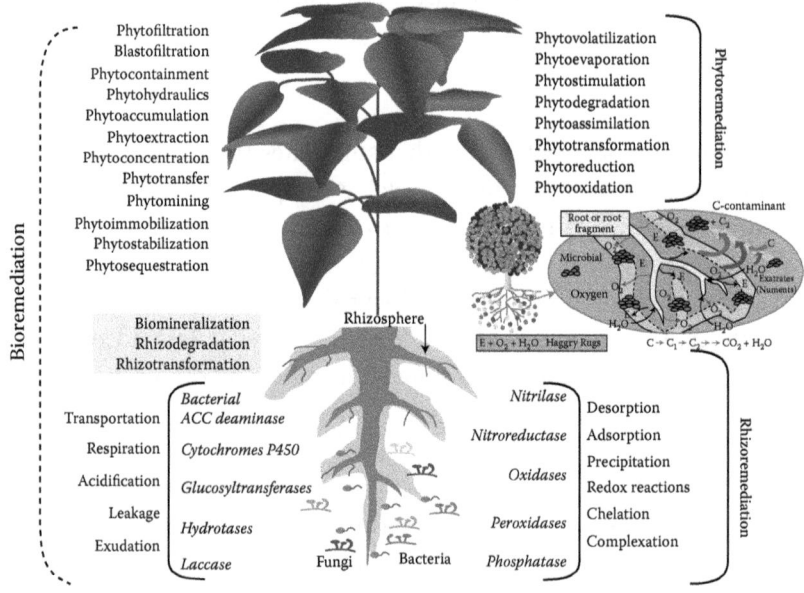

FIGURE 2.1   Different methods of phytoremediation.

Depending on the concentration and composition, plant species, and soil conditions, contaminants may be able to pass through the rhizosphere only partially or negligibly impeded by phytosequestration and/or rhizodegradation.

Phytovolatilization

It is the volatilization of contaminants from the plant either from the leaf stomata or from plant stems. Once volatilized, many chemicals that are recalcitrant in the subsurface environment react rapidly with the hydroxyl radicals in the atmosphere, forming an oxidant during the photochemical cycle.

Phytostabilization

It refers to the holding of contaminated soils and sediments in place by vegetation, and the immobilization of toxic contaminants in soils. Establishment of rooted vegetation prevents windblown dust, an important pathway for human exposure at hazardous waste sites. It is especially applicable for metal contaminants at waste sites where the best alternative is often to hold contaminants in place.

Phytoextraction

Phytoextraction refers to the ability of plants to take up contaminants into the roots and translocate them to the aboveground shoots or leaves. For contaminants to be extracted by plants, the constituent must be dissolved in the soil water and come into contact with the plant roots through the transpiration stream. The uptake may occur through vapor adsorption onto the organic root membrane in the vadose zone. Once adsorbed, the contaminant may dissolve into the transpiration water or be actively taken up through plant transport mechanisms.

Rhizofiltration

It can be defined as the use of plant roots to absorb, concentrate, and/or precipitate hazardous compounds, particularly heavy metals or radionuclides, from aqueous solutions. It is effective in cases where wetlands can be created, and all of the contaminated water is allowed to come in contact with roots. Roots of plants are capable of sorbing large quantities of lead and chromium from soil water or from water that is passed through the root zone of densely growing vegetation.

REMEDIATION BY MICROBES

Microbial bioremediation is the process by which microorganisms are stimulated to degrade rapidly the hazardous contaminants to environmentally safe levels in soil, subsurface materials, water, sludge, and residues. Microbes are adapted to thrive in "adverse conditions" of high acidity/alkalinity/ toxicity and high temperature. They can develop "biological resistance" against any toxic substance in the environment due to special "jumping genes." Microbes can biodegrade/biotransform the complex hazardous organic chemicals into simpler and reliable ones. Exposure to metals leads to the establishment of a tolerant/resistant microbial population. Microbial "resistance" is defined as the ability of a microorganism to survive toxic effects of metal exposure through a detoxification mechanism produced in direct response to the "heavy metal species" concerned. Microbial "tolerance" is defined as the ability of a microorganism to survive metal toxicity by means of intrinsic properties and/or environmental modification of toxicity. Microorganisms use a variety of mechanisms to resist and cope with toxic metals. Microbial activity is proved to play an important role in remediating metals in soil residues. Microbial metal uptake can either occur actively or passively. Microorganisms cannot destroy metals but they can alter their chemical properties via a surprising array of mechanisms. The principal mechanism of resistance of inorganic metals by microbes are metal oxidation, metal reduction, methylation, demethylation, enzymatic reduction, metal–organic complexion, metal–ligand degradation, intracellular and extracellular metal sequestration, metal efflux pumps, exclusion by permeability barrier, and production of metal chelators such as metallothioneins and biosurfactants (Figure 2.2) (Tabak et al., 2005).

METAL-MOBILIZING MICROORGANISMS

Metals can be extracted from contaminated environments by two potentially useful mechanisms: by mobilizing metals via the production of organic acids by heterotrophic bacteria and by metal leaching. This mechanism has been used to leach metals from low-grade ores and a lucrative global market in mineral extraction.

ENZYME-CATALYZED TRANSFORMATIONS

Microorganisms are ubiquitous and offer a potentially enormous gene pool to select enzymes that can help us to treat metal contamination. They have evolved a wide range of biochemical tricks to protect themselves from potentially toxic metals and these activities can be useful for

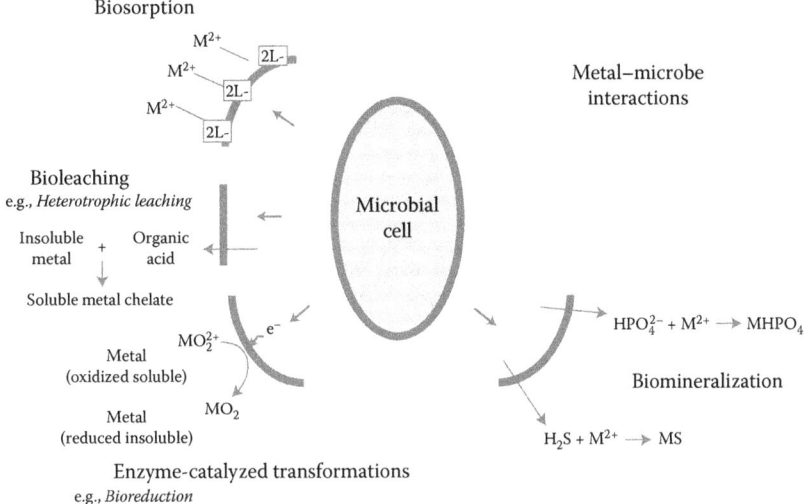

FIGURE 2.2   Microbial remediation of contaminates.

bioremediation applications. Microbial detoxification processes involve efflux or exclusion of metal ions from the cell, which result in high local concentrations of metals at the cell surface. Alternative mechanisms involve redox transformations.

INDIRECT MECHANISMS

Anaerobic bacteria are also able to reduce and precipitate a range of metals via indirect mechanisms. Once metals are in solution, one of the simplest ways to remove them is through "biosorption," which can be defined as the metabolism-independent sorption of heavy metals to biomass. The cell surface carries a net negative charge at neutral pH due to the presence of carboxyl, amine, hydroxyl, phosphate and sulfhydryl groups, and can adsorb appreciable quantities of positively charged cationic metals.

CASE STUDY

Indigenous bacterial isolates from the chromite mining waste of Sukinda Valley, Odisha, India, showed a considerable reduction of hexavalent chromium (Cr (VI)) through adaptation and consortia development. The bacterial isolates showed high tolerance at 500–1000 ppm of Cr (VI). The Gram positive strain, *Bacillus cereus* strain BAB-806 (GenBank Accession Number: KC250198.1) was identified to reduce 97% of hexavalent chromium at 37°C and pH 9.0. The strain also exhibited resistance to various antibiotics such as erythromycin, chloramphenical, and ofloxacin. Microbial reduction of toxic Cr (VI) may be very successful since the biological strategies provide cost-effective green technology and native bacterial strains have significant potential to be used as a tool for bioremediation of chromites mine sites.

CONCLUSION

Chromium contamination of soil and water is a major problem in the world. This heavy metal is a potential threat to the environment and to public health primarily because it is nonbiodegradable and environmentally persistent. Bioremediation is a multidisciplinary technology and successful application that requires a deep understanding of all the relevant scientific fields and attenuation processes. It provides a technique for cleaning up pollution by enhancing the natural biodegradation processes. So by developing an understanding of biological agents and their response to the natural environment and pollutants can be arisen as a cost effective technique for a long term. It is in the process of paving a way to greener pastures. This technology offers an efficient and eco-friendly way to treat contaminated ground water and soil. The technology is influenced by various factors such as soil type, pH, temperature, nutrient, amendments, and oxygen. It has been successfully employed in the field and is gaining more importance with increased acceptance of eco-friendly remediation solutions. Plant and microbe strains are capable of tolerating and reduce hexavalent chromium in soil and effluent that will be further used by the scientists and helpful to the environment as well as mankind. In conclusion, we believe that there are good prospects for developing cost-effective and eco-friendly bioremediation technologies that offer incredible opportunities for treating arsenic in affected areas and in industrial wastewater treatment.

FUTURE PERSPECTIVE

Bioremediation, which involves the use of naturally occurring organisms alone or in association with dead biomass/biomass residues, could be the best alternative and cost-effective technology as compared with conventional methods to mitigate environmental pollution. Biomolecular engineering can be successfully used to improve the capabilities of the biological agents (plants, algae, microbes, etc.) in bioremediation systems. Bioremediation of heavy metal has numerous advantages including eco-friendliness, adaptability, specificity, self-reproducibility, and recycling of by-products. The main drawbacks are that the process is slow and difficult in controlling. The safe removal of heavy metal is a high concern and therefore the presented process is the most logical, long-term solution for remediation.

BIBLIOGRAPHY

Akpor O B, and Muchie M. 2010. Remediation of heavy metals in drinking water and wastewater treatment systems: Processes and applications. *International Journal of Physical Sciences*; 5(12), 1807–1817.

Asha L P, and Sandeep R S. 2013. Review on bioremediation-potential tool for removing environmental pollution. *International Journal of Basic and Applied Chemical Sciences*; 3(3), 21–33. ISSN: 2073–2277.

Black Smith Institute Report. 2007. The world's worst polluted places. A project of Blacksmith Institute, pp. 16–17.

Black Smith Institute Report. 2007. The world's worst polluted places. A project of Blacksmith Institute, pp. 16-17.

Brar S K, Verma M, Surampalli R Y, Mishra K, Tyagi R D, Meunier N, and Blais J F. 2006. Bioremediation of hazardous waste—A review. *Practice Periodical of Hazardous, Toxic, and Radioactive Waste Management*; 10, 59–72.

Cervantes C, Campos-García J, Devars S, Gutiérrez-Corona F, Loza-Tavera H, Torres-Guzmán J C, and Moreno-Sanchez R. 2001. Interactions of chromium with microorganisms and plants. *FEMS Microbiology Reviews*; 25, 335–347.

Cervantes C, and Silver S. 1992. Plasmid chromate resistance and chromate reduction. *Plasmid*; 27, 65–71.

Codd R, Rillon C T, Levina A, and Lay P. A. 2001. Studies on the genotoxicity of chromium: From the test tube to the cell. *Coordination Chemistry Review*; 216,217, 537–582.

Costa M. 2003. Potential hazards of hexavalent chromium in our drinking water. *Toxicology and Applied Pharmacology*; 188, 1–5.

Das A. and Mishra S. 2010. Biodegradation of the metallic carcinogen hexavalent chromium Cr (VI) by an indigenously isolated bacterial strain. *Journal of Carcinogen*; 9(6), 19–24.

Francisco R, Moreno A, and Vasconcelos Morais P. 2010. Different physiological responses to chromate and dichromate in the chromium resistant and reducing strain *Ochrobactrum tritici* 5bvl1. *Bio Metals*; 23, 713–725.

Gadd G M. 1992. Metals and microorganisms: A problem of definition. *FEMS Microbiology Letters*; 100, 197–204.

Ghodsi H, Hoodaji M, Tahmourespour A, and Gheisari M. M. 2011. Investigation of bioremediation of arsenic by bacteria isolated from contaminated soil. *African Journal of Microbiology Research*; 5(32), 5889–5895.

Gibb H J, Lees P S, Pinsky P F, and Rooney B. C. 2000a. Clinical findings of irritation among chromium chemical production workers. *American Journal of Industrial Medicine*; 38, 127–131.

Gibb H J, Lees P S, Pinsky P F, and Rooney B. C. 2000b. Lung cancer among workers in chromium chemical production. *American Journal of Industrial Medicine*; 38, 115–126.

Jeyasingh J, and Philip L. 2005. Bioremediation of chromium contaminated soil: Optimization of operating parameters under laboratory conditions. *Journal of Hazardous Materials*; 118(1–3), 113–120.

Kumar A, Bisht B S, Joshi V D, and Dhewa T. 2011. Review on bioremediation of polluted environment: A management tool. *International Journal of Environmental Sciences*; 1(6), 1079–1093.

Lebedeva E V, and Lyalikova N. N. 1979. Reduction of chrocoite by *Pseudomonas chromatophila* sp. nov. *Mikrobiologya*; 48, 517–522.

Mathiyazhagan N, and Natrajan D. 2011. Bioremediation of Effluents from Magnetite and Bauxite Mines using *Thiobacillus* Spp and *Pseudomonas* Spp, Department of Biotechnology, Periyar University, Salem, Tamilnadu, India.

Mohanty M., and Patra, H. K. 2011. Attenuation of chromium toxicity in mine waste water hyacinth. *Journal of Stress Physiology and Biochemistry*; 7(4), 336–346.

Muhammad I, and Edyvean, R. G. J. 2007. Ability of loofa sponge—Immobilized fungal biomass to remote lead ions from aqueous solution. *Pakistan Journal of Botany*; 39(1), 231–238.

Ohta N, Galsworthy P R, and Pardee A. B. 1971. Genetics of sulfate transport by *Salmonella typhimurium*. *Journal of Bacteriology*; 105, 1053–1062.

Ohtake H, Cervantes C, and Silver S. 1987. Decreased chromate uptake in *Pseudomonas fluorescens* carrying a chromate resistance plasmid. *Journal of Bacteriology*; 169, 3853–3856.

Park J H, Lamb D, Paneerselvam P, Choppala G, Bolan N, and Chung J-W. 2011. Role of organic amendments on enhanced bioremediation of heavy metal(loid) contaminated soils. *Journal of Hazardous Materials*; 185(2–3), 549–574.

Pepi M, and Baldi F. 1992. Modulation of chromium (VI) toxicity by organic and inorganic sulphur species in yeast from industrial waste. *Bio Metals*; 5, 179–185.

Pepi M, and Baldi F. 1995. Chromate tolerance in strains of *Rhodosporidium toruloides* modulated by thiosulphate and sulfur amino acids. *Bio Metals*; 8, 99–109.

Pesti M, Gazdag Z, and Belágyi J. 2000. In vivo interaction of trivalent chromium with yeast plasma membrane as revealed by EPR spectroscopy. *FEMS Microbiology Letters*; 182, 375–380.

Plaper A, Jenko-Brinovec S, Premzl A, Kos J, and Raspor P. 2002. Genotoxicity of trivalent chromium in bacterial cells. Possible effects on DNA topology. *Chemical Research in Toxicology*; 15, 943–949.

Poopal A C, and Laxman R. S. 2009. Chromate reduction by PVA-alginate immobilized *Streptomyces griseus* in a bioreactor. *Biotechnology Letters*; 31, 71–76.

Prasad M. N., and Freitas H. 1999. Removal of Toxic Metals from Solution by Leaf, Stem and Root Phytomass of Quercus Ilex L. (Holly Oak). *Journal of Environment Pollution*, 110(2)277–283.

Raspor P, Batic M, Jamnik P, Josic D, Milacic R, Pas M, Recek M, Rezic-Dereani V, and Skrt M. 2000. The influence of chromium compounds on yeast physiology. *Acta Microbiology Immunology Hung*; 47, 143–173.

Ray S A, and Ray M. K. 2009. *Bioremediation of Heavy Metals Toxicity—With Special Reference to Chromium*; special, 57–63, ISSN 0974-1143.

Reynolds M F, Peterson-Roth E C, Bespalov I A, Johnston T, Gurel V M, Menard H L, and Zhitkovich A. 2009. Rapid DNA double-strand breaks resulting from processing of CrDNA cross-links by both MutS dimers. *Cancer Research*; 69, 1071–1079.

Romanenko V I, and Koren'kov V. N. 1977. A pure culture of bacteria utilizing chromates and bichromates as hydrogen acceptors in growth under anaerobic conditions. *Mikrobiologiya*; 46, 414–417.

Saxena D K, Murthy R C, Jain V K, and Chandra S. V. 1990. Fetoplacental-maternal uptake of hexavalent chromium administered orally in rats and mice. *Bulletin of Environmental Contamination and Toxicology*; 45, 430–435.

Shazia I, Uzma, Sadia G R, and Talat A. 2013. Bioremediation of heavy metals using isolates of filamentous fungus *Aspergillus fumigatus* collected from polluted soil of Kasur, Pakistan. *International Research Journal of Biological Sciences*; ISSN 2278–3202, 2(12), 66–73.

Shi W, Becker J, Bischoff M, Turco R F, and Konopka A. E. 2002. Association of microbial community composition and activity with lead, chromium, and hydrocarbon contamination. *Applied and Environmental Microbiology*; 68, 3859–3866.

Silver S, Schottel J, and Weiss A. 2001. Bacterial resistance to toxic metals determined by extrachromosomal R factors. *International Biodeterioration and Biodegradation*; 48, 263–281.

Tabak H. H., Lens P., van Hullebusch E. D. and Dejonghe W. 2005. Developments in bioremediation of soils and sediments polluted with metals and radionuclides—1. Microbial processes and mechanisms affecting bioremediation of metal contamination and influencing metal toxicity and transport. *Reviews in Environmental Science and Bio/Technology*; 4, 115–156.

Tiwary R. K., Dhakate, R., Rao, V. A., and Singh, V. S. 2005. Assessment and prediction of contaminant migration in ground water from chromite waste dump. *Environmental Geology*; 48, 420–429.

Valko M, Morris H, and Cronin M. T. 2005. Metals, toxicity and oxidative stress. *Current Medical Chemistry*; 12, 1161–1208.

Vidali M. 2001. Bioremediation. An overview. *Pure Applied Chemistry*; 73(7), 1163–1172.

Xu X R, Li H B, and Gu J-D. 2004. Reduction of hexavalent chromium by ascorbic acid in aqueous solutions. *Chemosphere*; 57, 609–613.

Xu X R, Li H B, Gu J-D., and Li X. Y. 2005. Kinetics of the reduction of chromium (VI) by vitamin C. *Environmental Toxicology and Chemistry*; 24, 1310–1314.

Zhitkovich Y, Song G, Quievryn V, and Voitkun V. X. 2001. Nonoxidative mechanisms are responsible for the induction of mutagenesis by reduction of Cr(VI) with cysteine: Role of ternary DNA adducts in Cr(III)-dependent mutagenesis. *Biochemistry*; 16, 549–560.

Zhou J, Xia B, Treves D S, Wu L-Y, Marsh T L, O'Neill R V, Palumbo A V, and Tiedje J. M. 2002. Spatial and resource factors influencing high microbial diversity in soil. *Applied and Environmental Microbiology*; 68, 326–334.

CHAPTER 3

Recent Developments in Food Biotechnology to Improve Human Health with Probiotics with Special Emphasis on Lowering Cholesterol

Renu Agrawal

CONTENTS

ABSTRACT

Recently, there has been a lot of discussions on the natural cure/inhibition of human diseases. Probiotics top the list as they come under GRAS category. These are beneficial microorganisms that kill or competitively inhibit the pathogenic microorganisms. They adhere to the intestinal wall and multiply there ultimately providing beneficial effects.

This chapter deals with the recent developments in the field of probiotics, probiotic foods, and the role of probiotics to reduce the level of cholesterol.

Among the probiotic foods, this chapter includes dairy, meat, and cereals.

Clinical studies of probiotic products in human beings are the correct yard stick to know about its usage. This chapter deals with the clinical studies done on human beings using various probiotic products. It also covers the market and the companies manufacturing these products.

INTRODUCTION

Today's consumer is very much aware of the side effects of dietary supplements. They are looking for natural/organic foods which could improve health and could also address specific diseases. In this context, the use of innovative probiotic fermented food products has taken an upper hand as they contain friendly bacteria. These healthier foods have been termed as probiotic functional foods.

The risk of coronary diseases has known to increase by the intake of chemical drugs. It has been found that this causes many side effects and also can cause an increase in cholesterol levels. It has been found that when cholesterol levels rise above 200 mg/dL (Nissen et al., 2004), the risk is high. Chemicals such as cholestyramine, nicotinic acid, neomycin, triparanol, clofibrate (CPIB), D-thyroxine, plant sterols, and estrogenic hormones were utilized for lowering cholesterol in the 1950s and 1960s (Mann et al., 1977; Weintraub et al., 1973).

SIDE EFFECTS OF COMMON CHEMICALS
TAKEN TO LOWER CHOLESTEROL

Nicotinic acid is known to reduce both cholesterol and also triglycerides in humans. This happens due to a decrease in lipoprotein synthesis which ultimately reduces low density lipoprotein (LDL) cholesterol. There are many side effects of nicotinic acid especially they cause skin-related infections. Other known side effects are rashes, gastrointestinal upset, hyperuricemia, hyperglycemia, and hepatic dysfunction (AMA Department of Drugs, 1977).

An anion-exchange resin namely cholestyramine is helpful in lowering cholesterol levels. During the action it binds with the bile acids in the intestinal lumen. This increases their fecal excretion as it interferes with the reabsorption of bile acids. This stimulates the bile acid synthesis. Therefore, high amounts of cholesterol are required by the liver, increasing the activity of hepatic HMG-CoA reductase. Cholestyramine has been found to be very effective in treating patients with high cholesterol levels. Due to its many side effects it is also not advisable.

In hyperlipoproteinemia it reduces LDL cholesterol. Pharmacologically it stimulates lipolysis by lipoprotein lipase. Biochemically more studies need to be done.

Neomycin is another chemical which has been found to be effective in reducing cholesterol. Its mode of action is by precipitating cholesterol inside the intestinal tract so that the absorption is inhibited. It causes side effects such as severe nausea and diarrhea and is therefore not recommended (Sedaghat et al., 1975).

Interference with the cholesterol absorption in the intestinal tract is also possible by using plant sterols. They are recommended as they do not cause any side effects. A lot of work is being carried out in this direction.

Another chemical namely, triparanol also reduces serum cholesterol. The inhibition is seen in the final stage(s) of the biosynthetic pathway along with the accumulation of other sterols. However, as it causes cataract it was banned in the 1960s (Goyette et al., 1960).

Euthyroid and hypothyroid patients were given D-thyroxine, and were found to reduce the level of LDL cholesterol. However, due to side effects such as ischemic heart diseases along with increased mortality rate in patients who had arrhythmias, angina pectoris, or multiple infarctions D-thyroxine is not much preferred (AMA Department of Drugs, 1977).

Hormones such as estrogens are helpful in treating hyperlipidemia. Due to the feminizing effect it is found to be unsuitable as hypolipidemic agents in men and it also elevates very low density lipoprotein (VLDL) and triglyceride levels (AMA Department of Drugs, 1977).

Therefore, if was found that most of these chemicals are not good cholesterol-lowering agents (Nissen et al., 2004).

Other diseases such as diarrhea, blood pressure, obesity, and lactose intolerance are causing great concern among the population due to the side effects of allopathic medicines. The population is looking toward natural and healthier foods.

With the emergence of a more health-conscious society, the role of probiotic food products in human health has gained much attention from both consumers and producers. Some of the benefits are inhibition of pathogens, stimulation of the immune system, synthesis of vitamins, decrease of lactose intolerance, and reduction of serum cholesterol.

Food development in the world is taking place mainly in three strategic directions as functional fermented foods, organic foods, and national foods. Every year the global demand is increasing at the rate of 15%–20%.

Biotechnological advances are being used to promote old and new industries. As a result, the microflora, manufacturing processes, and the healthy function of these foods products are coming to the forefront. The application and progress of biotechnological food products are different from the traditional fermented food. Traditionally, these are fermented milk products (yogurt, cheese), fermented sausages, fermented vegetables (kimchi, sauerkraut), fermented cereals (sourdough), and fermented beans (tempeh, natto). Promotion of microbial processes of traditional fermented foods and optimizing the level of microorganisms have helped different processes for industrialization with homogeneous batches.

Fermentation by newer technologies involves a specific strain of lactic acid bacteria (LAB). This brings specific fermentation under controlled conditions resulting in a specific fermented product with enhanced organoleptic, nutritional, and therapeutic qualities. For adequate growth an individual requires adequate nutrients such as proteins, carbohydrates, vitamins, and minerals to maintain a balanced diet. Today the industries are aware of the consumer interest in the role of enhancement of health by some specific foods or physiologically active food components, the so-called functional foods (Cochin, 2007).

> Functional foods are the processed foods, containing ingredients that aid specific bodily functions in addition to being nutritious.

The present society has the knowledge how to be healthy. It is important to provide nutritionally balanced foods with the goodness of healthy bacteria. Probiotic fermented foods seem to be the best option. LAB foods and supplements have been shown to aid in the treatment and prevention of hyperlipidemia, acute diarrhea, and in decreasing the severity and duration of rotavirus infections in children and travelers' diarrhea in adults (Guandalini, 2011).

The probiotic functional food concept was first developed in Japan in the 1990s. A lot of side effects of medicines were observed. It was then that the Ministry of Health and Welfare initiated a regulatory system to approve certain foods with documented health benefits to improve health. These are now recognized as Foods for Specialized Health Use (FOSHU). According to the American Dietetic Association (ADA) these foods are potential healthful food or food ingredients that may provide health benefits beyond the traditional nutrients. Probiotics (for life) are fermented foods which are cultured by beneficial microorganisms. Probiotic foods as yogurt and sauerkraut are very familiar. There are many more which have been enjoyed for centuries in different parts of the world. Probiotic fermented foods have values beyond their original states. Fermentation increases nutrients, gives a tasty zing, and increases the shelf life without the use of preservatives. Probiotic fermented foods improve the immune system. The beneficial bacteria in fermented foods help to keep the digestive tract healthy and protect them against virulent pathogens that cause food-borne illnesses. It has been observed that the typical American diet makes the body vulnerable to many infections. The problem increases with the use of antibiotics and our increasing resistance to them. Use of antibiotics not only kills the pathogens but also beneficial microorganisms.

Probiotic foods can prevent or alleviate many health disorders ranging from allergy and asthma to yeast infection and heart diseases that result from food-borne infections and antibiotic resistance. Probiotic foods provide special nutritional and therapeutic properties over traditional foods. For any probiotic food to be commercialized lot of modifications and parameter optimizations are needed. Most of the probiotic fermented foods have the basic raw material as milk, vegetables, beans, fruit juices, and cereal grains. These are not only nutritious but also add variety and can be consumed as the daily diet. These diets are nutritious and healthy. They carry microbes that reach the intestine. These beneficial microorganisms improve health (Russo et al., 2005; The benefits of probiotics for your pet, 2010; Timmermann et al., 2001). Evaluation of probiotics in food can be checked in the guidelines given (FAO/WHO 2002). FAO/WHO (2001) has also evaluated the health and nutritional properties of milk powder incorporated with live LAB.

PROBIOTIC FERMENTED FOODS AND HEALTH

The presence of high levels of biologically active components in probiotic fermented foods provides health benefits which are above the basic nutrition. On fermentation, dairy products, meat and fish products, and cereal products mostly contain LAB which can enhance gastrointestinal function. These bacteria can improve lactose digestion, reduce cholesterol, and prevent diarrhea. This has action on the immune system which helps the body to fight infection. Probiotics are taking an upper hand as they also prevent or lower the growth of cancer in the colon, infections in the urogenital regions, lowering constipation, and sometimes also food allergy. Downregulation of the allergic reaction by the probiotic food has been studied (Kalliomaki and Isolauri, 2004). The probiotic microorganisms should also have good hydrophobicity which enables them to adhere to the intestinal epithelium and colonize. Probiotics are helpful in the modulation of immune system, exclusion of pathogenic microorganisms, production of bacteriocins, anticarcinogenic, and in lowering cholesterol levels (Ziemer and Gibson, 1998). Various probiotic products such as meat and fish, and cereal products have been found to enhance gastrointestinal function. However, it is important that a LAB must survive the stomach acid (pH 1.5) and high bile acid concentration of the GIT tract in order to exert a probiotic effect. Adherence to the intestinal wall is another important characteristic to impart beneficial effect and should be able to grow. Each LAB strain is specific for a particular disease in improving human health. These probiotics and their beneficial effects can be utilized by supplementation in various foods (Parvez et al., 2006). The gut flora can be altered to maintain the natural balance and the animal/human would return to better nutrition, growth, and health status (Russo et al., 2005). These bacteria cause fermentation and coagulation in milk by producing different optical forms of lactic acid (Henneberg, 1904) derived by the reduction of pyruvic acid. As initially the experiments can be done in the laboratory for screening some *in-vitro* methods have been standardized which may determine the tolerance of the organism in the GIT (Charteris et al., 1998).

Biochemically, these are mostly Gram positive, non-spore formers, form lactic acid, tolerant to acid, aerobic or facultative anaerobic, nonmotile, and catalase negative.

LABs were first referred as probiotics by Lily and Stillwell (1965). These organisms are naturally available in many food materials (Rodriguez et al., 2000). The development of probiotic food incorporates LAB as starter cultures in which they perform fermentation by the production of lactic acid. Apart from lactic acid they also produce low molecular weight compounds such as acids, alcohols, hydrogen peroxide, diacetyls, esters, and other metabolites. They inhibit the growth of spoilage organisms in food, contributing to the flavor and textural properties of fermented foods. These properties have enabled these strains to be used for the production of a wide range of probiotic fermented foods. They come under GRAS (generally recognized as safe) status (Agrawal, 2005). *Lactobacillus* and *Bifidobacterium* spp. are the commonly used probiotic LAB in fermented foods. The names of the probiotic cultures and their source availability are given in Table 3.1. The recommended dose of lactobacilli to impart a functional effect is given in Table 3.2.

A probiotic functional food improves health beyond the basic nutrition. As these probiotic foods are natural and have no side effects the

TABLE 3.1   Probiotic Culture and Source

Lactobacillus acidophilus NCFM® Danisco (Madison, WI)	*Bifidobacterium infantis* 35264 Procter & Gamble (Mason, OH)
Lactobacillus fermentum VRI003 (PCC) Probiomics, Eveleigh, Australia	*Lactobacillus rhamnosus* R0011, *L. acidophilus* R0052 Institute Rosell (Montreal, Canada)
Lactobacillus acidophilus LA-1, *L. paracasei* Chr. Hansen (Milwaukee, WI)	CRL 431, *B. lactis* Bb-12
	Lactobacillus casei Shirota, *B. breve* strain Yakult Yakult (Tokyo, Japan)
Lactobacillus casei DN114001 DN173 010 ("*Bifidus regularis*™")	("*L. casei* Defensis™", *B. animalis*
Dannon (Tarrytown, NY)	Danone (Paris, France),
Lactobacillus rhamnosus GR-1™	*Lactobacillus reuteri* RC-14™,
Urex Biotech (London, Ontario, Canada) and formerly *L. acidophilus* La-1)	Chr. Hansens (Milwaukee, WI),
	Lactobacillus johnsonii Lj-1 (same as
Lactobacillus plantarum 299V, *L. rhamnosus* 271 Probi AB (Lund, Sweden)	NCC533 Nestlé (Lausanne, Switzerland)
Lactobacillus rhamnosus GG ("LGG") Valio Dairy (Helsinki, Finland)	*Lactobacillus reuteri* SD2112 Biogaia (Stockholm, Sweden)
Lactococcus lactis L1A	*Lactobacillus rhamnosus* LB21, Essum AB (Umeå, Sweden)
Lactobacillus salivarius UCC118 University College (Cork, Ireland)	*Bifidobacterium longum* BB536 Morinaga Milk Industry Co., Ltd.
(Zama-City, Japan)	*Bifidobacterium lactis* HN019 (DR10),
Lactobacillus rhamnosus HN001 (DR20)	Danisco (Madison, WI)

TABLE 3.2   Recommended Dose of Lactobacilli

Lactobacillus casei shirota	6.5×10^9	*Lactobacillus rhamnosus GG*	$10^9 \times 10^{10}$
Lactobacillus plantarum 299 v	5×10^8	*Lactobacillus acidophilus NCFB 1748*	3×10^{11}
Lactobacillus reuteri	$1 \times 10^8 - 10^{11}$	*Lactobacillus rhamnosus DSM 6594*	16×10^9

Source: Probiotics market needs more awareness. 2011. http://www.businessstandard.com/common/news_article.php?leftnm=5&autono 311953.

population around the globe prefers it. Depending on the purpose different terminologies have been given to these probiotic functional foods such as medicinal foods, nutraceuticals, designer foods, therapeutic foods, superfoods or even medifoods (www.dairyprocessingcaft.com).

Fuller (1989) explained probiotics as a live microbial food supplement which beneficially affects the host and improves the microbial balance in the intestine.

It was Charteris et al. (1997) who explained that when a probiotic lactic acid bacterium (PLB) is consumed it imparts positive effect as it prevents the growth of pathogens. The probiotic foods are being recognized globally. A lot of attention is now being given to use probiotic microorganisms in other food products such as ice creams, chocolates, and juices. These nondairy probiotic foods are becoming more popular in European countries (Stanton et al., 2001). An excellent review article has been published earlier (Sanders, 1998, 2008). Bioactive peptides or bacteriocins are also utilized in many commercial products (Table 3.3).

TABLE 3.3   Commercial Dairy Products and Ingredients with Health or Function Claims Based on Bioactive Peptides

Brand Name	Product	Functional Bioactive Peptide	Health/Function Claim
Calpis	Sour milk	Val–pro–pro, Ile–pro–pro from casein and k-casein	Reduction blood pressure
Biozate	Hydrolyzed whey protein isolate	β-lactoglobulin fragments	Reduction blood pressure
Product F200/ Lactium	Flavored milk, drink, confectionery, and capsule	As 1-casein (91–100) (Try–Leu–Gly–Tyr–Leu–Glu–Gln–Leu–Leu–Arg)	Reduction of stress effects
Capolac	Ingredient	Casein phosphor peptide	Helps mineral absorption

Source: www.aseanbiotechnology.info.

PROBIOTIC LACTIC ACID BACTERIA IN MEAT PRODUCTS

In the production of fermented sausages PLB have been used as starter cultures since a long time (Diebel and Niven, 1961). Similar roles have contributed flavors in cured meats which grew more than 80% after storage at 20°C for 7 days (Kitchell and Ingram, 1963).

DAIRY FOODS

Milk can be enriched further with LAB as it has nutritive compounds and is consumed by all ages. Milk fat has saturated monounsaturated and polyunsaturated fatty acids. Conjugated linoleic acid prevents many diseases. Similarly milk proteins and bioactive peptides protect against many risk factors. Lactose derivatives are utilized to relieve constipation and in the modulation of intestinal flora. Milk minerals are used in the replacement of sodium in salt which prevents hypertension (Shahani and Chandan, 1979).

The development of functional dairy products needs many expert groups working together. These should include food microbiologists, medical doctors, and nutrition experts. Marketing of any product requires skill and knowledge. As probiotic products are involved with nutrition and therapeutic uses it requires nutrimarketing along with health-care professionals to convince the general public.

Dairy products are the most common among probiotic foods. Milk is consumed by both children and the elderly. Lactose which is a disaccharide is the major carbohydrate in milk. Galactose and glucose combine to form lactose. Of the total non-fat milk solids 54% is lactose which provides 30% energy. Milk provides proteins, minerals, and many essential vitamins. As milk is obtained from the animals, variations due to seasons cause many changes in milk yield and composition. Hence, preservation and scaling up of milk was worked out (Leporanta, 2001). Among the dairy probiotic foods, cultured buttermilk and fermented milk products are the most common. Dairy-based ingredients are used in milk-derived products. These are milk products such as concentrates of milk, various chesses and their products, milk or khoya desserts. Many diary products have come into the market for the convenience of the consumers.

In the preparation of any form of product milk the properties of milk should be studied such as its composition and constitution; biophysical and biochemical properties such as color, flavor, density, specific gravity, surface tension, foaming, viscosity, specific heat, electrical conductivity, freezing point, boiling point, and refractivity (FAO/WHO, 2001, 2002). Procedures for basic milk processing, raw milk handling, separation, fat standardization, pasteurization, homogenization, packaging, and storage must be known.

Fermented Milk

Mesophilic LAB such as *Lactococcus lactis* and *Leuconostoc cremoris* strains are fermented at 30°C for 16–20 h (Walstra et al., 2005). Starter cultures other than mesophilic cultures such as *Lactobacilli* are also utilized in milk products. Kefir is a traditional fermented milk drink containing *Lactococcus, Leuconostoc, Lactobacillus, Acetobacter* and yeasts. These also provide a special flavor and aroma. Yogurts are prepared by fermenting *Streptococcus thermophilus* and *Lactobacillus delbruekii* sub sp. bulgaricus growing in synergy. The temperature of fermentation is in the range of 30–43°C and fermentation time is kept for 20 h (factors affecting the growth and survival of probiotic in milk. Studies in probiotic fermented milk have been done to check the levels of lipoproteins in plasma (Richelsen et al., 1996).

The strain selection is specific and so is the fermentation time which imparts a specific flavor to the product. Whey is separated after milk coagulates during fermentation using different starter cultures. Depending on the live starter culture the process of production varies. Probiotic fermented milk contains live microorganisms. After fermentation if they are pasteurized then they may be killed. In a survey it was found that the amount of fermented milk and yogurt consumed in the EU in 2001 was 6.35 million tons (Bulletin of the International Federation, 2001).

A natural way to enhance the functionality of the products was thus found by the addition of selected and well-documented probiotic culture strains for effective health.

Various probiotic foods prepared from milk are divided into fermented milk, cheese, ice cream, by changing milk composition and dairy products with added prebiotics. Probiotics are added either before the fermentation or culture mixed with milk and added during fermentation and mixed homogenously. It is also added after fermentation in some cases.

Many products are available in the market today. Yakult (*Lb. casei*; shirota), Danone Actimel (*Lb. casei* imunitass) buttermilk, sweet milk, yogurt, fermented whey-based drinks, fermented milks, cheeses, fermented juices are some of them. These products contain *Lactobacillus* species with different strains. As milk matrix is very protective for probiotics it improves the strain survival in the intestine. Probiotics desserts are also becoming common these days. Recently, a yogurt enriched with *Lactobacillus gasseri* was shown to inhibit *Helicobacter pylori* (Filippo et al., 2001; Wang et al., 2004).

Probiotic Fermented Dairy Beverages

Probiotic dairy beverages mixing with fruits such as mango or strawberry were evaluated by mixed with fruit juices (Tamime, 2002). Probiotic dairy beverages mixed with fruits such as mango or strawberry were evaluated by Kailasapathy et al. (2008). This showed a reduction in the levels of pathogenic microorganisms. Fortification of products with ascorbic acid, amylase, or inulin had demonstrated significant growth and higher viability of probiotic cultures (Donkor et al., 2007).

There are many fermented milk beverages in India such as chaas, lassi, and buttermilk. Probiotic drinks in the European market include A-fi 1, Actimel Aktifit, AB-Piima, Bella-Vita, Bifidus, Biofit, Biola, Casilus, Cultura, Emmifit, Aktiv, Fundo, Gaio, Gefilac, Gefilus, kaiku actif, Onaka, Proviva, Yakult, Yoco activit.

Now in India there is a lot of awareness regarding better health by probiotic drinks. Yakult Danone—India has launched "yakult"; Nestle India has launched a new low fat Nesvita dahi. Mother dairy has introduced a probiotic curd with dietary fiber "b-Active Plus" and "Nutrifit"; Amul has launched probiotic lassi and ice cream.

Successful attempts have been made for carbonated probiotic fermented milk. *St. thermophilus* (MTCC 5460) and *Lb. rhamnosus* were used to prepare fermented milk from double toned milk. This was carbonated at a pressure of 15 kg/cm^2 (Shah, 2009). Kailasapathy and Supraidi (1996) have studied the survival of *Lb. acidophilus* in yogurt after hydrolyzing lactose and also the effect on whey protein concentrate under refrigerated conditions.

Cheeses

Cheese making process consists of lactic acid fermentation of milk into a complex biological substrate containing enzymes and varying in its physical and chemical properties. During cheese maturation, break down of the curd due to proteolysis, lipolysis, and other enzymatic processes gives rise to changes in texture and flavor in the cheese (Kosikowski and Mistry, 1997).

Cheddar cheese when analyzed by GCMS led to the identification of a large variety of compounds as ketones, aldehydes, alcohols, fatty acids, and volatile sulfur compounds which contributed to its aroma and flavor. Flavor quality was consistent for a particular starter and any off-flavors present were reproducible and characteristic of the starter strain used. The starter culture has a clearly defined role in cheddar flavor development.

Souring of milk is a potent mechanism for the preservation of the growth of pathogens without heating the product. Channa and paneer whey are commonly used by all strata of the society. Lactic acid bacteria easily ferment lactose in whey forming lactic acid. Jindal et al. (2004) had worked on the metabolic pathway and showed the replacement of organic and inorganic salts with the extracellular fluid which can be used in the treatment of various diseases such as arthritis and liver dysfunction.

Bioactive Peptides in Fermented Milk

Bioactive peptides are well known to inhibit pathogenic microorganisms. These have been found in different fermented products of milk. These are yogurt, buttermilk, and dahi. Scientists have recognized them for the inhibition of ACE and immunomodulation. Opioid peptides have been found to be present in yogurt and fermented milk containing probiotic *L. casei* ssp. *rhamnosus* strain. Similarly, in human casein immune stimulating hexa-peptides have been detected (Parker et al., 1984). In the market, probiotic fermented foods such as Evolus and Calpis have become common due to their health benefits (Shah, 2007).

Clinical studies after administering probiotics have been shown to reduce eczema (Kalliomäki et al., 2001; Marteau et al., 2002). Diarrhea is very common after antibiotic consumption. However, if probiotics are taken along with antibiotics a reduction in antibiotic-associated diarrhea (AAD) has been observed (Plummer et al., 2004). Bioactive peptides tablets such as Evolus and Calpis are available in the market which impart health benefits (Demin et al., 1994; Fleming et al., 1973; Tables 3.2 and 3.3).

Fortification of products with ascorbic acid, amylase, or inulin had demonstrated significant growth and higher viability of probiotic cultures (Donkor et al., 2007). Probiotic products in the United States are available with targeted health benefits (Table 3.4).

Cereal-Based Probiotic Fermented Dairy Beverages

Mineral value and fiber in milk can be increased with the addition of cereals. Further, fermentation with probiotics enhances not only the nutritive

TABLE 3.4 Probiotic Products with Targeted Health Benefits Available in the United States

Strain[a]	Product containing strain[b]	Sold by
L. acidophilus NCFM B. lactis Bi-07 B. lactis HN019 (DR10) L. rhamnosus HN001 (DR20)	Sold as ingredient	DuPont Nutrition Biosciences ApS (Madison WI)
Saccharomyces cerevisiae boulardii	Florastor	Biocodex (Creswell OR)
B. infantis 35624	Align	Procter & Gamble (Mason OH)
L. rhamnosus R0011 L. acidophilus R0052	Sold as ingredient	Lallemand (Montreal, Canada)
B. lactis Bb-12 L. acidophilus LA5 L. paracasei CRL 431 L. fermentum VRI003 (PCC) L. reuteri RC-14 L. rhamnosus GR-1 L. paracasei F19	Sold as ingredient	Chr. Hansen (Milwaukee WI)
L. casei Shirota B. breve Yakult	Yakult	Yakult (Torrence, CA)
L. casei DN-114 001 B. animalis DN-173 010	DanActive fermented milk Activia yogurt	Dannon (Tarrytown, NY)
L. johnsonii Lj-1 (NCC533; L. acidophilus La-1)		Nestlé (Lausanne, Switzerland)
L. plantarum299V	Sold as ingredient Good Belly Nature Made Digestive Probiotic Health	Probi AB (Lund, Sweden) NextFoods (Boulder, CO) Pharmavite
L. reuteri ATCC 55730 ("Protectis")	BioGaia Probiotic chewable tablets or drops	Biogaia (Stockholm, Sweden)
L. rhamnosus GG ("LGG")	Culturelle	Valio Dairy (Helsinki, Finland)
L. rhamnosus LB21 Lactococcus lactis L1A	Sold as ingredient	Essum AB (Umeå, Sweden)
L. salivarius UCC118		University College (Cork, Ireland)
B. longum BB536	Sold as ingredient	Morinaga Milk Industry Co., Ltd. (Zama-City, Japan)
L. acidophilus LB	Sold as ingredient	Lacteol Laboratory (Houdan, France)

(Continued)

TABLE 3.4   Probiotic Products with Targeted Health Benefits Available in the
United States

Strain[a]	Product containing strain[b]	Sold by
Bacillus coagulans BC30	Sustenex, Digestive Advantage Ingredient for food use	Schiff Nutrition InternationalGaneden Biotech Inc. (Cleveland, OH)

Source: Adapted from US.probiotics.org; California Dairy research Foundation: products
with probiotics.

Note: Commercial strains sold as probiotics: Selected probiotic strains and products available in the US and Europe. This table does not constitute an endorsement of any of
these products, nor does it include all strains/mixtures currently available.

[a] Parenthetic entries indicate alternative strain designations; *B. lactis* is a shorthand designation for*Bifidobacterium animalis* subsp. *lactis*.

[b] Strains sold as ingredients are available in numerous consumer products; contact responsible company for product list. Products listed are examples and do not reflect a comprehensive list of available products containing the indicated strain.

value but also the palatability and functionality of cereals. Antinutritional
factors are also reduced. The commonly used millets are sorghum, ragi,
and jowar. Solids from the cereals get incorporated into milk before and
after fermentation (Gupta et al., 2007).

CEREALS IN PROBIOTIC FUNCTIONAL FOODS

Another alternative besides milk are cereals for the preparation of probiotic functional foods. The multiple beneficial effects of cereals can be
exploited in different ways leading to the design of novel cereal foods or
cereal ingredients that can target specific populations. After knowing the
composition and processing of cereal grains they can be utilized as substrates for probiotic fermentation. The formulation of the substrate and
pattern of growth of LAB is important. It is important to know the viability of the culture (colony forming units), shelf stability, aroma properties
(organoleptically and sensorily), and finally if the product has improved
on the nutritional value (Wood, 1997). The effect of processing treatments
and fermentation on pearl millet has been studied (Hassan et al., 2006).

With increasing intakes of refined food products, there are increasing
incidences of bowel diseases among the population. Water-soluble fibers
such as β-glucan and arabinoxylan, oligosaccharides such as glucooligo-
saccharides and fructooligosaccharides and resistant starch are in high
amounts in cereals which act as prebiotics.

This has helped in the development of nondairy probiotic products. This
has also brought a new light in the food industry to utilize these natural

resources in the preparation of functional products. In the market, baby foods and confectionary formulations top the list (Saarela et al., 2000). A review article on the probiotic potential of spontaneously fermented cereal-based foods gives good information (Christine et al., 2010).

Asia and Africa have commonly been producing different types of beverages, gruels, and porridge by lactic acid fermentation of cereals. During the preparation of grains such as maize, sorghum or millet are taken which are soaked in water for 1–2 days. The grains become soft and are then milled into slurry. Hull and bran are removed after sieving and then fermented. The pH decreases and the acidity increases due to acid production by PLB.

For sourdough preparation usually wheat and rye are utilized in the west. The culture in fully density is kept at 10^9 cfu/mL in fermented sourdoughs. The growth of PLB along with a prebiotic was found to be good in any cereal. The fermentation should be under controlled conditions with well defined and known characteristics of the probiotic culture which could impart health promoting properties. It is also very vital to emphasize on the technological aspects such as composition and processing of the cereal grain and LAB. If the cell growth rate is very high then acidification may reduce the fermentation time. This may increase the viability of the specific strain and prevent the growth of pathogenic microorganisms present in the raw material (Marklinder and Lonner, 1992). Probiotic culture should get adaptable to the substrate (Oberman and Libudzisz, 1998).

The culture strains have very particular nutritional requirements for carbohydrates, amino acids, peptides, fatty esters, salts, nucleic acid derivatives, and vitamins (Severson, 1998). Many homofermentative and heterofermentative lactic acid bacteria were tested and oats were found to be a suitable substrate for the growth of probiotic LAC. *Lb. acidophilus* was found to grow slowly. The highest viability was found with *Lb. plantarum* and *Lb. reuteri* (Mills et al., 2011). The recommended dose of lactobacilli is given in Table 3.2. Probiotic products with targeted health benefits available in the United States are given in Table 3.4.

Among the different media used for probiotic culture such as malt, barley, and wheat, the malt medium was found to be the best (Charalampopoulos et al., 2002). It was also found to be strain specific. *Lb. plantarum* had the ability to tolerate low pH by maintaining a proton and charge gradient.

The physicochemical properties of the food carrier are responsible for the survival of the probiotic strains during gastric transit. Most importantly the type of carrier medium, the capacity to resist (high buffering capacity), and the pH (ranging from 3.5 to 4.5) of the gastric tract are responsible for the stability of the probiotic strain (Kailaspathy and Chin, 2000).

A lot of crude fiber is present in e grains such as corn, sorghum, millet, barley, rye, and oats. They also lack gluten which is present in wheat. Traditionally, due to lack of flavor and aroma these foods are not much relished. However, when they are fermented with PLB an improvement was found enhancing the sensorial value; volatiles such as higher alcohols, aldehydes, ethyl acetate, and diacetyl were formed (Agrawal et al., 2000). In general, a single strain of LAB is not very acceptable and therefore in order to enhance the aromatic profile of the final product many strains are incorporated to bring out the desired flavor. A study of the metabolic pathway of all the culture strains in the cereals will help in the preparation of good products.

Cereals are the generally suitable substrates for the growth of human-derived probiotic strains. A systematic approach is needed in order to identify the intrinsic processing factors that could enhance the growth and survival of the probiotic strain *in vitro* and *in vivo*. The improvement in the organoleptic properties can be done using supporting cultures that act synergistically on the probiotic strains. The functionality of colonic strains could be improved by the presence of specific nondigestible components of the cereal matrix that could act as prebiotics. The possibility of separating specific fractions of nondigestible soluble fibers from different types of cereals or cereal by-products through processing technologies as pearling and sieving or enzymatic modifications will be helpful.

FERMENTATION OF SAUERKRAUT

New cultivars of cabbage are constantly being developed to confirm the ever-changing agricultural and ecological demands. Sauerkraut fermentation is a good model to know the growth of mixed cultures and their biological control. Environmental factors such as substrate composition, anaerobiosis, temperature, pH, and salt concentration serve as a means to modify and direct the interactions of a microbial flora.

For the commercial production a controlled fermentation and information on the role and limitations of the undesirable or abnormal end product formation is necessary to establish the conditions which are optimal for the growth of selective cultures (Stamer et al., 1971).

BRINED CUCUMBERS

The process is favored by the introduction of starter cultures with the control of undesirable microbial activity. Brining treatments, environmental conditions, and initial microbial populations are the primary factors that decide the course of microbial activity. The fermentation is mainly by homofermentative LAB.

PROBIOTICS IN THE PRODUCTION OF SOY SAUCE

Soy sauce and sourdough both utilize the interaction between yeasts and lactobacilli which contribute to the flavor of the final product (Yong and Wood, 1974).

In the sourdough bread fermentation lactic acid is produced by the bacterium which contributes to a distinctive flavor and texture of the finished bread. In sour fermentation there is coexistence of bacteria and yeast. Enhanced growth of *Lactobacilli* on immobilization has been studied in soy milk (Teh et al., 2010).

PROBIOTIC TEMPEH

Tempeh is a fermented soybean product covered with white mold, a compact cake produced by the fermentation of dehulled, hydrated, soaked, and precooked soybean cotyledons. The fermenting organism is *Rhizopus oligosporus*. The fungus improves the flavor and aroma of the product. Tempeh is a traditional fermented food in Indonesia. Stimulation of growth on *Rhizopus* with potassium ions is studied earlier (Pefialoza et al., 1991). It is rich in calories, proteins, aminoacids, vitamins, minerals, antioxidants, antimicrobials, and dietary fibers. Biotechnologically, it is nutritionally rich with nutty taste, odor of fresh mushroom, and nougat-like texture. It can be consumed from infants to aged people.

Today, it is very common in countries such as Malaysia, Holland, Canada, and West Indies. When the substrates grow along with microorganisms which are not only desirable but are also edible then they become resistant to the attack of pathogenic, food poison causing, and toxin producing microorganisms. It is beneficial for health as it decreases the risk of heart diseases, strokes, osteoporosis, cancer, digestive disorders, and reduces fat.

There are other cereals and legume substrates such as chickpea, horse gram, lupin, common bean, groundnut, wheat, and maize which are used for tempeh preparation. Soya milk has also been utilized for tempeh production. Tempeh when fresh is perishable due to sporulation by fungi.

Therefore, the practice is to blanch the tempeh product, dry, and powder it which can be stored for a longer time. However, tempeh contains small amounts of oxalates which if accumulated can cause health problems.

Fermentation of soybean improves digestibility by reducing antinutritional factors such as tannin and phytate in addition to the production of acids which inhibit pathogenic bacteria. They are low-cost nutritious food which can be consumed by all socioeconomic groups.

Marketing strategies need to be worked out in order to develop new foods. This will also involve large campaigns and the needs of the consumer should be kept in mind while adapting a new product. Today the consumer is very much aware of new and innovative products or reformulations. To meet this, the manufacturer needs to constantly reinvent the products. This will ultimately bring a change in the health-conscious society.

Probiotic cereal products have a good matrix for the culture to grow in the human gut along with prebiotics and are therefore gaining a lot of importance.

CLINICAL STUDIES WITH PROBIOTIC PRODUCTS

Beneficial effects of probiotics have been proved by many clinical studies in diseases like eczema (Kalliomäki et al., 2001; Marteau et al., 2002). Diarrhea caused due to the consumption of antibiotics (AAD) has shown to alleviate with probiotics (Plummer et al., 2004).

MARKET OF PROBIOTIC PRODUCTS

The sale of functional foods is high in countries like Brazil which amounts to US$ 500 million per annum. The maximum sale is in the products of dairy, nutraceuticals, and soy-based sector. ABIA had predicted a substantial growth of 4.5%–5% in this segment (Stanton et al., 2001) which has proved to be correct.

Any probiotic product can contribute for better health effects if a sufficient cell number of the culture is present in the product till the end of the shelf storage. Research microbiologists and food scientists are taking a lot of interest as this market is growing. The beneficial effects and inhibition of diseases by probiotics proved by clinical trials has led to the demand of these products by Regulators in the U.S. Food and Drug Administration, U.S. Department of Agriculture and health departments. Technologically, the probiotic cultures should be able to grow in the presence of oxygen and should resist the acid and bile concentration of the gut region to impart beneficial effects. Many scientists are working in this direction to get good probiotic cultures. The commonly used cultures are *Lactobacillus* sp. and *Bifidobacterium* sp. (Stanton et al., 2003a,b).

Baby health food and foods to improve chronic health diseases are of utmost priority. Recently, many changes have been found in food regulations and health-care costs are increasing. With the changing scenario in consumer demands innovative products are required (Agriculture: A vision for the future—Probiotics "What's the hype", 2002; Functional foods Japan: Product Report, 2006; The benefits of probiotics for your pet, 2010). Lot of research is being taken up to check the products clinically for health attributes. Biomedical studies with probiotics are taken up in animal models dealing with therapies as alternative medicines. Probiotic product development is in focus. Prebiotics and probiotics are an important source of micronutrients and macronutrients. Scientists are developing bio-modification aids to check dietary plant molecules which will enhance health. Each prebiotic and probiotic product is specific for a particular health benefit and can be consumed as per particular health needs (Probiotics market needs more awareness, 2011). It is assumed that global probiotics market will exceed US$28.8 billion by 2016. As the consumer today has become very health conscious it is important that the companies switch to preventive health care. Some of the recent developments are given in Table 3.4.

Lot of companies are investing in the preparation of probiotic products in India. With awareness among the general public and health-consciousness, demand for these products is increasing day by day.

COMPANIES PREPARING PROBIOTIC PRODUCTS

1. Probiotic curd: Heritage Foods (India) Ltd.

2. "b-Activ" probiotic curd (*L. acidophilus* and *B. lactis*, strain *BB12*): Mother Dairy.

3. "Nesvita" probiotic yoghurt: Nestle.

4. Probiotic ice creams, "Amul Prolife," "Prolite," and "Amul Sugarfree": Amul (brands of Gujarat Cooperative Milk Marketing Federation Ltd.).

5. Yakult, probiotic curd with *L. casei* strain *Shirota* Yakult: Danone India (YDI) Private Limited.

6. Probiotic drugs, Binifit: Ranbaxy.

7. Probiotic drugs: Dr. Reddy's Laboratories.

8. Probiotic drugs: Zydus Cadila.

9. Probiotic drugs: Unichem.

10. Probiotic drugs: JB Chem.

11. Probiotic drugs: GlaxoSmithKline.

12. Fructooligosaccharides: Probiotic drugs Glenmark Alkem Labs.

Some of the dairy beverage products can be enriched with active ingredients such as omega-3 fatty acids, cholesterol-lowering ingredients, and probiotics besides micronutrients. Changes in food regulations, market opportunities, increased health-care costs, changing consumer demands, and new opportunities to add value to the existing products with higher profits has made probiotic functional foods very important.

REFERENCES

Agrawal, R. 2005. Probiotics: An emerging food supplement with health benefits. *Food Biotechnol.* 19: 227–246.

Agrawal, R., Rati, E.R., Vijayendra, S.V.N. et al. 2000. Flavour profile of idli batter prepared from defined microbial starter cultures. *World J. Microbiol. Biotechnol.* 16: 687–690.

Agriculture: A vision for the future—Probiotics "What's the hype". 2002. Bioag Enews letter. http://www.Bio-ag.com/info/newsletters/enews/enews2. Html. Accessed 22nd Jan, 2008.

American Medical Association. 1977. AMA: American Medical Association, Council on Drugs. Department of Drugs AMA drug evaluations, 4th ed., http://trove.nla.gov.au/version/13821202.

Bulletin of the International Federation. 2001. Bulletin of the International Dairy Federation. 362/2001. Reviewed. Elad D.56 (3).

Charalampopoulos, D., Wang, R., Pandiella, S.S, and Webb, C. 2002. Application of cereals and cereal components in functional foods: A review. *Int. J. Food Microbiol.* 79: 131–141.

Charteris, W.P., Kelley, P.M., Morelli, L., and Collins, J.K. 1997. Selective detection, enumeration and identification of potentially probiotic *Lactobacillus* and *Bifidobacterium* species in mixed bacterial populations. *Int. J. Fd. Microbiol.* 35: 1–27.

Charteris, W.P., Kelly, P.M., Morelli, L., and Collins, J.K. 1998. Development and application of an *in vitro* methodology to determine the transit tolerance of potentially probiotic *Lactobacillus* and *Bifidobacterium* species in the upper human gastro intestinal tract. *J. Appl. Microbiol.* 84: 759–768.

Cochin, B. 2007. Editor-in-chief: Newsletter of the Syndifrais science committee, Scientific coordinator and copywriter: Dr. Denis Mater, Review board: The Syndifrais Science Committee, Science survey: Vanessa Bodot (CERIN).

Demin, A.A., Malinin, V.V., Shataeva, L.K., and Chernova, I.A. 1994. Chromatographic methods for isolation of immunostimulating peptides of milk whey. *Appl. Biochem. Microbiol.* 30: 255–259.

Diebel, R.H. and Niven, C.F. Jr. 1961. Comparative study of Gaffkya homari, *Aerococcus viridans*, tetrad—Forming cocci from meat curing brines and the genus *Pediococcus. J. bact.* 79: 175–180.

Donkor, O.N., Nilmini, S.L.I., Stolic, P., Vasiljevic, T., and Shah, N.P. 2007. Survival and activity of selected probiotic organisms in set-type yogurt during cold storage. *Int. Dairy J.* 17: 657–665.

Effect of *Bifidobacterium longum* BB 536 on Prevention of Influenza virus Infection of Elderly. Morinaga Probiotics News Release. 2006. Annual Meeting of Japan Society for Bioscience, Biotechnology, and Agrochemistry.

FAO/WHO. 2001. Evaluation of health and nutritional properties of powder milk with live lactic acid bacteria. Report FAO/WHO expert consultation, 1–4 October, 2001. Cordoba, Argentina.

FAO/WHO. 2002. Guidelines for the Evaluation of Probiotics in Food Report of a Joint FAO/WHO Working Group on Drafting Guidelines for the Evaluation of Probiotics in Food, London Ontario, Canada, April 30 and May 1, 2002.

Filippo, C., Di Caro, S., Santarelli, L., Armuzzi, A., Gasbarrini, G., and Gasbarrini, A. 2001. *Helicobacter pylori* treatment: A role for probiotics. *Dig Dis.* 19: 144–147.

Fleming, H.P., Walter, W.M. Jr., and Etchells, J.L. 1973. Antimicrobial properties of oleuropein and products of its hydrolysis from green olives. *Appl. Microbiol.* 26: 777–782.

Fuller, R. 1989. Probiotics in man and animals. *J. Appl. bacterial.* 66: 365–378.

Goyette, E.M. and Elder, J.C. 1960. The use of Triparanol to lower serum cholesterol. *Am. Heart J.* 60: 536–538.

Guandalini, S.J. 2011. Probiotics for prevention and treatment of diarrhea. *Clin. Gastroenterol.* 45(Suppl): S149–S153. doi: 10.1097/MCG.0b013e3182257e98.

Gupta, V., Sharma, A., and Nagar, R. 2007. Preparation, acceptability and nutritive value of rabadi—A fermented moth bean food. *J. Fd. Sci. Technol.* 44: 600–601.

Hassan, A.B., Ahmed, I.A.M., Osman, N.M., Eltayeb, M.M., Osman, G.A., and Babiker, E.E. 2006. Effect of processing treatments followed by fermentation on protein content and digestibility of pearl millet (*pennisetum typhoideum*) cultivars. *Pak. J. Nutr.* 5: 86–89. ISSN 1680-5194. © Asia Network for Scientific Information.

Henneberg, W.C. 1904. Zur Kenntris der Milchsaurebakterien der brennerei-Maische, chlichen Magens (Working with twenty-two strains, made an extensive morphological and cultural study of this group. The information contained in his orginal paper (Zeitschrift fur Spiritusindustrie) formed the basis for many subsequent systems of classification. He divided *L. leichmanni* into 3 sub-groups according to the fermentation reactions on 14 carbohydrates.). Zentbl. Bakt. *Parasitkde, Abt II*, 11, 154–170.

Jindal, A.R., Shore, M., Shukla, F.C., and Singh, B. 2004. Studies on the use of channa and paneer whey in the preparation of puras (pan cakes). *Int. J. dairy Technol.* 57: 221–225.

Kailaspathy, K. and Chin, J. 2000. Survival and therapeutic potential of probiotic organisms with reference to *Lactobacillus* and *Bifidobacterium* species. *Immunol. Cell Biol.* 78: 80–88.

Kailasapathy, K., Harmstorf, I., and Phillips, M. 2008. Survival of *Lactobacillus acidophilus* and *Bifidobacterium animalis* ssp. lactis in stirred fruit yogurts. *LWT Fd. Sci. Technol.* 41: 1317–132.

Kailasapathy, K. and Supraidi, D. 1996. Effect of whey protein concentrate on the survival of *L. acidophilus* in lactose hydrolysed yogurt during refrigerated storage. *Milchwissenschaft.* 51: 565–568.

Kalliomäki, M., Salminen, S., Arvilommi, H., Kero, P., Koskinen, P., and Isolauri, E. 2001. Probiotics in primary prevention of atopic disease: A randomised placebo-controlled trial. *Lancet* 357 (9262): 1076–1079.

Kalliomaki, M.A. and Isolauri, E. 2004. Probiotics and down regulation of the allergic response. *Immunol. Allergy Clin. North Am.* 24: 739–752.

Kalui, C.M., Mathara, J.M., and Kutima, P.M. 26 April, 2010. Probiotic potential of spontaneously fermented cereal based foods—A review. *Afr. J. Biotechnol.* 9: 2490–2498.

Kitchell, A.G. and Ingram, M. 1963. Vacuum packed sliced Wiltshire becon. *Fd. Process Packag.* 32: 1–7.

Kosikowski, F.V. and Mistry, V.V. 1997. Bakers, neufchatel, cream, Quark and Ymer. *Cheese and Fermented Milk Foods-Origins and Principles.* Vol. 2, 3rd ed. pp. 147–161. F V Kosikowksi, L.L.C., Wesport, USA.

Leporanta, K. 2001. Developing fermented milks into functional foods. *Innov. Food Technol.* 10: 46–47.

Lily, D.M. and Stillwell, R.H. 1965. Probiotics-growth promoting factors produced by microorganisms. *Science.* 147: 747–748.

Mann, J.I., Harding, P.A., Turner, R.C., and Wilkinson, R.H. 1977 .A comparison of cholestyramine and nicotinic acid in the treatment of familial type II hyperlipoproteinaemia. *Br J Clin. Pharmacol.* 26(2):249–53.

Marklinder, J. and Lonner, C. 1992. Fermentation properties of intestinal strains of *Lactobacillus* of a sourdough and of a yogurt starter culture in an oat-based nutritive solution. *Fd. Microbiol.* 9: 197–205.

Marteau, P., Seksik, P., and Jian, R. 2002. Probiotics and health: New facts and ideas. *Curr. Opin. Biotechnol.* 13(5): 486–489.

Mills, S., Catherine, S., Gerald, F.F., and Ross, R.P. 2011. Enhancing the stress responses of probiotics for a lifestyle from gut to product and back again. *Microb. Cell Fact.* 10(Suppl 1):S19. doi: 10.1186/1475-2859-10-S1-S19.

Nissen, S.E., Murat Tuzcu, E., and Schoenhagen, P. et al. 2004. Effect of intensive compared with moderate lipid-lowering therapy on progression of coronary atherosclerosis. *JAMA.* 291(9): 1071–1080. doi: 10.1001/jama.291.9.1071.

Oberman, H. and Libudzisz, Z. 1998. Fermented milks. In: *Microbiology of Fermented Foods.* Wood, B.J.B. (Ed), Vol. I, Blackie Academic and Professional, London, UK. 308–349.

Parker, F., Migliore-Samour, D., Floch, F., Zerial, A., Werner, G.H., Jolles, J., Casaretto, M., Zahn, H., and Jolles, P. 1984. Immunostimulating hexapeptide from human casein: Amino acid sequence, synthesis and biological properties. *Euro. J. Biochem.* 145: 677–682.

Parvez, S., Lee, H.C., Kim, D.S., and Kim, H.Y. 2006. Bile salt hydrolase and chlolesterol removal effect by *Bifidobacterium bifidum* NRRL 1976. *World J. Microbiol. Biotechnol.* 22: 455–459.

Pefialoza, W., Davey, C.L., Hedger, J.N., and Kell, D.B. 1991. Stimulation by potassium ions of the growth of *Rhizopus oligosporus* during liquid- and solid-substrate fermentations. *World J. Microbiol. Biotechnol.* 7: 260–268.

Plummer, S., Weaver, M.A., Harris, J.C., Dee, P., and Hunter, J. 2004. *Clostridium difficile* pilot study: Effects of probiotic supplementation on the incidence of *C. difficile* diarrhea. *Int. Microbiol.* 7: 59–62.

Probiotics market needs more awareness. 2011. http://www.businessstandard.com/common/news_article.php?leftnm=5&autono 311953.

Richelsen, B., Kristensen, K., and Pedersen, S.B. 1996. Long term (96months) effect of a new fermented milk product on the level of plasma lipo proteins: A placebo-controlled and double–blind study. *Eur. J Clin. Nutr.* 50: 811–815.

Rodriguez, E., Gonzalis, B., Gaya, P., Nunez, M., and Medina, M. 2000. Diversity of bacteriocins production by lactic acid bacteria isolated from raw milk. *Intl. Dairy J.* 10: 7–15.

Russo, J. et al. 2005. Nutraceuticals and functional foods. *Encyclopedia of Physical Pharmacy*, 1st ed. 6: 32–40.

Saarela, M., Mogensen, G., Fonden, R., Matto, J., and Mattila-Sandholm, T. 2000. Probiotic bacteria: Safety functional and technological properties. *J Biotechnol.* 84: 197–215.

Sanders, M.E. 1998. Overview of functional foods; emphasis on probiotic bacteria. *Int. Dairy. J.* 8: 341–347.

Sanders, M.E. 2008. Products with probiotics. www.usprobiotics.org Sciences University of Wales Institute, Cardiff, United Kingdom.

Sedaghat, A., Samuel, P., Crouse, J.R., and Ahrens, E.H. Jr. 1975. Effects of neomycin on absorption, synthesis, and/or flux of cholesterol in man. *J. Clin. Invest.* 55(1): 12–21. doi: 10.1172/JCI107902.PMCID: PMC301712.

Severson, D.K.1998. Lactic acid fermentations. In: *Nutritional Requirements of Commercially Important Microorganisms.* T.W. Nagodawithana and G. Reed (Eds.), Esteekay Assoc., Milwaukee, 258–297.

Shah, N.P. 2007. Functional cultures and health benefits. *Int. Dairy J.* 17: 1262–1277.

Shah, N.P. 2009. Development of artificially carbonated fermented milk. MSc thesis, Anand Agricultural Univ., Anand, Gujarat, India.

Shahani, K.M. and Chandan, R.C. 1979. Nutritional and healthful aspects of cultured and culture-containing dairy foods. *J Dairy Sci.* 62: 1685–1694.

Stamer, J.R., Stoyla, B.O., and Dunckel, B.A. 1971. Growth rates and fermentation patterns of lactic acid bacteria associated with the sauerkraut fermentation. *J. milk Food Technol.* 34:521–525.

Stanton, C., Desmond, C., Coakley, M., Collins, J.K., Fitzgerald, G., and Ross, R.P. 2003a. Challenges facing development of probiotic containing functional foods. 27–58. In: *Handbook of Functional Fermented Foods.* Chap. 11. Farnworth, E.R. (Ed.), CRC press, Boca Raton, FL.

Stanton, C., Desmond, C., Fitzgerald, G., and Ross, R.P. 2003b. probiotic health benefits—reality or myth? *Aust. J. Dairy Technol.* 58: 107–113. Leporanta 2001.

Stanton, C., Gardiner, G., Meehan, H., Collins, K., Fitzgerald, G., Lynch, P.B., and Ross, R.P. 2001. Market potential for probiotics. *Am. J. Clin. Nutr.* 73: 476–483.

Tamime, A.Y. 2002. Fermented milks: A historical food with modern applications—A review. *Eur. J. Clin. Nutr.* 56: S2–S15.

Teh, S.S., R. Ahmad et al. 2010. Enhanced growth of lactobacilli in soymilk upon immobilization on agrowastes. *J. Food Sci.* 75 (3): 155–164.

The benefits of probiotics for your pet. 2010. http://www.FlintriverCom/product Info. Asppi.probiotics-overview. Html. Accessed 8th Jan, 2010.

Timmermann, R., Huys, G., Pot, B., and Swings, J. 2001. Identification and antibiotic resistance of isolates from probiotic products. Abstracts of 101st ASM general meeting. The American Society for Microbiology, Washington, DC, USA, p.c-289.

Walstra, P., Walstra, P., Wouters, J. T. M., and Geurts, T. J. 2005. *Dairy science and technology. Food Science and Technology.* (2nd ed.), CRC Press, UK, 808 pp.

Wang, K.Y., Li, S.N., Liu, C.S., Perng, D.S., Su, Y.C., Wu, D.C., Jan, C.M., Lai, C.H., Wang, T.N., and Wang, W.M. 2004. Effects of ingesting Lactobacillus and Bifidobacterium containing yogurt in subjects with colonized Helicobacter pylori. *Am. J. Clin. Nutr.* 80: 165–172.

Weintraub, M., Breckenridge, R.T., and Griner, P.F. 1973. The effects of dextrothyroxine on the kinetics of prothrombin activity: Proposed mechanism of the potentiation of warfarin by D-thyroxine. *J. Lab. Clin. Med.* 81: 273–279.

Wood, P.J. 1997. Functional foods for health: Opportunities for novel cereal processes and products. In: *Cereals: Novel Uses and Processes.* Campbell, M.G., Mckee, L.S., and Webb, C. (Eds.), Plenum, NY. 233–239.

Yong, F.M. and Wood, B.J.B. 1974. Microbiology and biochemistry of the soy sauce fermentation. *Adv. Appl. Microbiol.* 17: 157–194.

Ziemer, C.J. and Gibson, G.R. 1998. An overview of probiotics, prebiotics and synbiotics in the functional food concept: Perspectives and future strategies. *Int. Dairy J.* 8: 473–479.

Molecular Characterization and Quantification of Microbial Communities in Wastewater Treatment Systems

Jashan Gokal, Oluyemi Olatunji Awolusi, Abimbola Motunrayo Enitan, Sheena Kumari, and Faizal Bux

CONTENTS

ABSTRACT

Understanding the biological nature of any given environment through characterization of the microbial populations therein is of critical importance to illuminate that particular ecological system. This need for characterization is particularly essential for complex wastewater environment systems, wherein both the distribution and quantity of the native microbial population can determine the success or failure of the system. Although the microbial groups found in many biological wastewater treatment systems are analogous, their relative quantities and metabolic activity can vastly differ between systems. In systems that are directly mediated by microbial activity, even small changes in the microbiota can produce instability, thereby affecting the performance of the system as a whole. Since much of the microbiota found within the wastewater system cannot be quantified through traditional microbiological methods, highly specialized molecular methods based on the screening of group-specific genomic biomarkers have been developed. Owing to the sheer complexity of environmental samples, many of these techniques were initially developed for medical research, and later adapted for environmental microbiological analysis. As such, many of these techniques require significant modification and optimization to suit each type of environmental sample type. This chapter reviews the diverse molecular methods specifically developed to analyze microbial populations within the wastewater system, including a practical approach to detect and quantify specific microbial community, a guideline for obtaining unquestionable data, and a comprehensive list of the primers and probes used for the key functional group in AS microbiota.

INTRODUCTION

Over the course of the preceding millennium, the Industrial Revolution coupled with a human population explosion has resulted in a manifold increase in the amount of wastewater produced. Consequently, the polluted wastewater resulting from anthropogenic activities alone has far exceeded any natural capacity for remediation. As such, the human endeavor to reclaim and reuse some of this wastewater forms a large body of scientific inquiry, facilitates a thriving industry, and inspires vast political discourse. Current technologies for the treatment of wastewater utilize physical, chemical, and biological treatment methods, or some combination thereof. Despite the success of chemical and physical treatment methods, their cost makes them prohibitive for large-scale application. Thus, the biological approach is still the most widely used and preferred method for efficient, economical, and environmentally low-impact wastewater remediation (Akpor and Muchie, 2010; Amenaghawon and Obahiagbon, 2014).

The principles underlying biological wastewater treatment have made it a suitable and sustainable treatment approach. All biological systems constitute a complex microbial community that works synergistically to degrade organic and inorganic components in the influent wastewater. These biological systems manifest themselves in a multitude of designs and process configurations, often as combinations of aerobic- and anaerobic-based technologies (Akpor et al., 2014). The main aerobic technologies applied in wastewater treatment include the AS process, rotating biological contractors, trickling filters, constructed wetlands, stabilization ponds, and membrane bioreactors. In contrast, anaerobic treatment involves the treatment of wastewater in a closed system in the absence of oxygen, to reduce the pollutant load and promote the production of biogas through the activities of different types of microorganisms (Enitan et al., 2014a).

Although biological treatment has significant advantages over other processes, the efficiency of any type of biological treatment approach is directly dependent on the microbial diversity, abundance, and activity within the system. As such, a proper analysis of this microbial community is critical for optimum performance of the system, and by extension, the efficiency of the system in removing organic and inorganic compounds.

Many of the microorganisms within the wastewater treatment plants (WWTPs) are currently non-culturable and cannot be studied with the conventional microbiological techniques, thus making a full characterization difficult. Contemporary advanced molecular technologies, especially in the realm of high-throughput DNA metagenome sequencing are thus the best alternative to fully understand the functions, structure, activities, and dynamics of the microbial community in their natural environments.

The ecophysiology of the major microbial communities present within the WWTP and the standard molecular methods that are used to characterize and quantify these microorganisms are discussed in this chapter. New advancements in molecular techniques such as the next-generation sequencing (NGS) that are useful for creating unique microbial community fingerprints are also described in order to provide a comprehensive overview of the current available detection methods.

AS SYSTEMS

The AS system represents a component of the largest biotechnology industry by footprint worldwide (Seviour et al., 1998). In its most basic form, the AS process has been in existence for more than 100 years, and its versatility has lent itself to multiple process configurations and iterations as treatment needs and technologies have changed over time. The AS itself refers to a metabolically active biomass of microorganisms coated in a thick extracellular polymeric layer. This biomass consists of a mixed population of synergistically complementary microorganisms that remediate wastewater by actively metabolizing the biodegradable organic and inorganic compounds entering the system. The conventional activated sludge (CAS) system was originally designed to remove carbonaceous organic compounds and ammonia (Jeppsson, 1996). When left unchecked, these nutrients in excess could elicit eutrophication events within the receiving water bodies which could result in serious disruption of the aquatic ecosystem. Consequently, safeguarding our aquatic environment by focusing on the degradation of the excessive pollutant load in an environmentally manageable fraction is a primary goal of any wastewater treatment process.

To improve both flexibility and efficiency of this task in the AS process, several modifications in operational features and design have been carried out over the years. Many AS plants are now designed to achieve nitrification, denitrification, and enhanced biological phosphorous removal (Akpor and Muchie, 2010). In many cases, nutrient removal is a multistep process depending on the physiology, function, and microbial diversity in

AS treatment systems. Optimum and efficient nutrient removal not only hinges on proficient process control, but also on a better understanding of the structure and dynamics of the microbial community structure within (Xia et al., 2008; Hu et al., 2012). The microbial community composition of the constituent microbes responsible for the treatment of wastewater is determined primarily by the type of wastewater treated, the constituents of that particular type of wastewater, the treatment plant configuration, the type of treatment process, and a whole array of other factors that contribute to its distinctive fingerprint (Vieno et al., 2007; Vandenberg et al., 2012). While the composition of the microbial community is dependent on a number of abiotic factors acting within each treatment system, there are some common bacterial clades across all conditions. These particular bacterial populations often have very specific nutritional requirements and are key players in nutrient removal, and will thus be outlined further.

THE KEY PLAYERS WITHIN THE AS MICROBIOTA
Nitrifiers and Denitrifiers

The nitrifying bacteria are an extremely important group of microbes that are especially critical within the wastewater treatment system as they mediate the nitrification process. Termed nitrifiers, this group includes both bacterial and archaeal members, which collectively function to mineralize organic nitrogen. Nitrification capacity by other organisms (protozoa, algae, and fungi) has been also reported; however, it is usually at a relatively low rate (about 1,000–10,000 times less) as compared with rates that have been reported with true nitrifiers (Gerardi, 2002; Nicol and Schleper, 2006). Nitrification is a biological process that involves two sequential steps: the oxidation of ammonia (NH_3) to nitrite (NO_2^-) and subsequently to nitrate (NO_3^-) (Figure 4.1). A consortium of ammonia-oxidizing bacteria (AOB) is usually involved in the first rate-limiting ammonia-oxidizing step, whereas the nitrite-oxidizing bacteria (NOB) subsequently oxidize nitrite (NO_2^-) to nitrate (NO_3^-) (Ramond et al., 2015).

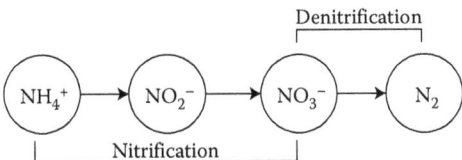

FIGURE 4.1 Nitrogen removal pathway.

Molecular characterization of the key organisms involved in nitrifica-tion through comparison of 16S rDNA sequences has shown that these two groups, AOB and NOB, are phylogenetically distinct (Daims and Wagner, 2010). All AOB can be classified in the β-subclass of proteobacteria with the sole exception of *Nitrosococcus*, which belongs to a distinct branch of the γ-subclass. The NOB can be found within the α- and γ-subclasses of pro-teobacteria, with the exceptions of *Nitrospira*, which has its own distinct phylum (Duan et al., 2013), and *Nitrospina* which belongs to the δ-subclass of proteobacteria (Zeng et al., 2012). For over a century, it was assumed that ammonia oxidation was limited to the β- and γ-proteobacteria and to some extent the heterotrophic nitrifying bacteria. The recent discovery of a homologous archaeal amoA gene has shown that there are still many organisms capable of nitrification that have yet to be discovered (Francis et al., 2005; Pester et al., 2011). As an example, *Candidatus Nitrosopumilus maritimus* belonging to the phylum *Thaumarchaeota* was the first AOA to be isolated (Stahl and de la Torre, 2012), exhibiting the same growth and cell production rates as AOB (Figure 4.2), and similarly using ammonia as

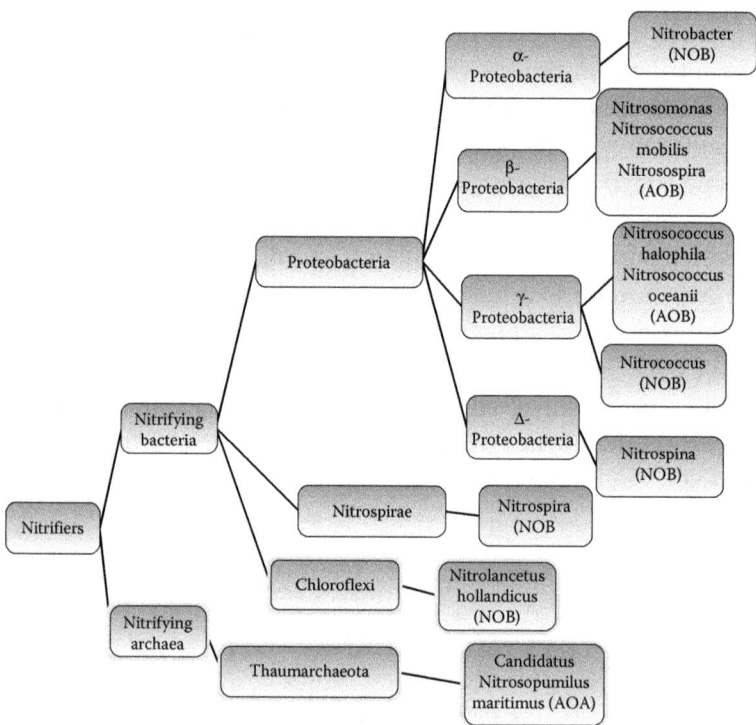

FIGURE 4.2   The schematic representation of nitrifiers in AS microbiota.

its sole energy source (You et al., 2009). The discovery of these new species is not limited to only the AOB. They reported the discovery of *Nitrolancetus hollandicus*, a brand new NOB that belongs to the phylum Chloroflexi—a phylum not previously known to contain any nitrifier (Sorokin et al., 2012) (Figure 4.2). During the last decade, groundbreaking findings in microbial ecology, such as bacteria capable of oxidizing ammonia anaerobically called anaerobic ammonia oxidation (anammox) (Terada et al., 2011) and the existence of above-mentioned ammonia-oxidizing archaea (AOA) (Biller et al., 2012), have proved that our knowledge regarding the nitrogen cycle and related microbial key players is still limited.

Denitrifying bacteria which reduce nitrate or nitrite to gaseous nitrogen compounds, i.e., NO, N_2O, and N_2 are collectively referred to as denitrifiers. Recently, more studies are focusing on denitrifiers since they have large potential of contributing to nitrous oxide (N_2O) emissions (a potent greenhouse gas) that to the climate change burden and stratospheric ozone destruction (Black et al. 2016). The denitrifiers are distributed across more than 50 phylogenetic bacteria genera, which includes Proteobacteria, Firmicutes, Actinobacteria, Bacteroides, and Planctomyces (Yu et al., 2014; Kim et al., 2013). Due to their high taxonomic diversity, molecular techniques targeting functional genes that encode key enzymes involved in the denitrification process have been established as molecular markers. These targeted gene clusters include nitrate reductase (*NarG*), nitrite reductase (*NirK*), nitric oxide reductase (*NorB*), and nitrous oxide reductase (*NosZ*).

Anammox Bacteria

As previously mentioned, cycling of nitrogen compounds through the biosphere is made up of a number of successive catabolic and anabolic processes, which are carried out by a huge variety of microorganisms. Over the past decade, as molecular and analytical techniques have improved, previously unknown or overlooked microorganisms have shown themselves to be significant players in the global nitrogen cycle.

The anammox species is one such bacterial clade that has completely shifted the previously understood paradigm of the nitrogen cycle. Until the latter half of the twentieth century, ammonium was only considered metabolically active under highly oxygenated conditions, and thus the sole pathway for ammonia removal lay through the route of aerobic nitrification (Jetten et al., 2009). Thermodynamic calculations by Broda (1977), however, theorized the existence of a process whereby ammonia could be oxidized via a nitrate- or nitrite-mediated pathway. The actual organism that executed this pathway was only elucidated 22 years later by Strous et al. (1999), who first described the anammox organism from a laboratory-enriched anaerobic digester sample (Jetten et al., 1998). Simply termed "anammox" bacteria—as an abbreviation of anaerobic ammonium oxidation, the anammox clade is uniquely able to convert ammonia to dinitrogen gas under anoxic conditions using nitrite as the electron acceptor and through a hydrazine intermediate (Strous et al., 1999; Schmidt et al., 2003; Kartal et al., 2004; Jetten et al., 2009). The full reaction for anammox is represented in the following equation:

$$NH_4^+ + NO_2^- \rightarrow N_2 + NO_3^- + Biomass$$

The anammox bacterial species is a fairly recent discovery within the relatively ubiquitous phylum Planctomyceteales. These Planctomycetes themselves have been divided into eight culturable genera: *Pirellula*, *Gemmata*, *Planctomyces*, *Isosphaera*, *Blastopirellula*, *Rhodopirellula*, *Schlesneria*, and *Singulisphaera*, and five unculturable *Candidatus* genera—to which the anammox species belongs (Table 4.1) (Jetten et al., 2009; Shu et al., 2011).

TABLE 4.1   List of Five Unculturable *Candidatus* Genera in the Phylum Planctomyceteales

Genus	Species	Electron Acceptors	References
Brocadia	Candidatus Brocadia anammoxidans	NO_2^-	Strous et al. (1999)
	Candidatus Brocadia fulgida	NO_2^-	Kartal et al. (2007)
	Candidatus Brocadia sinica	NO_2^-	Hu et al. (2012)
Kuenenia	Candidatus Kuenen stuttgartiensis	NO_2^-	Schmid et al. (2000)
Scalindua	Candidatus Scalindua brodae	NO_2^-	Schmidt et al. (2003)
	Candidatus Scalindua wagneri	NO_2^-	Schmidt et al. (2003)
	Candidatus Scalindua sorokinii	NO_2^-	Kuypers et al. (2003)
	Candidatus Scalindua arabica	NO_2^-	Woebken et al. (2008)
	Candidatus Scalindua sinooifield	NO_2^-	Li et al. (2010)
	Candidatus Scalindua zhenghei	NO_2^-	Hong et al. (2011)
	Candidatus Scalindua richardsii	NO_2^-	Fuchsman et al. (2012)
Jettenia	Candidatus Jettenia asiatica	NO_2^-	Tsushima et al. (2007)
Anammoxoglobus	Candidatus Anammoxoglobus propionicus	NO_2^-	Kartal et al. (2007)
	Candidatus Anammoxoglobus sulfate	SO_4^{2-}	Liu et al. (2012)

TABLE 4.2   Anammox Oddities

Feature	Common Prokaryotes	Anammox	References
Cell wall composition	Peptidoglycan	Protein	Strous et al. (1999), Jetten et al. (2003)
Intracellular compartmentalization	Absent	Presence of a defined anammoxasome and nuclear envelope	Jetten et al. (2003)
Membrane structure	Polar or nonpolar lipids	Additional ladderane lipid bilayer	Rattray et al. (2008)
Sterol synthesis	Absent	Present	Fuerst and Sagulenko (2011)
Hydrazine oxidation	Unable	Capable	Fuerst and Sagulenko (2011)

Of the currently elucidated species, two *"Candidatus"* genera Brocadia and Kuenenia are the most relevant for wastewater treatment, while the others are more commonly found in freshwater or marine ecosystems where they play a central role in nitrogen cycling (Bagchi et al., 2012; Lotti et al., 2012).

Anammox bacterial species are extremely slow growing, strictly anaerobic chemoautotrophs. They have a doubling time of between 11 and 20 days under reactor conditions; however, this may be even slower *in situ* due to suboptimal growth conditions, inhibition, or competition (Jetten et al., 2009). As traditional isolation techniques are inadequate at dealing with slow growing or as yet nonculturable bacterial species, the study of anammox bacteria is often centered on indirect measurements of N-species transformation within a system, or through the use of specific molecular techniques. The lack of pure cultures of anammox bacteria has made the genomic approach slightly less straightforward; however, the anammox species, like the other members of its parent phylum (the Planctomycetes), have several peculiarities that define them (Table 4.2). It is these oddities that are often targeted as unique biomarkers for their detection and quantification.

Denitrification

The denitrification process completes the nitrogen cycle and acts as a complementary metabolic process to that of conventional nitrification. Denitrifcation is a heterotrophic process carried out in anaerobic conditions, and is responsible for the final conversion of NO_3 to N_2 gas, in the presence of organic carbon. It was found that many organisms actually possess genes for denitrification as a secondary metabolic pathway, and thus denitrifiers as a group will not be discussed in this chapter.

PAOs and GAOs

The polyphosphate accumulating organisms (PAOs) also form a major clade within the enhanced biological phosphate removal (EBPR) pathway. They form a competitive relationship with the glycogen accumulating organisms (GAOs), with the PAOs and GAOs being in competition for carbon substrates in the anaerobic phase (Ong et al., 2014). Molecular-based studies have revealed the Candidatus "*Accumulibacter phosphatis*" as the significant/dominating PAOs, while "*Competibacter phosphatis*" have been revealed to be the dominating GAOs (Crocetti et al., 2002; Oehmen et al., 2007). Molecular analysis of PAOs focus primarily on the polyphosphate kinase functional gene (*ppk1*), but culture independent studies for these groups are still limited and will thus not be covered in detail within this chapter (He et al., 2007).

Filamentous Bacteria

The filamentous bacteria represent a diverse group of nonculturable bacteria, which play a critical role in the functioning of the wastewater treatment process. Filamentous bacteria serve as a "backbone" to the floc structure, allowing the formation of larger, stronger flocs that settle well resulting in a low turbid supernatant (Jenkins et al., 2004). Though filamentous bacteria are beneficial for typical floc formation, excessive growth of these organisms can lead to sludge settling problems such as bulking or foaming incidents that cause poor settling of AS within the secondary clarifier (Tandoi et al., 2006). The proliferation of these filamentous organisms is influenced by operational factors that include: mean cell residence time (MCRT), food to microorganism (F/M) ratio, presence of an unaerated zone (either anaerobic or anoxic) preceding the aeration basin, dissolved oxygen (DO) concentrations, nitrogen and phosphorus concentrations, pH, sulfide concentrations, and the nature of organic substrates (soluble or particulate and readily or slowly biodegradable) in the WWTP (Blackbeard et al., 1986; Seviour et al., 1994; Eikelboom et al., 1998). Table 4.3 presents the parameters commonly correlated to the dominant bulking or foaming filamentous bacteria that are conventionally identified, but not characterized and classified.

A large diversity of different types of filamentous bacterial species have been observed in domestic WWTW, with an even higher diversity present in industrial WWTW. In fact, more than 30 different hydrophobic filament morphotype growths that cause sludge bulking have been observed in AS systems treating primarily municipal wastewater (Vanysacker et al., 2014). Most frequently observed bulking cases include *Gordonia*,

TABLE 4.3 Operational Parameters Favoring the Preferential Proliferation of Certain Filamentous Bacteria

Operational Factor	Filaments
Low DO concentration (DO < 1 mg/L)	*Sphaerotilus natans* Type 1701* *Haliscomenobacter hydrossis*
Low F/M ratio (the minimum and maximum F/M ratio range varies due to the process design, retention times, and influent type of any given system)	Type 0041 Type 0675 Type 1851 Type 0803
Septicity	Type 021N *Thiothrix* I and II *Nostocoida limicola* I, II, III Type 0914 Type 0411 Type 0961 Type 0581 Type 0092
Grease and oil	*Nocardia* spp. *Microthrix parvicella* Type 1863
Nutrient deficiency 　Nitrogen	Type 021N *Thiothrix* I and II *Nostocoida limicola* III
Phosphorus	*Haliscomenobacter hydrossis* *Sphaerotilus natans*

Source: Jenkins, D., Richard, M. G., and Daigger, G. T. 1993. *Manual on the Causes and Control of Activated Sludge Bulking and Foaming.* 2nd ed. Chelsea, MI: Lewis Publisher.
* NB. The designation type indicates filamentous morphotypes that have been conventionally identified but not characterized and classified.

Nostocodia, Microthrix, Thiothrix, and *Sphaerotilus* (Marrengane et al., 2011; Kim et al., 2013). *Microthrix parvicella, Gordonia amarae,* Type 0041, Type 0092, *Sphaerotilus natans,* and Type 021N are the most commonly reported ones (Kappeler and Gujer, 1992; Madoni et al., 2000; Jenkins et al., 2004; Kumari et al., 2009). Although these bacteria have been widely characterized morphologically, the absence of a pure culture has hampered attempts at a more thorough molecular characterization.

MOLECULAR IDENTIFICATION AND QUANTIFICATION OF MICROBIOTA WITHIN A WWTP

Identifying and quantifying the complex microbial community in wastewater treatment systems are essential to understand and effectively exploit the biochemical transformation pathways offered by these microbes.

Traditional culture-dependent techniques require isolation of the organism of interest using selective media, and this approach has historically had its merits. Problematically, a great majority of the Key organisms in AS remain nonculturable in pure form, and this has made the traditional culture-dependent methods highly biased (Awolusi et al., 2015). Most alarmingly, these traditional methods severely underestimate the amount of microorganisms in a system as many of the AS communities are difficult to culture *ex situ*, due to slow growth rate, restricted environmental conditions, and selective nutritional requirements (Briones and Raskin, 2003). Furthermore, conventional techniques severely limit the ability to study the functional gene regulation and the population dynamics that contribute to creating an effective system.

To overcome these limitations, advanced molecular biology methods have been developed (Figure 4.3) that allow for rapid, accurate, specific, and direct identification and quantification of target microbial consortia present in complex environmental samples (Amann et al., 1990; Akpor et al., 2014; Awolusi et al., 2016). These methods can be divided into two main types: quantitative and qualitative molecular techniques, which have all been successfully adopted to study the complex microbial populations in WWTPs (Pernthaler et al., 2002; McHugh et al., 2003; Bialek et al., 2011; Ziganshin et al., 2011). The qualitative techniques include polymerase chain reaction (PCR), PCR-based denaturing gradient gel electrophoresis (PCR-DGGE), temperature gradient gel electrophoresis (TGGE), terminal-restriction or amplification fragment length polymorphism (T-RFLP/AFLP), and cloning.

Quantitative/real-time PCR (qPCR), fluorescence *in-situ* hybridization (FISH), and flow cytometry are the most common quantitative techniques available (Figure 4.3). Gene amplification techniques including real-time PCR (RT-PCR), competitive quantitative PCR (QC-PCR), and most recently droplet digital PCR (dPCR) have also been used for the quantification of microbial ecology in WWTPs with some degree of success (Rački et al., 2014) (Figure 4.3). The techniques and composite outline of the MIQE guidelines for qPCR, RT-qPCR, and dPCR will also be mentioned subsequently. Furthermore, these amplified nucleic acid fragments need to be sequenced, and compared with known sequences in the GenBank database for identifying related microorganisms. Sanger sequencing is the method of choice for this, however it is limited to small sequence reads from highly purified gene products. The sheer quantity of reads generated from complex environmental samples makes Sanger sequencing too expensive and inefficient for wastewater community analysis. Recently, the application of NGS

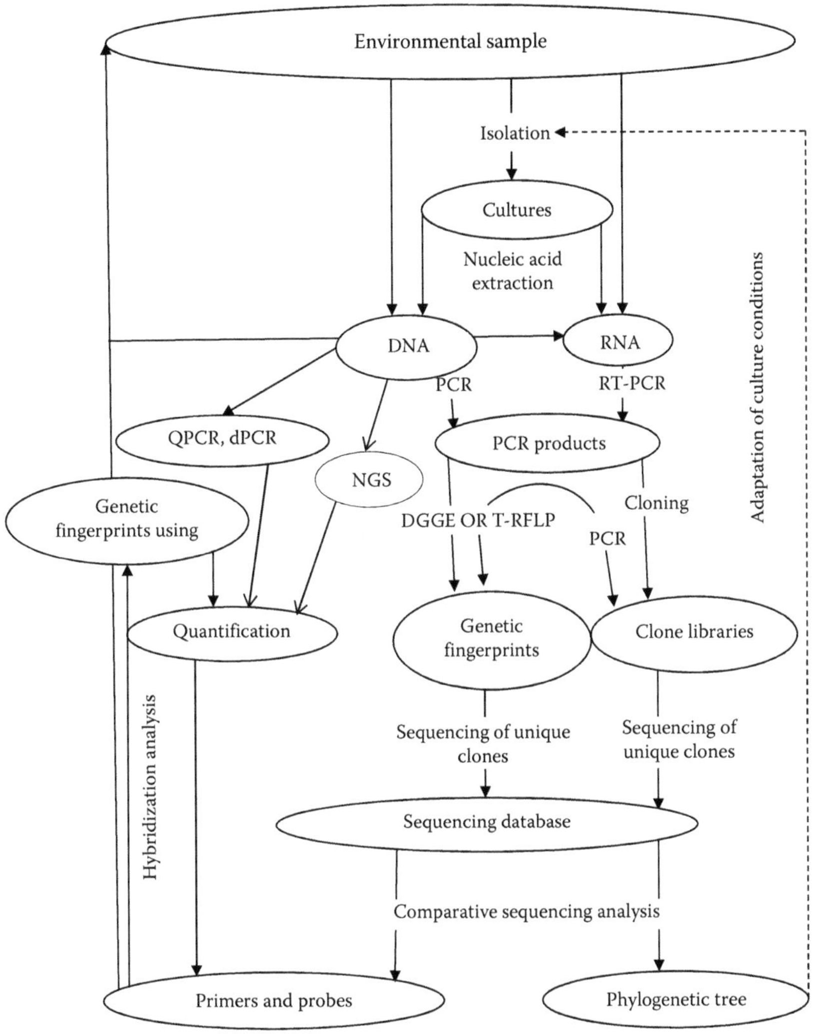

FIGURE 4.3 Flow diagram of different techniques used in studying the structure and function of microbial communities in environmental samples.

technologies to determine the composition and gene content of the microbial population in WWTPs have become the most innovative platform in environmental microbiology (Pernthaler and Pernthaler, 2007; Krober et al., 2009; Enitan et al., 2014c; Nakasaki et al., 2015).

Quantitative PCR for Detection of Wastewater Microbes

In molecular biology, DNA or RNA extraction is often used as a starting point for all downstream molecular genomic techniques (apart from flow cytometry and FISH which require whole cells). Wastewater samples are notoriously difficult to extract pure, clean DNA/RNA due to the inherent nature of the sample itself. Many laboratories develop their own DNA/RNA extraction methods that are specific to their samples, while others use commercial kits. The topic of DNA/RNA extraction and optimization steps, and the synthesis of a complementary DNA (cDNA) strand from the RNA extracts are too broad to be discussed in the context of this chapter. Instead the focus will be on PCR and its derivatives. In traditional (end point) PCR, detection and quantification of amplified sequences of a cDNA or gDNA template are carried out at the end of the PCR assay. Post-PCR analysis is often qualitative and the success of a reaction is often judged by comparing the amplified product with a known size standard. An approximate quantification of the initial target may be carried out by extrapolating from the final amplicon concentration; however, due to biases of conventional PCR and some inherent limitations of the technique, this end-point concentration cannot be used to accurately quantify the initial concentration of template DNA or RNA.

To improve the specificity, sensitivity, and speed of detecting PCR-amplified products, qPCR was developed (Tenover and Moellering, 2007). The principle of qPCR is similar to that of conventional PCR in that the target gene is amplified over a defined number of typical PCR. However, unlike the conventional PCR technique that uses only end-point detection, real-time assays measure the amount of amplified DNA after each cycle as fluorescence resonance emission using a fluorescent dye or probe. Since the accumulating fluorescent signal is measured by the instrument in direct proportion to the number of amplified PCR amplicons generated, a detected change in fluorescence intensity reflects the concentration of amplified gene in real time (Alvarado et al., 2014; Musa, 2014). The qPCR thus shows a great benefit over endpoint PCR analysis since the data being collected during the exponential amplification phase in real time allows users to determine and obtain quantitative information on the initial starting quantity of the amplified target with great precision (Kim et al., 2013).

The real advantage of qPCR is that it can be used to amplify and simultaneously quantify a gene target by using a PCR-based technique that enables one to quantify the number of gene copies or relative number of gene copies in a DNA sample. Quantification of either a DNA or RNA target using qPCR is performed in two ways: absolute and relative quantification.

- Relative quantification is based on a comparison between changes in the expression of a specific functional gene in relation to a constitutively expressed housekeeping gene (Yu et al., 2013; Alvarado et al., 2014). This method is rarely used for total microbial quantification, but more routinely for monitoring changes in gene expression levels within cells (Wong and Medrano, 2005). The two most widely used relative quantification methods are the *Pfaffl* mathematical model and the comparative $\Delta\Delta CT$ method for evaluating the mean normalized gene level from the obtained qPCR results (Pfaffl et al., 2004; Ahiamadu, 2007; Onwughara et al., 2011; Uyom et al., 2014). A more comprehensive guide to these relative quantification methods in qPCR can be easily found elsewhere (Livak and Schmittgen, 2001; Ma et al., 2001).

- Absolute quantification involves the construction of a linear standard curve for direct quantification along this curve. The instrument generates an amplification plot by delimiting fluorescence signals from each sample against the cycle number to create accumulation plots of products (Figure 4.4a). The determination of any unknown concentrations of the target gene in a sample is based on the relationship between the defined copy number of the standard and their corresponding fluorescence intensity, over the duration of RT-PCR assay (Figure 4.3) (Dhanasekaran et al., 2010).

Absolute quantification is the most commonly applied method and is widely used to quantify the microbial population in natural environments, especially for the monitoring of microbial dynamics in WWTPs (Yu et al., 2006; Stams et al., 2012; Kim et al., 2013). In fact, the absolute quantification method allows for the inference of many other important qualitative observations: an early detection of amplification in the cycle signifies the abundance of target RNA or DNA in environmental samples; while amplification is observed much later in the cycle if the target sequence is scarce. The standard curve can be prepared from single-stranded DNA, cDNA/double-stranded DNA, genomic DNA from pure culture strains, recombinant purified plasmid of representative strains, PCR fragments of the gene of interest or commercially synthesized DNA fragments (Figure 4.4b) (Wong and Medrano, 2005). Once prepared, these standards are easy to handle, reproducible, and can be prepared in high amounts for a large number of quantification assays (Yu et al., 2005).

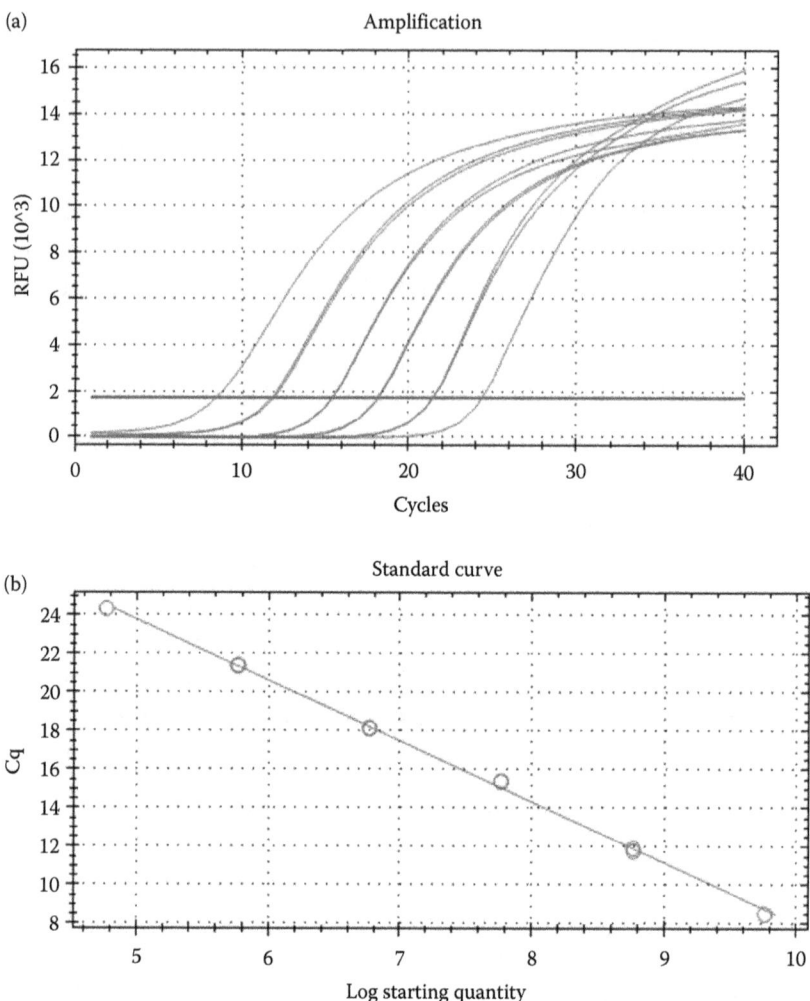

FIGURE 4.4 Principle of a qPCR application using the standard curve method for absolute quantification. (a) Fluorescence intensity changes during amplification of the target gene by using seven standard solutions from 10^1 to 10^7 target gene copy numbers per reaction. (b) Cq values of the standard curve for absolute quantification.

The most widely used and economical method for standard preparation is through the use of purified plasmid genes as a template for the standard curve. Equation 4.1 can be used to calculate the target 16S rDNA gene copy numbers in each standard plasmid DNA (Yu et al., 2006; Enitan et al., 2014b). An average molecular weight of 660 Da was assumed per double-stranded copy, and 6.02×10^{23} (Avogadro's number) was assumed as the average number of DNA copies per milliliter of DNA in a standard plasmid (He et al., 2003). For standard curve preparation, the plasmid could be diluted in a 10-fold series using PCR grade water and analyzed in triplicate with its corresponding primer set:

$$DNA\ (copy/mL) = \frac{DNA\ concentration\ (g/mL) \times 6.02 \times 10^{23}\ (copy/mole)}{DNA\ amplicon\ size\ (bp) \times 660\ (g\ DNA/mol/bp)}$$

$$(4.1)$$

For each qPCR assay, the value of the logarithmic starting quality would be plotted against the threshold cycle (Cq) values (Figure 4.4b) and the linear range of the standard curves should be selected based on the slope ($R^2 > 0.950$). The concentration of the target sequence in the unknown sample could then be estimated by interpolation from linear regression of the standard curve, and the amplified gene copy number from the original DNA reflects the relative abundance of the microorganisms in the community (Enitan et al., 2014b).

Detection Chemistry and Sample Preparation for qPCR Assay

Various detection chemistries have been developed that involve different fluorescent molecules, hybridization probes, nonspecific DNA-binding dyes, light-up probes, scorpion primers, sunrise primers, and molecular beacons (Ma et al., 2006; Lim et al., 2013). Each detection method has its own unique mode of operation, amounts of amplified targets, and sample-specific advantages and disadvantages. The most widely used detection chemistries among these for investigating the microbial ecology in environmental samples are based on either the incorporation of DNA intercalating fluorescent dyes or fluorescence resonance emission tagging (FRET) based assays. The most common commercial examples of these are in the form of the SYBR Green 1 dye and TaqMan probe brands, respectively, and will henceforth be referred to as such.

SYBR Green 1 is a fluorescent DNA-binding dye that intercalates with the minor groove of double-stranded DNA (dsDNA) to produce a >1000-fold fluorescent signal over the unbounded dye in a sequence-independent way (Kim et al., 2013). The SYBR Green intercalating dye is easy to use, sensitive, and inexpensive, but it does suffer from some major disadvantages: primer dimers and nonspecific binding that could cause false-positive results and overestimation of the target concentration (Ma et al., 2006).

Specificity of the qPCR assay when the SYBR green dye is used must be confirmed by melting curve analysis (Bustin et al., 2009). This melting curve analysis will identify the presence of primer dimers or any nonspecific products in no-template controls (NTCs) due to nucleic acid contamination of reagent components. It will also differentiate the DNA fragments based on the difference in melting temperature (T_m) into separate melting peaks (Kim et al. 2013).

Conversely, FRET chemistry involves the attachment of a fluorescent TaqMan probe to the primers used in the amplification reaction. Using fluorescent probes in combination with the two oligonucleotide primers greatly lowers the chances of false signals and greatly improves the quantification sensitivity of the reaction when compared with SYBR Green dye (Yu et al., 2005). This enhanced sensitivity, however, requires custom designing of primer/probe sets that need to be specific to a gene or target sequence. Not only is this a costly affair, but can also sometimes cause complication due to the interfering regions on the primer, and probe sequences that may both target the sequence of interest (Yu et al., 2005). Molecular beacons are also a popular quantification chemistry that functions in a manner similar to other FRET-based assays. The molecular beacons consist of a single-stranded oligonucleotide probe, with the sequence designed to allow for the formation of a hairpin structure such that the fluorescent dye and the quencher are in close proximity. Upon annealing to a target sequence, the loop opens and the fluorescent reporter and quencher are spatially separated, resulting in a fluorescent signal upon excitation. The amount of signal is proportional to the amount of target sequence, and is measured in real time to allow quantification of the target sequence. Molecular beacons are generally used to detect waterborne pathogens, and are a great deal more sensitive than the TaqMan probes (Ma et al., 2006). They are routinely used to detect single-nucleotide polymorphisms (SNPs) within a target; however, they are often more costly than the other methods and are thus preferentially not used for routine population quantification analysis.

Despite the advantages of TaqMan probes, SYBR Green dye is still the most widely used chemistry due to its easy application, simplicity, low cost, and flexibility compared with other options (Malinen et al., 2003; Kim et al., 2013). However the high degree of specificity of the TaqMan assay has proven an attractive option for investigating mixed culture in environmental samples (O'Reilly et al., 2009; Lee et al., 2010; Kim et al., 2013). A new technology referred to as droplet dPCR has been recently developed and is seen as a more advanced version of qPCR and RT-qPCR as it obviates the need for the preparation of a standard curve. Although it has not yet significantly been used in environmental or wastewater microbial analysis, it shows remarkable potential and is thus outlined below.

Digital PCR: A Brief Overview

dPCR is a recently developed quantitative PCR method that provides a highly sensitive and reproducible way of measuring the amount of nucleic acid in a sample. This method is similar to qPCR with regard to the reaction components, assembly, and amplification protocols, but differs in the way the target is measured. With dPCR, the machine first partitions your sample into hundreds or even thousands of separate reaction chambers/individual wells prior to the amplification step, resulting in either 1 or 0 targets being present in each well (Rački et al., 2014). The current generation of dPCR machines achieve this with micropumps and microvalves that distribute and seal PCR fluid into isolated reaction chambers; with microfluidics plates that funnel each reaction into its own separate well; or by streaming each sample into thousands of nanoliter-sized droplets (Morley, 2014). This step is unique to dPCR and relies on the assumption that sample partitioning will follow a Poisson distribution resulting in 0 or 1 target per well. On completion of sample partitioning, PCR amplification reactions are run to end point. The incidence of presence or absence of fluorescence in the amplified reaction wells is then used to calculate the absolute number of targets present in the original sample. Wells with fluorescent signal are positives and scored as "1"; wells with the background signal are negatives and scored as "0"—hence the term "digital." Algorithms calculate the total number of reactions, giving the precise number of target molecules in your sample the number of positive versus negative reactions (Rački et al., 2014). Poisson statistical analysis is then used to determine the absolute concentration of target present in the initial and final samples.

Although qPCR is currently the molecular quantification benchmark, dPCR offers some significant advantages. These include a higher degree of sensitivity, the fact that it does not rely on a user-generated calibration curve for sample target quantification, nor does it require any reference standards or endogenous controls. Furthermore, both qPCR and dPCR are used to amplify, detect, and count individual nucleic acid molecules; however dPCR is more precise, making it better for quantifying rare genetic mutations, deletions, and duplications in DNA. For example, with dPCR, it is possible to distinguish samples containing 10 copies of a gene from those with 11, while in contrast with qPCR, it is difficult to distinguish even two copies from three.

dPCR for Wastewater and Environmental Samples

Being a relatively new and slightly more expensive technology, dPCR has not yet been widely applied to wastewater research. While qPCR has been used extensively in wastewater analysis for the detection and quantification of the many specific groups within the wastewater system, it is still limited by the reliance on standard reference material for quantification. The reliability and consistency of the utilized standards therefore greatly affect the accuracy of qPCR quantification of the unknown. Recent studies have found a great degree of variation even between commercial standards in that the standard material was responsible for approximately half a log difference in results between vendors (Cao et al., 2013) and twofold between batches within a single vendor (Sivaganesan et al., 2011). As such, lack of access to reliable and consistent standard material has been identified as the biggest obstacle to use qPCR for water monitoring (Cao et al., 2013).

Other limitations of qPCR are also problematic for environmental applications. Many microbiological targets are present in environmental waters only at very low concentrations; thus it is often difficult to detect the target molecules through qPCR alone, as they may fall well below the detection limits of conventional methods. Furthermore, qPCR is susceptible to inhibition from common constituents found in environment samples, which are complex and often contain substances that interfere with PCR amplification (Cao et al., 2013). Since dPCR counts the frequency of positives in small volume partitions, droplet dPCR is less affected by PCR inhibitors that could reduce the amplification efficiency and therefore, more robust against inhibition (Huggett et al., 2013). Since only a small amount of target or nontarget DNA is present in each partition, PCR interference between DNA molecules and substrate competition during amplification of different DNA targets is minimized. This feature could enable cost-saving strategies such as multiplexing to simultaneously quantify multiple targets (Morisset et al., 2013), which may be particularly advantageous for environmental monitoring applications with limited budgets. For accuracy, relevance, repeatability, correct interpretation, and most importantly standardization of qPCR, Bustin et al. (2009) has recently described a set of minimum criteria for qPCR experiments called the minimum information for publication of quantitative real-time experiments (MIQE). This guideline will help in experimental transparency, promote consistency between laboratories, improve the reliability of qPCR analysis and assay characteristics, and standardize the technique for other researchers to be able to reproduce results (Bustin et al., 2009). This will hopefully make comparison of quantitative data between different studies on environmental microbial ecology simpler and more accurate.

IMPORTANCE AND OVERVIEW OF THE MIQE GUIDELINES FOR STANDARDIZATION OF PUBLISHED DATA

The MIQE guidelines represent all the necessary information for standardization and publication of any quantitative PCR method. The original MIQE guidelines were published in 2009 by Bustin et al., who noticed that the lack of standardization between published materials was often incomparable due to insufficient information provided by the authors on diverse reagents, protocols, analysis methods, and reporting formats.

The MIQE checklist consists of 42 points that cover the most serious technical deficiencies including

- DNA extraction
- Experimental design
- Sample storage
- Sample preparation
- Sample quality
- Choice of primers and probes
- RNA
- Target information
- Inappropriate data and statistical analysis

Since its publication, the MIQE aims to

- Instill confidence in the reliability of qPCR
- Standardize the qPCR technique
- Encourage better experimental and reporting practice
- Interpretation of qPCR results

The MIQE guidelines for qPCR publications are divided into the following categories: essential information and desirable information. The essential information must be included in the submitted manuscript or accompanying supplementary material. Desirable information should be included in the submitted manuscript and is intended to help the reader understand the study. A recently published update now provides a checklist for dPCR-minimum information for publication of quantitative digital PCR experiments (dMIQE). The requirements in the MIQE checklist are specific to dPCR-related reaction partitioning and data analysis. Table 4.4 gives a summary of both dMIQE and MIQE as applied to environmental samples.

TABLE 4.4   Summary of dMIQE and MIQE as Applied to Environmental Samples

Parameter	Details
Sample source	Is it a single point source sample, a composite sample? Has the sampling location and sampling time been recorded and is the sampling point consistent across all similar samples?
Method of sample preservation	Was the sample stored or preserved before DNA/RNA extraction What type of preservation protocol was used?
Sample storage	Conditions of storage and time in storage before extraction.
Extraction method	Was DNA extraction performed with a kit or a custom method? If a custom method was used, list: reagents used. Accurate timings, speeds, temperatures, and equipment specifications
Analysis of extract	Nucleic acid analysis with a Nanodrop/fluorometer and an agarose/polyacrylamide gel. Note DNA or RNA contamination/shearing extent. Is the extract good enough to use downstream?
Amplicon details	Size and sequence of the amplicons
Primer sequence/ probe sequence	Protocol reference, nucleotide sequence, accession number, supplier name, purification grade of the primers.
Empirical data	Concentrations of reaction components, annealing temperature, thermo cycling conditions and protocol reference, reagent name, and supplier
PCR efficiency	Curve regression, no secondary products/dimers
Limits of detection	Include both upper and lower detection limits
Intra-assay variation	Copy numbers and standard variation
Duplicate RT	Cq values
NAC	Change in Cq at the beginning and end of the run
Controls	Type of positive control used, use of a negative control, use of a no template control, Cq and melt curve analysis for all
Data analysis	Software name and version, statistical analysis used, normalization procedures

Fluorescent *In-Situ* Hybridization

Nucleic acid hybridization in terms of quantification is primarily based on FISH and derivations thereof. This method targets ribosomal RNA to precisely detect the distribution and abundance of the bacterial population of interest (Shu et al., 2011). Due to its relative sensitivity and speed, FISH has been an essential tool for the elucidation of wastewater microbes. It allows visualization of individual bacterial cells within an environmental sample by combining the specificity of molecular techniques with the visual information obtained from microscopy (Amann et al., 1990). In FISH, a specific fluorescently labeled oligonucleotide probe detects and hybridizes with its complementary target nucleic acid sequence, within the intact cell of interest, which then can be viewed using a specialized epifluorescent microscope or quantified using flow cytometry or microscopic image analysis. The most commonly used target nucleic acid sequence is the 16S rRNA which has a reasonably high copy number and is genetically stable, containing a domain structure with conserved and variable regions (Amann et al., 1990).

The probes developed for FISH consist of between 15 and 30 nucleotide bases and are covalently linked at the 5′ end with a fluorescent dye, for example, fluorescein, tetramethylrhodamine, texas red, and carbocyanine (Amann et al., 2001). Ideal FISH probe offers both high degree of sensitivity and specificity and allow for a fairly high fluorescent signal as opposed to the background fluorescence of nontargeted cells. This will give accurate identification and quantification of the organism of interest within the mixed microbial population at a desired taxonomy level. The specificity of the FISH probe allows for the differentiation between the target organism and nontargeted organisms including those that are closely related which could skew data interpretation (Yilmaz et al., 2010). Many researchers have designed FISH probes for the *in-situ* detection of wastewater microbes from domain level to different species, which can be accessed online using probeBase (http://www.microbial-ecology.net/probebase (Loy et al., 2003).

There are three key steps involved in FISH analysis namely

- Stabilization and permeabilization (cell fixation)

- Hybridization of whole cells with target-specific probes

- And microscopic analysis of tagged cells by epifluorescence microscopy, flow cytometry, or scanning electron microscopy (Bokonyte et al., 2003; Ma et al., 2006) (Figure 4.4)

Cell fixation is an important first step as it assures inactivation of bacterial cells and increases cell permeability, while still preserving the rRNA content and cell membrane integrity (Amann et al., 1990; Enitan et al., 2014a) (Figure 4.5).

Hybridization or annealing of the FISH probe to the target molecule is then carried out in a suitable hybridization buffer under conditions that favor duplex formation. This hybridization step is often the most critical step of the FISH procedure, and is the step where most optimization can

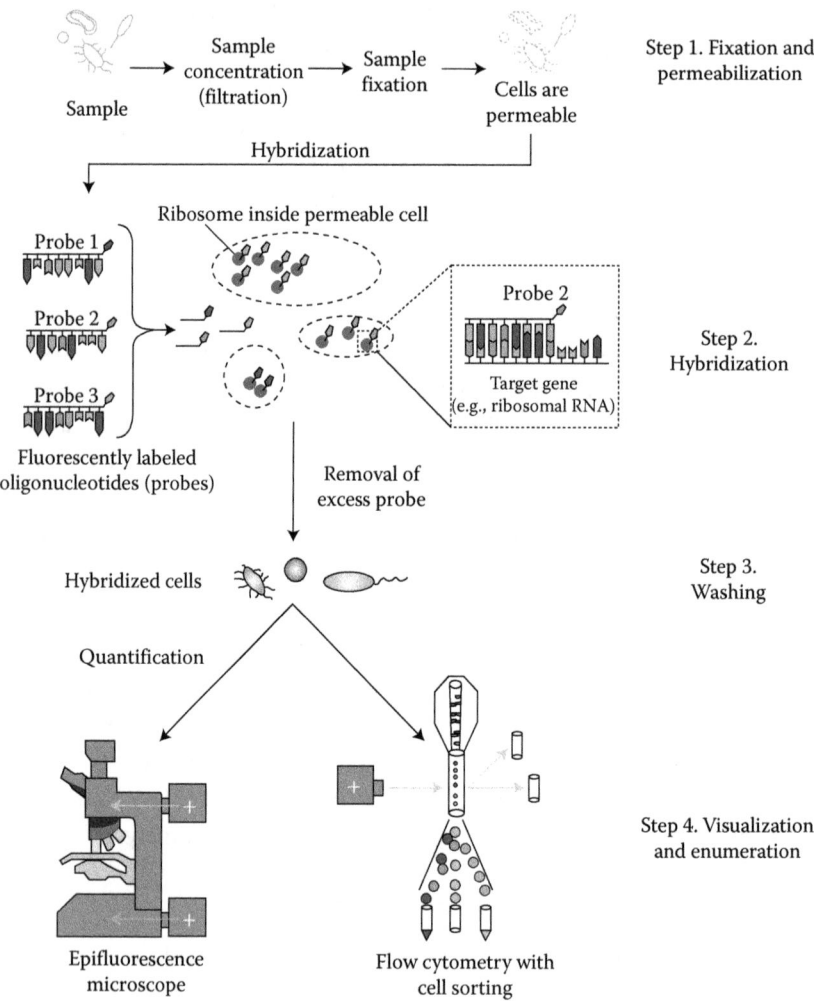

FIGURE 4.5   Flow diagram depicting the steps involved in the FISH procedure. (Interstate Technology and Regulatory Council, 2013. www.itrcweb.org.)

be carried out. The hybridization conditions, probe sequence, and concentrations of the various hybridization buffer components must be stringent such that probe–rRNA duplexes contain minimal mismatches, and results in a successful FISH analysis (Amann et al., 1990; Yilmaz et al., 2010). Some of the key optimization steps carried out at this stage include

- Modification of salt concentrations: Nucleic acids possess an overall negative charge; sodium chloride is incorporated into the hybridization buffer and thus provides the Na^+ anion which neutralizes this charge, thereby facilitating oligonucleotide probe–rRNA duplex formation.

- Enhancing cell permeability: The sequences of the labeled probe recognize the 16S rRNA sequences in fixed cells, sodium dodecyl sulfate facilitates the disruption of the cell membrane, and denatures proteins, thus increasing probe accessibility to the target site (*in situ* DNA–RNA matching).

- Optimized formamide concentrations: The stringency of the probe–rRNA duplex formation is often strongly controlled by the formamide concentration. Thus the formamide concentrations must often be optimized to increase stringency of the duplex formation.

- Optimization of probe concentration: The probe concentration within the buffer is typically 1.5–5 ng/μL, since lower concentrations could result in inefficient probe hybridization and thus reduced sensitivity, while higher concentrations could result in nonspecific binding and thus reduces specificity (Fuchs et al., 1998).

- Optimizing hybridization times: Increased hybridization times can increase probe binding efficiency (Amann et al., 1990; Yilmaz et al., 2010) and typical hybridization times can range from a minimum of 3 h to overnight incubation.

On completion of the hybridization protocol, a washing step is recommended to remove the excess unhybridized probes. Furthermore, an antifade agent that prevents photobleaching of the fluorescent dye is often used to enhance the longevity of the prepared slides. The slides are then viewed using an epifluorescent microscope (Amann et al., 1990; Enitan et al., 2014b). Using epifluorescent microscopy, researchers are able to visualize the spatial arrangement of positive cells by observing the fluorescent signal emitted from fluorescent probe bounded cells. However, it must be noted that the fluorescent dye will only be excited at a wavelength specific to the dye; thus, the type of filters and light source of the epifluorescent microscope must be considered before selecting the fluorophore. The fluorescent intensities of individual cells have also been used to infer changes in activity (Poulsen et al., 1993), but this is not always appropriate for all organisms, such as the AOB that appear to maintain high ribosomal copy numbers despite their physiological status.

Quantification using FISH is a tedious and highly user-biased task. Semiautomated and fully automated analyses are available, but they require the use of expensive epifluorescent or confocal microscopes with the appropriate image analysis software (Daims et al., 2001). The FISH targeted populations are often expressed as percentages of the total bacterial population detected by domain level oligonucleotide probes (EUB mix probes) or by DNA intercalating dyes such as 4',6-diamidino-2-phenylindole (DAPI). For statistical analysis of FISH quantifications via microscopy, a minimum of 10 random images per probe combination must be captured and processed.

The following advantages make the FISH technique the most widely used one for investigating microbial ecology of wastewater treatment:

- The technique allows direct visualization of noncultured microorganisms

- Preferential or differentiation of active microorganisms and dead cells

- In contrast to conventional techniques such as plate counts, most probable number, quantification of specific microbial group is possible

- Relatively basic knowledge of microscopy and laboratory experience are required

However, the FISH technique is relatively time consuming and sometimes difficult to use in complex environmental samples due to low contents of rRNA molecules per microbial cell and the formation of dense microbial clusters. Other shortcomings of FISH technique include

- Nonavailability of probes targeting some bacterial group and background or *a priori* knowledge of expected microorganisms in environmental samples is often required.

- Quantification could be tedious, complex (image analysis), and subjective (manual counting).

- FISH technique may require the use of other confirmatory techniques.

- Detection and quantification of particular organisms require that the rRNA sequence must be known.

- The design of probes that share the metabolic properties of interest is not always easy or possible (e.g., halo-respiring bacteria, nitrifying bacteria).

- Optimization of hybridization conditions is a difficult process that requires time, dedication, and experience.

- Pretreatment and dilution of samples may be required for easy penetration of probes, to avoid fluorescent background, nonspecific binding, and spatial distribution of cells for quantification (especially in environmental and sludge samples).

- Image analysis required expensive confocal microscope and skilled personnel.

- Inadequacy or variability in ribosomal content can result in low signal intensity.

- Loop and hairpin formation of rRNA structure, as well as rRNA–protein interactions can inhibit hybridization.

- Unlike fast-growing microorganisms, the cellular rRNA content of anammox and β-proteobacterial ammonia oxidizers do not really reflect the physiological activity of these organisms, especially during starvation and inhibition periods. Thus, correlation of the nitrifier population to its physiological activity can be biased. Witzig et al. (2002) observed that due to the low food-to-microorganism conditions in membrane bioreactors with the resultant low rRNA molecules for the organisms, less than half of the population were detectable by FISH whereas 80% in conventional AS systems.

Despite the disadvantages of FISH techniques, upgrades and new ideas to overcome some of the shortcomings have been widely published by researchers. These include Spike FISH (Daims et al., 2001), catalyzed reporter deposition-FISH (CARD-FISH) first described by Bobrow et al. (1989) and introduced to the study of microbial ecology of WWTP in 2002 by Pernthaler et al. (2002). An improved CARD-FISH that considered simultaneous hybridization of both rRNA and messenger-RNA (mRNA) in living cells that are present in environmental samples is reported by the same author (Pernthaler, 2005). This improved CARD-FISH technique is extremely important in studying the microbial population of WWTPs because it allows the detection and differentiation of actively participating bacteria in the system excluding any bacteria in the dormant form. Other optimized FISH protocol for quantitative detection of bacteria include the increase of probe signal intensity by polynucleotide FISH (Zwirglmaier, 2005); minimization of probe penetration problems, and increasing hybridization efficiencies with different probes chemistries such as peptide nucleic acid FISH (Perry-O'Keefe et al., 2001) and locked nucleic acid FISH (Kubota et al., 2006). Some newer FISH methods, such as Mar-FISH and CLASI-FISH focus on the multiplexing of probes and the utilization of a CLSM for accurate viewing. Some of the recent publications on the application of FISH for microbial analysis of AS, nitrogen, and phosphorus removal systems will be reviewed in this text.

Terminal-Restriction Fragment Length Polymorphism

T-RFLP is one of the molecular techniques which have been used by the researchers for microbial community structure shift. The technique is based on the restriction banding pattern of an amplified gene, or specifically of the 16S rRNA gene for microbial community analysis. The restriction enzymes are used to cleave the PCR-amplified genes that are already labeled at the terminals (Sanz and Kochling, 2007; Gao and Tao, 2012). It can be used in investigating spatial and temporal shifts in microbial community composition of a given natural or artificial ecosystem (Yang et al., 2011). The technique is highly sensitive and useful for semi-quantitative analysis of microbial populations in a particular ecosystem as an alternative to DGGE (Liu et al., 2010). The microbial fingerprints obtained from T-RFLP are usually insufficient for identification of individual taxonomic units (Yang et al., 2011). However, comparing the generated fragments with a sequence from a public database or a related clone library can be employed in order to sequence and identify the dominant organisms (Yang et al., 2011). However, similar to other PCR-based techniques, the biases related to DNA isolation steps and amplification also can affect the accuracy of this method (Sanz and Kochling, 2007). Some notable examples include Liang et al. (2010) who used the T-RFLP technique successfully to investigate the difference in nitrifier population from two different

MBR systems. Likewise, Amenaghawon and Obahiagbon (2014) designed a T-RFLP technique in monitoring the changes in relative abundance of Accumulibacter clades and other key members of bacterial community in enhanced biological phosphorus removal systems. The bacterial communities of the sequencing batch biofilm reactors having different Anammox start-up inoculum were investigated using TRFLP by Seviour et al. (1998) and they discovered that different genera of anammox bacteria became dominant after 20 and 52 days of operation.

MICROBIAL COMMUNITY ANALYSIS USING NGS-BASED HIGH-THROUGHPUT TECHNIQUE

The traditional Sanger/dideoxy-sequencing approach to process complex environmental samples has shown to be grossly inadequate, due to the hundreds or thousands of important sequences that go unnoticed when employing this method (Shokralla et al., 2012). Due to thousands of potential DNA templates usually present in wastewater samples, there is a strong need for a technique that is capable of simultaneous detection of diverse microbial communities in different DNA templates (Shokralla et al., 2012). The Sanger or dideoxy sequencing method, though useful in its own right, is limited in the quantity of targets that can be sampled because of the read length limitations, purity requirements and expense involved (Mardis, 2008). A major shortcoming of this technique is that it requires *in vivo* amplification of DNA fragments in bacterial hosts prior to sequencing.

Contrarily, the NGS approach offers a speedy, relatively inexpensive alternative with a vastly improved amount of data production. This allows for the investigation of microbial ecology on a larger scale and with more detail than was possible with previously used sequencing technology (Ju and Zhang, 2015). NGS offers the advantage of direct sequencing from environmental samples without a prior cloning step in a bacterial host as in the traditional Sanger approach. Undoubtedly, NGS has revolutionized environmental metagenomic research, with stiff competition between manufacturers for ever improving platforms and technologies allowing for a variety of NGS options at an ever decreasing cost. The platforms that are gaining widespread usage include (Table 4.5): 454/Roche FLX system, the Illumina/Solexa Genome Analyzer, Applied Biosystems SOLiD system, Helicos Heliscope, Ion Torrent Personal Genome Machine (PGM), and Pacific Biosciences (PacBio) SMRT instruments (Mardis, 2008; Liu et al., 2012; Quail et al., 2012; Shokralla et al., 2012).

TABLE 4.5   Advances in DNA Sequencing Technologies and Comparison of Different Next-Generation DNA Sequencers

	Platform		
	Roche(454)	Illumina/Solexa	ABI/SOLiD
Sequencing chemistry	Pyrosequencing	Polymerase-based SBS	Ligation-based sequencing
Amplification approach	Emulsion PCR	Bridge amplification	Emulsion PCR
Read length	200–300 bp	30–40 bp	35 bp
Time/run (paired ends)	7 h	4 days	5 days
Paired ends/separation	Yes/3 kb	Yes/200 bp	Yes/3 kb
Mb/run	100 Mb	1300 Mb	3000 Mb
Cost per run (total directa)	$8439	$8950	$17,447
Cost per Mb	$84.39	$5.97	$5.81

454/Roche FLX System

The 454 platform marketed by Roche Applied Science was the first NGS technology that was made commercially available in 2004, and can be credited for starting the metagenomics era (Mardis, 2008). When it was first released it employed an innovative approach that was termed pyrosequencing—a real-time DNA sequencing technique that monitors DNA synthesis through a series of linked enzymatic processes (Ronaghi, 2001). This technique was revolutionary as it allowed for detecting, identifying, and typing bacteria at a previously unprecedented scale and resolution (Clarke, 2005; Sanapareddy et al., 2009). The 454 technology is capable of generating 80–120 Mb of sequence in 200 to 300 bp reads in a 4 h run. Ye et al. (2011) noted that the traditional molecular techniques do not give a complete profile of the community structure present in the wastewater; however, pyrosequencing has the potential of a truer estimation and more detailed reflection of such communities. Although the technology is slightly dated, it still ranks among the most published next-generation technologies with the main application in WWTPs to investigate the plasmid metagenome and the antimicrobial resistance pattern of the AS (Clarke, 2005; Morozova and Marra, 2008; Hu et al., 2012).

Illumina/Solexa Genome Analyzer

The Illumina/Solexa Genome Analyzer is based on "DNA clusters" or "DNA colonies," which involves the amplification of DNA that has been attached onto a flow cell. After the amplification step, more than 40 million clusters of a flow cell are produced. Each of these clusters contains approximately one million copies of the original fragment, which generates an adequately strong signal density to indicate bases incorporated during sequencing (Mardis, 2008; Morozova and Marra, 2008). The Illumina system adopts the sequencing-by-synthesis (SBS) technology, where DNA polymerase and the four nucleotides are added at the same time to the flow cell channels for incorporation into the oligo-primed cluster fragments (Mardis, 2008; Liu et al., 2012). This SBS technology uses an exclusive, reversible terminator-based method to detect single bases as they are incorporated into DNA template strands and nonincorporated nucleotides are washed away. The fluorescently labeled nucleotides images are captured by the camera, after which the fluorescent dye/terminal 3′ blocker is chemically removed for the next cycle of incorporation to begin (Quail et al., 2012). Unlike pyrosequencing, the DNA chains are elongated one nucleotide at a time, thus image capturing can be done at a delayed time, giving room for very large arrays of DNA colonies to be captured by successive images taken from a single camera (Parmar et al., 2014). The Illumina platform is more effective at sequencing homopolymeric regions than pyrosequencing, due to shorter sequence reads, however the accuracy is still comparable to that of pyrosequencing (Varshney et al., 2009; Ju and Zhang, 2015). Typically, in 2–3 days at least 1 Gb of sequences are produced per run using 1G genome Illumina analyzer, Inc. (capable of generating 35 bp reads). One major disadvantage to this method is that substitution errors have been observed in Illumina sequencing data due to the use of reversible terminators and modified DNA polymerases (Morozova and Marra, 2008).

Ion Torrent Personal Genome Machine (PGM)

Ion Torrent PGM is relatively new technology that was released into the market in 2010. As such, its usage for metagenomic research in wastewater is still limited compared to other established NGS platforms. The Ion Torrent uses semiconductor sequencing technology that works on the principle of detecting hydrogen ions produced during synthesis (Liu et al., 2012). Ion Torrent PGM detects nucleotide addition based on the pH change that occurs as proton is released during this reaction. During sequencing, each of the four nucleotide bases is added in sequence and there is proportional voltage signal detection when matching base is incorporated (Quail et al., 2012; Shokralla et al., 2012). This technology is unique in that unlike other sequencing platforms, it does not rely on modification of nucleotides or optical detection of its reaction process (Liu et al., 2012). Moreover when compared to other platforms it has the shortest run time of between 3.5 and 5.5 h (Shokralla et al., 2012). However, presence of homopolymeric chains or a sequence of identical bases (GGGG) on a DNA template during nucleotide incorporation results in a larger pH change and in turn proportionally large electronic signal production. This usually causes difficulty in differentiating the signals from one high repeat number and the other that is close in terms of number of repeats (Churko et al., 2013). So far this platform has only found few applications in wastewater (Cao et al., 2016).

APPLICATION OF THESE TECHNIQUES IN WASTEWATER TREATMENT AS MOLECULAR TOOLBOX

Although the techniques outlined above are all widely utilized for environmental microbial analysis, accurate quantification of particularly complex wastewater samples often requires the application of many of these techniques in unison. Each technique offers its own set of advantages and disadvantages, and combining these techniques for the analysis of a single sample greatly minimizes the error margin on any single technique alone. Several key wastewater bacterial populations have been identified and quantified using FISH techniques employing different oligonucleotides probes (Table 4.6) (Matsumoto et al., 2010; Ramdhani, 2012; Zielińska et al., 2012; Benakova and Wanner, 2013; Awolusi et al., 2015). Principally among wastewater samples, FISH has been particularly effective for elucidation of filamentous samples. Many researchers have applied fluorescence *in-situ* hybridization for identifying and quantifying filamentous bacteria in different types of WWTPs (Wagner et al., 1994; Kragelund et al., 2007; Mielczarek et al., 2012; Deepnarain et al., 2015; Miłobędzka and Muszyński, 2015), and recently Mielczarek et al. (2012) applied FISH technique to quantify filamentous bacteria, specifically type 0092 and type 0803, in different Danish WWTPs. In this study a full-scale BNR plant treating primarily domestic wastewater with bulking problem was investigated for filamentous bacterial growth under various plant operating parameters using FISH techniques by Deepnarain et al. (2015). Eikelboom Type021N, *Thiothrix* spp., Eikelboom Type 1851 and Eikelboom Type 0092 were found to be dominant in the investigated plant. Another study on the investigation of population dynamics of filamentous bacteria in five Polish full-scale municipal WWTPs was performed using quantitative FISH technique by Miłobędzka and Muszyński (2015). Of the total bacterial community found in the plants, filamentous bacteria constituted about 28% with significant abundant of Chloroflexi

TABLE 4.6   rRNA: Targeted Oligonucleotides Probes for Detecting Nitrifiers in Environmental Samples

Probe Name	Target	Sequence (5'–3')	FA[a](%)	References
AOB				
Nso1225	β-proteobacterial ammonia-oxidizing bacteria	CGCCATTGTATTACGTGTGA	35	Mobarry et al. (1996)
Nso 190	β-proteobacterial ammonia-oxidizing bacteria	CGATCCCCTGCTTTTCTCC	55	Mobarry et al. (1996)
Nsc825	β-proteobacterial ammonia-oxidizing bacteria	CCCTCCCAACGTCTAGTT	ND	Siripong et al. (2006)
Nse1472	*Nitrosomonas europea, N. halophila, N. eutropha,* Kraftisried-Isolat Nm103	ACCCAGTCATGACCCCC	50	Juretschko et al. (1998)
Nsm 156	*Nitrosomonas sp., Nitrosococcus mobilis*	TATTAGCACATCTTTCGAT	5	Mobarry et al. (1996)
NEU	Most halophilic and halotolerant *Nitrosomonas sp.,*	CCCCTCTGCTGCACTCTA	35/40	Wagner et al. (1996)
NmIV	*Nitrosomonas cryotolerans* lineage	TCTCACCTCTCAGCGAGCT	35	Pommerening-Rösera, (1996)
NmII	*Nitrosomonas communis* lineage	TTAAGACACGTTCCGATGTA	25	Pommerening-Rösera, (1996)
Cluster 6a192	*Nitrosomonas oligotropha lineage* (Cluster 6a)	CTTTCGATCCCCTACTTTCC	35	Adamczyk et al. (2003)
Nmn657	*Nitrosomonas* spp.	TGGAATTCCACTCCCCTCTG	20	Araki et al. (1999)
Nmo218	*Nitrosomonas oligotropha*	CGGCCGCTCCAAAAGCAT	35	Gieseke et al. (2001)
Nsv443	*Nitrosospira spp., Nitroso vibrio, Nitrosolobus*	CCGTGACCGTTTCGTTCCG	30	Mobarry et al. (1996)
NSMR34	*Nitrosospira tenuis-like*	TCCCCACTCGAAGATACG	20	Burrell et al. (2001)
NmV	*Nitrosococcus mobilis*	TCCTCAGAGACTACGCGG	35	Juretschko et al. (1998)
Ntcoc206	*Nitrosococcus mobilis*	CGGTGGGAGCTTGCAAGC	10	
NOB				
Ntspa712	Phylum *Nitrospitae*	CGCCTTCGCCACCGGCCTTCC	35/50	Daims et al. (2001), Lopez Vazquez (2009)
Ntspa662	Genus *Nitrospira*	GGAATTCCGCGCTCCTCT	35	Daims et al. (2001)

(*Continued*)

TABLE 4.6 (*Continued*) rRNA: Targeted Oligonucleotides Probes for Detecting Nitrifiers in Environmental Samples

Probe Name	Target	Sequence (5'–3')	FA[a](%)	References
Ntspa1431	*Nitrospira* sub-linage I	TTGGCTTGGGCGACTTCA	35	Maixner et al. (2006)
Ntspa1151	*Nitrospira* sub-linage II	TTCTCCTGGGCAGTCTCTCC	35	Maixner et al. (2006)
Ntspa 1026	*Nitrospira moscoviensis*, activated sludge clones A4 and A11	AGCACGCTGGTATTGCTA	20	Irvin et al. (2007)
NSR826	*Nitrospira moscoviensis*	GTAACCCGCCGACACTTA	20	Schramm et al. (1998)
Nsr1156	*Nitrospira moscoviensis*, Freshwater *Nitrospira sp.*	CCCGTTCTCCTGGGCAGT	30	Schramm et al. (1998, Mota et al. (2012)
Nspmar62	*Nitrospira marina*-related *Nitrospira*	GCCCCGGATTCTCGTTCG	40	Foesel et al. (2008)
NTG840	*Nitrotogaarctica*	CTAAGGAAGTCTCCTCCC	10–20	Alawi et al. (2007)
NSR447	*Nitrospira* spp.	GGTTCCCGTTCCATCTT	30	Schramm et al. (1998)
NIT3	Genus *Nitrobacter*	CCTGTGCTCCATGCTCCG	40	Lin (2003)
Nb1000	*Nitrobacter* spp.	TGCGACGGGTCATGG	–	Mobarry et al. (1996)
Ntbl169	*Nitrobacter* spp.	TTGCTTCCCATTGTCACC	10	Araki et al. (1999)
Ntlc439	*Nitrolancetus hollandicus*	TTGCTTCGTCCCCACAA	40	Sorokin et al. (2012)
Ntlc804	*Nitrolancetus hollandicus*	CAGCGTTTACTGCTCGGA	20	Sorokin et al. (2012)
AOA				
CREN499	Most Crenarchaeota	CCAGRCTTGCCCCCGCT	0	Burggraf et al. (1994)
CREN537	Crenarchaea	TGACCACTTGAGGTGCTG	20	Teira et al. (2004)
CREN569	Most environmental Crenarchaeota	GCTACGGATGCTTTAGG	0	Radax et al. (2012)

[a] FA, Formamide; ND, Not determined.

with types 0803, 1851, and *Microthrix* strains, with similar results being reported by other researchers (Vanysacker et al., 2014; Wang et al., 2014a). Quantification of anammox bacteria has also been widely attempted with FISH and real-time PCR (Schmidt, 2005; Isaka et al., 2006; Tsushima et al., 2007; Li and Gu, 2011). Quantification of anammox bacteria through FISH is often challenging at low enrichment states as the anammox bacteria are strict anaerobes, they are often found deeply imbedded within the floc—especially when found in granular sludge. This makes direct quantification through pixel counts or colored area difficult and inaccurate.

PCR-based methods also help to monitor functional groups in wastewater systems using certain functional genes such as amoA for AOB, or the hzo or hzs gene clusters for hydrazine oxidase or hydrazine synthase in anammox bacteria, respectively (Matsumoto et al., 2010; Ramdhani, 2012). Among the phylogenetic biomarkers, the 16S rRNA gene using qPCR is the most commonly used target for detection of bacteria from various ecosystems. Quantitative PCR is a suitable and effective method for the quantification of anammox bacterial 16S DNA or RNA gene copies in WWTPs (Bae et al., 2010). Some of the challenges of using qPCR for anammox quantification include high sequence divergence among different genera of anammox bacteria (<87.1% identity); low abundance of anammox bacteria in most ecological niches; and difficulty in selecting 16S rRNA gene-based PCR primers that will amplify all known genera of anammox bacteria with a high specificity from any given sample (Jetten et al., 2009; Junier et al., 2010; Kartal et al., 2011). To overcome these limitations, a suite of functional genes was elucidated based on the unique characteristics that are universal traits to all anammox bacteria (Duan et al., 2013). Genomic data analyses have shown some unique pathways in Planctomycetes, including ladderane lipid biosynthesis, biological hydrazine metabolism, and carbon fixation. These relatively unique characteristics have shown potential as biomarkers for both detection and quantification (Strous et al., 2006). Owing to poor quantification of filamentous bacteria especially *M. parvicella* by FISH caused by low metabolic activity and possibly incomplete permeability of the cell wall by the probe has led to quantification using qPCR. Many recent studies have investigated the microbial community in foaming AS using qPCR (Kaetzke et al., 2005; Kumari et al., 2009; Vanysacker et al., 2014). Phylogeny and distribution of *M. parvicella* based on the 16S rRNA gene abundance was investigated in three different WWTPs in South Africa by Kumari et al. (2009). The distribution of *Gordonia* and *M. parvicella* populations in three AS plants experiencing foaming in South Africa was also performed by Marrengane et al. (2011) using qPCR combined with nested PCR restriction fragment length polymorphism (PCR-RFLP).

Vanysacker et al. (2014) primers targeting 16S rDNA genes of *Candidatus Microtrhix parvicella* and *Microthrix calida* were designed and used to quantify 29 DNA samples originated from different WWTPs using qPCR. The results were compared by using conventional microscopy, FISH, and an existing SYBR Green-based assay. Both *M. parvicella* and *M. calida* have shown 100% specificity and sensitivity (2.93×10^9 to 29 copy numbers/reaction), with PCR inhibition at 93% amplification efficiency. *Microthrix* concentrations, recovery rates of 65% to 98% were obtained for *Microthrix* concentrations when spiking was performed. Most of the samples were reported to be in accordance with the microscopic observation and qPCR assay using SYBR Green reaction. The high degree of accuracy has resulted in the qPCR method being recommended as an early warning tool for a reliable and fast detection of filamentous bacteria in sludge bulking incidences by researchers who study the ecology of AS.

In studying nitrifying communities in WWTPs, Ye et al. (2011) used pyrosequencing to identify *Nitrosomonas* spp., *Nitrospira* spp., *Nitrosospira* spp., *Nitrosococcus* spp., and *Nitrobacter* spp. They observed that apart from *Nitrosomonas* spp. and *Nitrospira* spp., other nitrifiers did not have significant contribution in the nitrification process. In another study by Zhang et al. (2011) the majority of the nitrifiers identified with pyrosequencing were related to *Nitrosomonas* spp. There was incongruity in the results when nitrifying communities in different wastewater bioreactors were analyzed using quantitative PCR and pyrosequencing. The abundance and diversity of AOB and NOB were equally investigated in tannery sludge samples by Wang et al. (2014b). Choi and Liu (2014), Hai et al. (2014), and Zhu et al. (2013) have applied the technique successfully in studying the microbial community shift in different wastewater treatment systems. Ju et al. (2014) used Illumina in investigating the seasonal dynamics of AS over a period of 4 years. A combination of Illumina and pyrosequencing has been used by Sorokin et al. (2012) for genomic study of NOB and an entirely novel nitrifier named *Nitrolancetus hollandicus* (Tables 4.7 through 4.9).

CONCLUSIONS AND FUTURE PERSPECTIVES

As molecular techniques and the technology that supports them improve, many more unique and interesting microbial species are bound to be discovered, even within previously explored ecological niches. The complex

TABLE 4.7 Specific Primers for the Identification and Quantification of AOB and NOB Communities in Environmental Samples

Primers	Target	Sequence (5′ → 3′)	Annealing Temperature	References
amoA-1F	Ammonia	GGGGTTTCTACTGGTGGT	55°C	Rotthauwe et al. (1997)
amoA-2R	monooxygenase	CCCCTCKGSAAAGCCTTCTTC		
FGPS872	*Nitrobacter*	CTAAAACTCAAAGGAATTGA	50°C	Degrange and Bardin (1995)
FGPS1269		TTTTTGAGATTTGCTAG		
NSR1113F	*Nitrospira*	CCTGCTTTCAGTTGCTACCG	65°C	Dionisi et al. (2002)
NSR1264R	*Nitrospira*	GTTTGCAGGCGTTTGTACCG		
CTO189fAB	β-*Proteobacteria* ammonia oxidizers	GGAGRAAAGCAGGGGATCG	57°C	Kowalchuk et al. (1997)
CTO189fC	β-*Proteobacteria* ammonia oxidizers	GGAGGAAAGTAGGGGATCG		
CTO654r	β-*Proteobacteria* ammonia oxidizers	CTAGCYTTGTAGTTTCAAACGC		
Arch-amoAF	Archaeal ammonia monooxygenase	STAATGGTCTGGCTTAGACG	53°C	Francis et al. (2005)
Arch-amoAR	Archaeal ammonia monooxygenase	GCGGCCATCCATCTGTATGT		
CRENamo_F	Archaeal ammonia monooxygenase	ATGGTCTGGCTAAGACGMTGTA	55°C	Jin et al. (2010)
CRENamo_R	Archaeal ammonia monooxygenase	CCCACTTTGACCAAGCGGCCAT		

TABLE 4.8 Summary of Some of the Primers for the Identification of Anammox

Primer Name	Primer Sequence	Target	Annealing Temperature	References
Pla46F	GACTTGCATGCCTAATCC	Planctomycetes	58	Neef et al. (1998)
An7f	GGCATGCAAGTCGAACGAGG	Ca. Kuenenia, Ca. Brocardia		Penton et al. (2006)
Amx368F	CCTTTCGGGCATTGCGAA	All anammox bacteria	56	Schmidt et al. (2003)
A438F	GTCRGGAGTTADGAAATG	All anammox bacteria	55	Humbert et al. (2012)
Brod541F	GAGCACGTAGGTGGGTTTGT	Ca. Scalindua	60	Penton et al. (2006)
A684r	ACCAGAAGTTCCACTCTC	All anammox bacteria	55	Humbert et al. (2012)
Amx820R	AAAACCCCTCTACTTAGTGCCC	Ca. Kuenenia, Ca. Brocardia	56	Schmidt et al. (2003)
BS820R	TAATTCCCTCTACTTAGTGCCC	Ca. Scalindua	56	Kuypers et al. 2003
Anl388r	GCTTGACGGGCGGTGTG	Ca. Scalindua	56	Penton et al. (2006)
Brod1260R	GGATTCGCTTCACCTCTCGG	Ca. Scalindua	60	Penton et al. (2006)
Amx667R	ACCAGAAGTTCCACTCTC	All anammox bacteria		Van der Star et al. (2007)
hzsB-396F	ARGGHTGGGGHAGYTGGAAG	Hydrazine synthase B subunit		Harhangi et al. (2012), Park et al. (2010), Schmid et al. (2005)
hzsB-742R	GTYCCHACRTCATGVGTCTG	Hydrazine synthase B subunit		Harhangi et al. (2012), Park et al. (2010), Schmid et al. (2005)
hzsA_1597F	WTYGGKTATCARTATGTAG	Hydrazine synthase subunit A	55	Harhangi et al. (2012)
hzsA_1859R	AAABGYGAATCATARTGGC	Hydrazine synthase subunit A	55	Harhangi et al. 2013
hzoAB1F	GAAGCNAAGGCNGTAGAAATTATCAC	Hydrazine oxidase		Hirsch et al. (2011)
hzoAB1R	CTCTTCNGCAGGTGCATGATG	Hydrazine oxidase		Hirsch et al. (2011)
hzoAB4F	TTGARTGTGCATGGTCTAWTGAAAG	Hydrazine oxidase		Hirsch et al. (2011)
hzoAB4R	GCTGACCTGACCARTCAGG	Hydrazine oxidase		Hirsch et al. (2011)
hzocl1F1	TGYAAGACYTGYCAYTGG	Hydrazine oxidase		Schmid et al. (2005)
hzocl1R2	ACTCCAGATRTGCTGACC	Hydrazine oxidase		Schmid et al. (2005)
hzoF1	TGTGCATGGTCAATTGAAAG	Hydrazine oxidase		Li and Gu (2011)
hzoR1	CAACCTCTTCWGCAGGTGCATG	Hydrazine oxidase		Li and Gu (2011)

TABLE 4.9 Probes Used for the Detection of Anammox Bacteria

Probe Name	Sequence 5′-3′	Gene Target	Specificity Group	Formamide %	mM NaCl	References
S-P-Planc-0046-a-A-18	GACTTGCATGCCTAATCC	16S rRNA	Planctomycetes	25	159	Neef et al. (1998)
S-P-Planc-0886-a-A-19	GCCTTGCGACCATACTCCC	16S rRNA	Isosphaera, Gemmata, Pirellula, Planctomycetales	30	112	Neef et al. (1998)
S-D-Bact-0338-b-A-18	GCAGCCACCCGTAGGTGT	16S rRNA	Specific for Planctomycetales	0	900	Daims et al. (1999)
S-*-Amx-0368-a-A-18	CCTTTCGGGCATTGCGAA	16S rRNA	All anammox organisms	15	338	Schmidt et al. (2003)
S-*-Amx-0820-a-A-22	AAAACCCCTCTACTTAGTGCCC	16S rRNA	Brocadia anammoxidans Kuenenia stuttgartiensis	40	56	Schmidt et al. (2000)
S-G-Sca-1309-a-A-21	TGGAGGCGAATTTCAGCCTCC	16S rRNA	Scalindua	5	675	Schmidt et al. (2003)
S-*-Scabr-1114-a-A-22	CCCGCTGGTAACTAAAAACAAG	16S rRNA	Scalindua brodae	20	225	Schmidt et al. (2003)
S-*-BS-820-a-A-22	TAATTCCCTCTACTTAGTGCCC	16S rRNA	Scalindua wagneri Scalindua sorokinii	40	56	Kuypers et al. (2003)
S-S-Kst-0157-a-A-18	GTTCCGATTGCTCGAAAC	16S rRNA	Kuenenia stuttgartiensis	25	159	Schmid et al. (2001)
S-*-Kst-1275-a-A-20	TCGGCTTTATAGGTTTCGCA	16S rRNA	Kuenenia stuttgartiensis	25	159	Schmid et al. (2000)

(Continued)

TABLE 4.9 (Continued) Probes Used for the Detection of Anammox Bacteria

Probe Name	Sequence 5'-3'	Gene Target	Specificity Group	Formamide %	mM NaCl	References
S-S-Ban-0162(B. anam.)-a-A-18	CGGTAGCCCCAATTGCTT	16S rRNA	Brocadia anammoxidans	40	56	Schmid et al. (2000)
S-*-Amx-0156-a-A-18	CGGTAGCCCCAATTGCTT	16S rRNA	Brocadia anammoxidans	40	56	Schmid et al. (2000)
S-*-Amx-0223-a-A-18	GACATTGACCCCTCTCTG	16S rRNA	Brocadia anammoxidans	40	56	Schmid et al. (2000)
S-*-Amx-0432-a-A-18	CTTAACTCCCGACAGTGG	16S rRNA	Brocadia anammoxidans	40	56	Schmid et al. (2000)
S-*-Amx-0613-a-A-22	CCGCCATTCTTCCGTTAAGCGG	16S rRNA	Brocadia anammoxidans	40	56	Schmid et al. (2000)
S-*-Amx-0997-a-A-21	TTTCAGGTTTCTACTTCTACC	16S rRNA	Brocadia anammoxidans	20	225	Schmid et al. (2000)
S-*-Amx-1015-a-A-18	GATACCGTTCGTCGCCCT	16S rRNA	Brocadia anammoxidans	60	14	Schmid et al. (2000)
S-*-Amx-1154-a-A-18	TCTTGACGACAGCAGTCT	16S rRNA	Brocadia anammoxidans	20	112	Schmid et al. (2000)
S-*-Amx-1240-a-A-23	TTTAGCATCCCTTTGTACCAACC	16S rRNA	Brocadia anammoxidans	60	56	Schmid et al. (2000)
S-*-Bfu-0613-a-A-24	GGATGCCGTTCTTCCGTTAAGCGG	16S rRNA	Brocadia fulgida	30	112	Kartal et al. (2008)
S-*-Apr-0820-a-A-21	AAACCCCTCTACCGAGTGCCC	16S rRNA	Anammoxoglobus propionicus Jettenia asiatica	40	450	Kartal et al. (2007)

(Continued)

TABLE 4.9 (Continued) Probes Used for the Detection of Anammox Bacteria

Probe Name	Sequence 5′-3′	Gene Target	Specificity Group	Formamide %	mM NaCl	References
L-*-Amx-1900-a-A-21	CATCTCCGGCTTGAACAA	23S rRNA	Brocadia and Kuenenia	30	450	Schmid et al. (2001)
I-*-Ban-0071(B.anam.)-a-A-18	CCCTACCACAAACCTCGT	ISR	Brocadia anammoxidans	10	450	Schmid et al. (2000)
I-*-Ban-0108(B.anam.)-a-A-18	TTTGGGCCCGCAATCTCA	ISR	Brocadia anammoxidans	10	450	Schmid et al. (2000)
I-*-Ban-0222(B.anam.)-a-A-19	GCTTAGAATCTTCTGAGGG	ISR	Brocadia anammoxidans	10	450	Schmid et al. (2000)
I-*-Ban-0389(B.anam.)-a-A-18	GGATCAAATTGCTACCCG	ISR	Brocadia anammoxidans	10	450	Schmid et al. (2000)
I-*-Kst-0031(K.stutt.)-a-A-18	ATAGAAGCCTTTTGCGCG	ISR	Kuenenia stuttgartiensis	10	450	Schmid et al. (2001)
I-*-Kst-0077(K.stutt.)-a-A-18	TTTGGGCCACACTCTGTT	ISR	Kuenenia stuttgartiensis	10	450	Schmid et al. (2001)
I-*-Kst-0193(K.stutt.)-a-A-19	CAGACCGGACGTATAAAAG	ISR	Kuenenia stuttgartiensis	10	450	Schmid et al. (2001)
I-*-Kst-0288(K.stutt.)-a-A-20	GCGCAAAGAAATCAAACTGG	ISR	Kuenenia stuttgartiensis	10	450	Schmid et al. (2001)

environs of the wastewater AS biota represent an especially rich ecosystem with a hugely diverse microbial population, many of which have not yet been fully characterized. In fact, the high degree of variability created by the microenvironments among different types of wastewater treatment systems creates considerable permutations in the microbial fingerprint of the system, which has a direct influence on the functioning of the system as a whole. Utilizing the techniques outlined in this chapter, a growing number of research groups worldwide are trying to characterize and catalog the unique microbial fingerprints that may exist within their biological wastewater treatment systems. As the available information on environmental microbial fingerprinting increases globally, many more interesting and possibly paradigm-shifting observations are expected to be made. The eventual goal of fingerprint cataloguing lies in identifying the optimum microbial community configuration based on the analysis of the core groups and their cumulative interspecies and intraspecies interactions, and engineering systems that optimally exploit these interactions.

REFERENCES

Adamczyk, J., Hesselsoe, M., Iversen, N., Horn, M., Lehner, A., Nielsen, P. H., Schloter, M., Roslev, P. and Wagner, M. 2003. The isotope array, a new tool that employs substrate-mediated labeling of rRNA for determination of microbial community structure and function. *Applied and Environmental Microbiology*, 69, 6875–6887.

Ahiamadu, N. M. 2007. The Challenges of Municipal Solid Waste Management In Nigeria. *8th International Conference "Waste Management, Environmental Geotechnology and Global Sustainable Development (ICWMEGGSD'07— GzO'07)*. Ljubljana, Slovenia, August 28–30.

Akpor, O. B. and Muchie, M. 2010. Bioremediation of polluted wastewater influent: Phousphorus and nitrogen removal. *Science Research and Essays*, 5, 3222–3230.

Akpor, O. B., Ogundeji, M. D., Olaolu, T. D. and Aderiye, B. I. 2014. Microbial roles and dynamics in wastewater treatment systems: An overview. *International Journal of Pure and Applied Bioscience*, 2, 156–168.

Alawi, M., Lipski, A., Sanders, T., Pfeiffer, E. M. and Spieck, E. 2007. Cultivation of novel cold-adapted nitrite oxidizing betaproteobacterium from the Siberian Arctic. *International Society for Microbial Ecology Journal*, 1, 256–264.

Alvarado, A., Montañez-Hernández, L. E., Palacio-Molina, S. L., Oropeza-Navarro, R., Luévanos-Escareño, M. P. and Balagurusamy, N. 2014. Microbial trophic interactions and mcrA gene expression in monitoring of anaerobic digesters. *Frontiers in Microbiology*, 5, 597.

Amann, R., Fuchs, B. M. and Behrens, S. 2001. The identification of microorganisms by fluorescence *in situ* hybridisation. *Current Opinion in Biotechnology*, 12, 231–236.

Amann, R. I., Binder, B. J., Olson, R. J., Chisholm, S. W., Devereux, R. and Stahl, D. A. 1990. Combination of 16S rRNA-targeted oligonucleotide probes with flow cytometry for analyzing mixed microbial populations. *Applied and Environmental Microbiology*, 56, 1919–1925.

Amenaghawon, A. N. and Obahiagbon, K. O. 2014. Wastewater treatment by bioremediation methods. *Wastewater Engineering: Advanced Wastewater Treatment Systems*, 108.

Araki, N., Ohashi, A., Machdar, I. and Harada, H. 1999. Behaviors of nitrifiers in a novel biofilm reactor employing hanging sponge-cubes as attachment site. *Water Science and Technology*, 39, 23–31.

Awolusi, O.O., Nasr, M., Kumari, S. and Bux, F., 2016. Artificial Intelligence for the Evaluation of Operational Parameters Influencing Nitrification and Nitrifiers in an Activated Sludge Process. Microbial ecology, 72, pp 49–63.

Awolusi, O. O., Kumari, S. K. S. and Bux, F. 2015. Ecophysiology of nitrifying communities in membrane bioreactors. *International Journal of Environmental Science and Technology*, 12, 747–762.

Bae, H., Park, K.-S., Chung, Y.-C. and Jung, J.-Y. 2010. Distribution of anammox bacteria in domestic WWTPs and their enrichments evaluated by real-time quantitative PCR. *Process Biochemistry*, 45, 323–334.

Bagchi, S., Biswas, R. and Nandy, T. 2012. Autotrophic ammonia removal processes: Ecology to technology. *Critical Reviews in Environmental Science and Technology*, 42, 1353–1418.

Benakova, A. and Wanner, J. 2013. Application of fluorescence *in situ* hybridization for the study and characterization of nitrifying bacteria in nitrifying/denitrifying wastewater treatment plants. *Environmental Technology*, 34, 2415–2422.

Bialek, K., Kim, J., Lee, C., Collins, G., Mahony, T. and O'Flaherty, V. 2011. Quantitative and qualitative analyses of methanogenic community development in high-rate anaerobic bioreactors. *Water Research*, 45, 1298–1308.

Biller, S. J., Mosier, A. C., Wells, G. F. and Francis, C. A. 2012. Global biodiversity of aquatic ammonia-oxidizing archaea is partitioned by habitat. *Frontiers in Microbiology*, 3, 1–15.

Black, A., Hsu, P. C. L., Hamonts, K. E., Clough, T. J. and Condron, L. M. 2016. Influence of copper on expression of *nirS*, *norB* and *nosZ* and the transcription and activity of NIR, NOR and N_2OR in the denitrifying soil bacteria Pseudomonas stutzeri. *Microbial Biotechnology*, 9, 381–388.

Blackbeard, J., Ekama, G. and Marais, G. 1986. A survey of filamentous bulking and foaming in activated-sludge plants in South Africa. *Water Pollution Control*, 85, 90–100.

Bobrow, M. N., Harris, T. D., Shaughnessy, K. J. and Litt, G. J. 1989. Catalyzed reporter deposition, a novel method of signal amplification application to immunoassays. *Journal of Immunological Methods*, 125, 279–285.

Bokonyte, D., Baranauskaie, A., Gcenaite, J., Sosnovkaja, A. and Stakenas, P. 2003. Molecular characterization of isoniazid-resistant Mycobacterium tuberculosis clinical isolates in Lithuania. *Antimicrob Agents Chemother*, 47, 2009–2011.

Briones, A. and Raskin, L. 2003. Diversity and dynamics of microbial communities in engineered environments and their implications for process stability. *Current Opinion in Biotechnology,* 14, 270–276.

Broda, E. 1977. Two kinds of lithotrophs missing in nature. *Zeitschrift für Allgemeine Mikrobiologie,* 17, 491–493.

Burggraf, S., Mayer, T., Amann, R., Schadhauser, S., Woese, C. R. and Stetter, K. O. 1994. Identifying members of the domain archaea with rRNA-targeted oligonucleotide probes. *Applied and Environmental Microbiology,* 60, 3112–3119.

Burrell, P. C., Phalen, C. M. and Hovanec, T. A. 2001. Identification of bacteria responsible for ammonia oxidation in freshwater aquaria. *Applied and Environmental Microbiology,* 67, 5791–800.

Bustin, S. A., Benes, V., Garson, J. A., Hellemans, J., Huggett, J., Kubista, M., Mueller, R., Nolan, T., Pfaffl, M. W. and Shipley, G. L. 2009. The MIQE guidelines: Minimum information for publication of quantitative real-time PCR experiments. *Clinical Chemistry,* 55, 611–622.

Cao, C., Lou, I., Huang, C. and Lee, M.-Y. 2016. Metagenomic sequencing of activated sludge filamentous bacteria community using the Ion Torrent platform. *Desalin Water Treatment,* 57, 2175–2183.

Cao, Y., Sivaganesan, M., Kinzelman, J., Blackwood, A. D., Noble, R. T., Haugland, R. A., Griffith, J. F. and Weisberg, S. B. 2013. Effect of platform, reference material, and quantification model on enumeration of Enterococcus by quantitative PCR methods. *Water Research,* 47, 233–241.

Choi, J. and Liu, Y. 2014. Biodegradation of oil sands process affected water in sequencing batch reactors and microbial community analysis by high-throughput pyrosequencing. *International Biodeterioration and Biodegradation,* 92, 79–85.

Churko, J. M., Mantalas, G. L., Snyder, M. P. and Wu, J. C. 2013. Overview of high throughput sequencing technologies to elucidate molecular pathways in cardiovascular diseases. *Circulation Research,* 112, 1613–1623.

Clarke, S. C. 2005. Pyrosequencing: Nucleotide sequencing technology with bacterial genotyping applications. *Expert Review of Molecular Diagnostics,* 5, 947–953.

Crocetti, R. G., Banfield, F., Keller, J., Bond, P. L. and Blackall, L. L. 2002. Glycogen-accumulating organisms in laboratory-scale and full-scale wastewater treatment processes. *Microbiology,* 148, 3353–3364.

Daims, H., Brühl, A., Amann, R., Schleifer, K.-H. and Wagner, M. 1999. The domain-specific probe EUB338 is insufficient for the detection of all Bacteria: Development and evaluation of a more comprehensive probe set. *Systematic and Applied Microbiology,* 22, 434–444.

Daims, H., Nielsen, J. L., Nielsen, P. H., Schleifer, K. H. and Wagner, M. 2001. In situ characterization of Nitrospira-like nitrite-oxidizing bacteria active in wastewater treatment plants. *Applied and Environmental Microbiology,* 67, 5273–5284.

Daims, H. and Wagner, M. 2010. The microbiology of nitrogen removal. *In:* Seviour, R. J. and Nielsen, P. H. (eds.), *Microbial Ecology of Activated Sludge.* London: IWA.

Deepnarain, N., Kumari, S., Ramjith, J., Swalaha, F. M., Tandoi, V., Pillay, K. and Bux, F. 2015. A logistic model for the remediation of filamentous bulking in a biological nutrient removal wastewater treatment plant. *Water science and Technology*, 72, 391–405.

Degrange, V. and Bardin, R. 1995. Detection and counting of *Nitrobacter* populations in soil by PCR. *Applied Environmental and Microbiology*, 61, 2093–2098.

Dhanasekaran, S., Doherty, T. M., Kenneth, J. and Group, T. T. S. 2010. Comparison of different standards for real-time PCR-based absolute quantification. *Journal of Immunological Methods*, 354, 34–39.

Dionisi, H. M., Layton, A. C., Harms, G., Gregory, I. R., Robinson, K. G. and Sayler, G. S. 2002. Quantification of Nitrosomonas oligotropha-like ammonia-oxidizing bacteria and Nitrospira spp. from full-scale wastewater treatment plants by competitive PCR. *Applied and Environmental Microbiology*, 68, 245–253.

Duan, L., Song, Y., Xia, S. and Hermanowicz, S. W. 2013. Characterization of nitrifying microbial community in a submerged membrane bioreactor at short solids retention times. *Bioresource Technology*, 149, 200–207.

Eikelboom, D. H., Andreadakis, A. and Andreasen, K. 1998. Survey of filamentous populations in nutrient removal plants in four European countries. *Water Science and Technology*, 37, 281–289.

Enitan, A. M., Kumari, S., Swalaha, F. M., Adeyemo, J., Ramdhani, N. and Bux, F. 2014a. Kinetic modelling and characterization of microbial community present in a full-scale UASB reactor treating brewery effluent. *Microbial Ecology*, 67, 358–368.

Enitan, A. M., Kumari, S., Swalaha, F. M. and Bux, F. 2014b. Real-time quantitative PCR for quantification of methanogenic Archaea in an UASB reactor treating brewery wastewater. *Conference of the International Journal of Arts and Sciences*, 07, 103–106.

Enitan, A. M., Swalaha, F. M., Adeyemo, J. and Bux, F. 2014c. Assessment of brewery effluent composition from a beer producing industry in KwaZulu-Natal, South Africa. *Fresenius Environmental Bulletin*, 23, 693–701.

Foesel, B. U., Gieseke, A., Schwermer, C., Stief, P., Koch, L., Cytryn, E., De-La-Torre, J. R. et al. 2008. Nitrosomonas Nm 143-like ammonia oxidizers and Nitrospira marina-like nitrite oxidizers dominate the nitrifier community in a marine aquaculture biofilm. *FEMS Microbiology Ecology*, 63, 192–204.

Francis, C. A., Roberts, K. J., Beman, J. M., Santoro, A. E. and Oakley, B. B. 2005. Ubiquity and diversity of ammonia-oxidizing archaea in water columns and sediments of the ocean. *Proceedings of the National Academy of Sciences of the United States of America*, 102, 14683–14688.

Fuchs, B. M., Wallner, G., Beisker, W., Schwippl, I., Ludwig, W. and Amann, R. 1998. Flow cytometric analysis of the *in situ* accessibility of *Escherichia coli* 16S rRNA for fluorescently labeled oligonucleotide probes. *Applied and Environmental Microbiology*, 64, 4973–4982.

Fuchsman, C. A., Staley, J. T., Oakley, B. B., Kirkpatrick, J. B. and Murray, J. W. 2012. Free-living and aggregate-associated Planctomycetes in the Black Sea. *FEMS Microbiology Ecology*, 80, 402–416.

Fuerst, J. A. and Sagulenko, E. 2011. Beyond the bacterium: Planctomycetes challenge our concepts of microbial structure and function. *Nature Reviews Microbiology*, 9, 403–413.

Gao, D. and Tao, Y. 2012. Current molecular biologic techniques for characterizing environmental microbial community. *Frontiers of Environmental Science and Engineering*, 6, 82–97.

Gerardi, M. 2002. *Nitrification and Denitrification in the Activated Sludge Process*. New York, NY: John Wiley and Sons, Inc.

Gieseke, A., Purkhold, U., Wagner, M., Amann, R. and Schramm, A. 2001. Community structure and activity dynamics of nitrifying bacteria in a phosphate-removing biofilm. *Applied and Environmental Microbiology*, 67, 1351–1362.

Hai, R., Wang, Y., Wang, X., Li, Y. and Du, Z. 2014. Bacterial community dynamics and taxa-time relationships within two activated sludge bioreactors. *PLoS One*, 9, 1–8.

Harhangi, H. R., Le Roy, M., Van Alen, T., Hu, B.-L., Groen, J., Kartal, B., Tringe, S. G., Quan, Z.-X., Jetten, M. S. and Den Camp, H. J. O. 2012. Hydrazine synthase, a unique phylomarker with which to study the presence and biodiversity of anammox bacteria. *Applied and Environmental Microbiology*, 78, 752–758.

He, S., Gall, D. L., and McMahon, K. D. 2007. "Candidatus accumulibacter" population structure in enhanced biological phosphorus removal sludges as revealed by polyphosphate kinase genes. *Applied and Environmental Microbiology*, 73, 5865–5874.

He, J., Ritalahti, K. M., Aiello, M. R. and Löffler, F. E. 2003. Complete detoxification of vinyl chloride by an anaerobic enrichment culture and identification of the reductively dechlorinating population as a Dehalococcoides species. *Applied and Environmental Microbiology*, 69, 996–1003.

Hirsch, M. D., Long, Z. T. and Song, B. 2011. Anammox bacterial diversity in various aquatic ecosystems based on the detection of hydrazine oxidase genes (hzoA/hzoB). *Microbial Ecology*, 61, 264–276.

Hong, Y.-G., Yin, B. and Zheng, T.-L. 2011. Diversity and abundance of anammox bacterial community in the deep-ocean surface sediment from equatorial Pacific. *Applied Microbiology and Biotechnology*, 89, 1233–1241.

Hu, M., Wang, X., Wen, X. and Xia, Y. 2012. Microbial community structures in different wastewater treatment plants as revealed by 454-pyrosequencing analysis. *Bioresource Technology*, 117, 72–79.

Huggett, J. F., Foy, C. A., Benes, V., Emslie, K., Garson, J. A., Haynes, R., Hellemans, J., Kubista, M., Mueller, R. D. and Nolan, T. 2013. The digital MIQE guidelines: Minimum information for publication of quantitative digital PCR experiments. *Clinical Chemistry*, 59, 892–902.

Humbert, L., Weinand, Y. and Buri, H. U. 2012. Curved Origami Beams. SAH Statusseminar 2012.

Interstate Technology and Regulatory Council, I. 2013. *Environmental Molecular Diagnostics, New Site Characterization and Remediation Enhancement Tools. EMD-2*. Washington, DC: Interstate Technology and Regulatory Council, Environmental Molecular Diagnostics Team. www.itrcweb.org.

Irvin, S., Baker, K. and Li, B. 2007. The variation of nitrifying bacterial popula-tion sizes in a sequencing batch reactor (SBR) treating low, mid, high con-centrated synthetic wastewater. *Journal of Environmental Engineering and Science*, 6, 651–663.

Isaka, K., Date, Y., Sumino, T., Yoshie, S. and Tsuneda, S. 2006. Growth char-acteristic of anaerobic ammonium-oxidizing bacteria in an anaerobic biological filtrated reactor. *Applied Microbiology and Biotechnology*, 70, 47–52.

Jenkins, D., Richard, M. G. and Daigger, G. T. 1993. *Manual on the Causes and Control of Activated Sludge Bulking and Foaming*. 2nd ed. Chelsea, MI: Lewis Publisher.

Jenkins, D., Richard, M. G. and Daigger, G. T. 2004. *Manual on the Causes and Control of Activated Sludge Bulking, Foaming, and Other Solids Separation Problems*. UK: IWA Publishing.

Jeppsson, U. 1996. *Modelling Aspects of Wastewater Treatment Processes*. Sweden: Lund University.

Jetten, M. S., Niftrik, L. V., Strous, M., Kartal, B., Keltjens, J. T. and Op Den Camp, H. J. 2009. Biochemistry and molecular biology of anammox bac-teria. *Critical Reviews in Biochemistry and Molecular Biology*, 44, 65–84.

Jetten, M. S., Sliekers, O., Kuypers, M., Dalsgaard, T., Van Niftrik, L., Cirpus, I., Van De Pas-Schoonen, K., Lavik, G., Thamdrup, B. and Le Paslier, D. 2003. Anaerobic ammonium oxidation by marine and freshwater planctomycete-like bacteria. *Applied Microbiology and* Biotechnology, 63, 107–114.

Jetten, M. S., Strous, M., Van De Pas-Schoonen, K. T., Schalk, J., Van Dongen, U. G., Van De Graaf, A. A., Logemann, S., Muyzer, G., Van Loosdrecht, M. C. and Kuenen, J. G. 1998. The anaerobic oxidation of ammonium. *FEMS Microbiology Reviews*, 22, 421–437.

Jin, T., Zhang, T. and Yan, Q. 2010. Characterization and quantification of ammo-nia-oxidizing archaea (AOA) and bacteria (AOB) in a nitrogen-removing reactor using T-RFLP and qPCR. *Applied Microbiology and Biotechnology*, 87, 1167–1176.

Ju, F., Guo, F., Ye, L., Xia, Y. and Zhang, T. 2014. Metagenomic analysis on sea-sonal microbial variations of activated sludge from a full-scale wastewater treatment plant over 4 years. *Environmental Microbiology*, 6, 80–89.

Ju, F. and Zhang, T. 2015. 16S rRNA gene high-throughput sequencing data mining of microbial diversity and interactions. *Applied Microbiology and Biotechnology*, 99, 4119–4129.

Junier, P., Molina, V., Dorador, C., Hadas, O., Kim, O. S., Junier, T., Witzel, J. P. and Imhoff, J. F. 2010. Phylogenetic and functional marker genes to study ammonia-oxidizing microorganisms (AOM) in the environment. *Applied Microbiology and Biotechnology*, 85, 425–440.

Juretschko, S., Timmermann, G., Schmid, M., Schleifer, K.-H., Pommerening-Roser, A., Koops, H.-P. and Wagner, M. 1998. Combined molecular and conventional analyses of nitrifying bacterium diversity in activated sludge: Nitrosococcus mobilis and Nitrospira-like bacteria as dominant popula-tions. *Applied and Environmental Microbiology*, 64, 3042–3051.

Kaetzke, A., Jentzsch, D. and Eschrich, K. 2005. Quantification of Microthrix parvicella in activated sludge bacterial communities by real-time PCR. *Letters in Applied Microbiology*, 40, 207–211.

Kappeler, J. and Gujer, W. 1992. Estimation of kinetic parameters of heterotrophic biomass under aerobic conditions and characterization of wastewater for activated sludge modelling. *Water Science and Technology*, 25, 125–139.

Kartal, B., Geerts, W. and Jetten, M. S. 2011. 4 Cultivation, detection, and ecophysiology of Anaerobic ammonium-oxidizing bacteria. *Methods in Enzymology*, 486, 89–108.

Kartal, B., Keltjens, J. T. and Jetten, M. S. 2008. The metabolism of anammox. *Encyclopedia of Life Sciences*. Doi: 10.1002/9780470015902.a0021315.

Kartal, B., Rattray, J., Van Niftrik, L. A., Van De Vossenberg, J., Schmid, M. C., Webb, R. I., Schouten, S., Fuerst, J. A., Damsté, J. S. and Jetten, M. S. 2007. Candidatus "Anammoxoglobus propionicus" a new propionate oxidizing species of anaerobic ammonium oxidizing bacteria. *Systematic and Applied Microbiology*, 30, 39–49.

Kartal, B., Van Niftrik, L., Sliekers, O., Schmid, M. C., Schmidt, I., Van De Pas-Schoonen, K., Cirpus, I., Van Der Star, W., Van Loosdrecht, M. and Abma, W. 2004. Application, eco-physiology and biodiversity of anaerobic ammonium-oxidizing bacteria. *Reviews in Environmental Science and Bio/Technology*, 3, 255–264.

Kim, J., Lim, J. and Lee, C. 2013. Quantitative real-time PCR approaches for microbial community studies in wastewater treatment systems: Applications and considerations. *Biotechnology Advances*, 31, 1358–1373.

Kowalchuk, G. A., Stephen, J. R., De Boer, W., Prosser, J. I., Embley, T. M. and Woldendorp, J. W. 1997. Analysis of ammonia-oxidizing bacteria of the beta subdivision of the Class Proteobacteria in coastal sand dunes by denaturing gradient gel electrophoresis and sequencing of PCR-amplified 16S ribosomal DNA fragments. *Applied and Environmental Microbiology*, 63, 1489–1497.

Kragelund, C., Caterina, L., Borger, A., Thelen, K., Eikelboom, D., Tandoi, V., Kong, Y. et al. 2007. Identity, abundance and ecophysiology of filamentous Chloroflexi species present in activated sludge treatment plants. *FEMS Microbiology Ecology*, 59(3), 671–682.

Krober, M., Bekel, T., Diaz, N., Goesmann, A., Jaenicke, S., Krause, L., Miller, D. et al. 2009. Phylogenetic characterization of a biogas plant microbial community integrating clone library 16S-rDNA sequences and metagenome sequence data obtained by 454-pyrosequencing. *Journal of Biotechnology*, 142, 38–49.

Kubota, K., Ohashi, A., Imachi, H. and Harada, H. 2006. Improved *in situ* hybridization efficiency with locked-nucleic-acid-incorporated DNA probes. *Applied and Environmental Microbiology*, 72, 5311–5317.

Kumari, S. S., Marrengane, Z. and Bux, F. 2009. Application of quantitative RT-PCR to determine the distribution of *Microthrix parvicella* in full-scale activated sludge treatment systems. *Applied Microbiology and Biotechnology*, 83, 1135–1141.

Kuypers, M. M., Sliekers, A. O., Lavik, G., Schmid, M., Jørgensen, B. B., Kuenen, J. G., Damsté, J. S. S., Strous, M. and Jetten, M. S. 2003. Anaerobic ammonium oxidation by anammox bacteria in the Black Sea. *Nature*, 422, 608–611.

Lee, C., Kim, J., Shin, S. G., O'Flaherty, V. and Hwang, S. 2010. Quantitative and qualitative transitions of methanogen community structure during the batch anaerobic digestion of cheese-processing wastewater. *Applied Microbiology and Biotechnology*, 87, 1963–1973.

Li, M. and Gu, J.-D. 2011. Advances in methods for detection of anaerobic ammonium oxidizing (anammox) bacteria. *Applied Microbiology and Biotechnology*, 90, 1241–1252.

Li, M., Yiguo H., Martin G. K. and Ji-Dong G. 2010. A comparison of primer sets for detecting 16S rRNA and hydrazine oxidoreductase genes of anaerobic ammonium-oxidizing bacteria in marine sediments. *Applied Microbiology and Biotechnology*, 86, 781–790.

Liang, Z., Das, A., Beerman, D. and Hu, Z. 2010. Biomass characteristics of two types of submerged membrane bioreactors for nitrogen removal from wastewater. *Water Research*, 44, 3313–3320.

Lim, J., Chen, C.-L., Ho, I. and Wang, J.-Y. 2013. Study of microbial community and biodegradation efficiency for single-and two-phase anaerobic co-digestion of brown water and food waste. *Bioresource Technology*, 147, 193–201.

Lin, S. 2003. Probing for the diversity of Nitrobacter species rDNA in a wastewater treatment bioreactor using molecular analysis. *Journal of Experimental Microbiology and Immunology*, 4, 27–32.

Liu, L., Li, Y., Li, S., Hu, N., He, Y., Pong, R., Lin, D., Lu, L. and Law, M. 2012. Comparison of next-generation sequencing systems. *BioMed Research International*, 2012. Article ID 251364, pp. 1–11. Doi:10.1155/2012/251364.

Liu, Y., Shi, H., Xia, L., Shen, T., Wang, Z., Wang, G. and Wang, Y. 2010. Study of operational conditions of simultaneous nitrification and denitrification in a Carrousel oxidation ditch for domestic wastewater treatment. *Bioresource Technology*, 101, 901–906.

Livak, K. J. and Schmittgen, T. D. 2001. Analysis of relative gene expression data using real-time quantitative PCR and the 2-$\Delta\Delta$CT method. Methods, 25, 402–408.

Lopez Vazquez, C. M. 2009. *The competition between polyphosphate-accumulating organisms and glycogen-accumulating organisms: Temperature effects and modelling.* TU Delft, Delft University of Technology.

Lotti, T., Van Der Star, W., Kleerebezem, R., Lubello, C. and Van Loosdrecht, M. 2012. The effect of nitrite inhibition on the anammox process. *Water Research*, 46, 2559–2569.

Loy, A., Horn, M. and Wagner, M. 2003. ProbeBase: An online resource for rRNA-targeted oligonucleotide probes. *Nucleic Acids Research*, 31, 514–516.

Ma, H., Shieh, K.-J., Chen, G., Qiao, T. and Chuang, M.-Y. 2006. Application of real time polymerase chain reaction (RT-PCR). *Journal of American Science*, 2, 1–5.

Ma, J. F., Goto, S., Tamai, K. and Ichii, M. 2001. Role of root hairs and lateral roots in silicon uptake by rice. *Plant Physiology*, 127, 1773–1780.

Madoni, P., Davoli, D. and Gibin, G. 2000. Survey of filamentous microorganisms from bulking and foaming activated-sludge plants in Italy. *Water Research*, 34, 1767–1772.

Maixner, F., Noguera, D. R., Anneser, B., Stoecker, K., Wegl, G., Wagner, M. and Daims, H. 2006. Nitrite concentration influences the population structure of Nitrospira-like bacteria. *Environmental Microbiology*, 8, 1487–1495.

Malinen, E., Kassinen, A., Rinttilä, T. and Palva, A. 2003. Comparison of real-time PCR with SYBR Green I or 5′-nuclease assays and dot-blot hybridization with rDNA-targeted oligonucleotide probes in quantification of selected faecal bacteria. *Microbiology*, 149, 269–277.

Mardis, E. R. 2008. Next-generation DNA sequencing methods. *Annual Review of Genomics and Human Genetics*, 9, 387–402.

Marrengane, Z., Kumar, S. K. S., Pillay, L. and Bux, F. 2011. Rapid quantification and analysis of genetic diversity among Gordonia populations in foaming activated sludge plants. *Journal of Basic Microbiology*, 51, 415–423.

Matsumoto, S., Katoku, M., Saeki, G., Terada, A., Aoi, Y., Tsuneda, S., Picioreanu, C. and Van Loosdrecht, M. 2010. Microbial community structure in autotrophic nitrifying granules characterized by experimental and simulation analyses. *Environmental Microbiology*, 12, 192–206.

Mchugh, S., O'Reilly, C., Mahony, T., Colleran, E. and O'Flaherty, V. 2003. Anaerobic granular sludge bioreactor technology. *Rev. Environmental Science and Biotechnology*, 2, 225–245.

Mielczarek, A. T., Kragelund, C., Eriksen, P. S. and Nielsen, P. H. 2012. Population dynamics of filamentous bacteria in Danish wastewater treatment plants with nutrient removal. *Water Research*, 46, 3781–3795.

Miłobędzka, A. and Muszyński, A. 2015. Population dynamics of filamentous bacteria identified in Polish full-scale wastewater treatment plants with nutrients removal. *Water Science and Technology*, 71, 675–684.

Mobarry, B. K., Wagner, M., Urbain, V., Rittmann, B. E. and Stahl, D. A. 1996. Phylogenetic probes for analyzing abundance and spatial organization of nitrifying bacteria. *Applied and Environmental Microbiology*, 62, 2156–2162.

Morisset, D., Štebih, D., Milavec, M., Gruden, K. and Žel, J. 2013. Quantitative analysis of food and feed samples with droplet digital PCR. *PLoS One*, 8, e62583.

Morley, A. A. 2014. Digital PCR: A brief history. *Biomolecular Detection and Quantification*, 1, 1–2.

Morozova, O. and Marra, M. A. 2008. Applications of next-generation sequencing technologies in functional genomics. *Genomics*, 92, 255–264.

Mota, C. R., So, M. J. and De Los Reyes III, F. L. 2012. Identification of nitrite-reducing bacteria using sequential mRNA fluorescence *in situ* hybridization and fluorescence-assisted cell sorting. *Microbial Ecology* 64(1), 256–267.

Musa, J. J. 2014. Effect of domestic waste leachates on quality parameters of groundwater. *Leonardo Journal of Sciences*, 24, 28–38.

Nakasaki, K., Kwon, S. H. and Takemoto, Y. 2015. An interesting correlation between methane production rates and archaea cell density during anaerobic digestion with increasing organic loading. *Biomass and Bioenergy*, 78, 17–24.

Neef, A., Amann, R., Schlesner, H. and Schleifer, K.-H. 1998. Monitoring a widespread bacterial group: *In situ* detection of planctomycetes with 16S rRNA-targeted probes. *Microbiology*, 144, 3257–3266.

Nicol, G. W. and Schleper, C. 2006. Ammonia-oxidising Crenarchaeota: Important players in the nitrogen cycle?. *Trends in Microbiology*, 14, 207–212.

Oehmen, A., Lemos, P. C., Carvalho, G., Yuan, Z., Keller, J., Blackall, L. L. and Reis, M. A. M. 2007. Advances in enhanced biological phosphorus removal: From micro to macro scale. *Water Research*, 41, 2271–2300.

Ong, Y. H., Chua, A. S. M., Fukushima, T., Ngoh, G. C., Shoji, T. and Michinaka, A. 2014. High-temperature EBPR process: The performance, analysis of PAOs and GAOs and the fine-scale population study of Candidatus "*Accumulibacter phosphatis.*" *Water Research*, 64, 102–112.

Onwughara, N. I., Umeobika, U. C., Obianuko, P. N. and Iloamaeke, I. M. 2011. Emphasis on effects of storm runoff in mobilizing the heavy metals from leachate on waste deposit to contaminate nigerian waters: Improved water quality standards. *International Journal of Environmental Science and Development*, 2, 55–63.

O'Reilly, J., Lee, C., Collins, G., Chinalia, F., Mahony, T. and O'Flaherty, V. 2009. Quantitative and qualitative analysis of methanogenic communities in mesophilically and psychrophilically cultivated anaerobic granular biofilims. *Water Research*, 43, 3365–3374.

Park, H., Rosenthal, A., Ramalingam, K., Fillos, J., and Chandran, K. 2010. Linking community profiles, gene expression and N-removal in anammox bioreactors treating municipal anaerobic digestion reject water. *Environmental Science and Technology*, 44, 6110–6116.

Parmar, A. M., Patel, K. D., Doshi, N. S., Kapadiya, G. M., Patel, B. S. and Sen, D. J. 2014. Correlation approach between shotgun sequencing with DNA sequencing in molecular genomics. *World Journal of Pharmacy and Pharmaceutical Science*, 3, 963–995.

Penton, C. R., Devol, A. H. and Tiedje, J. M. 2006. Molecular evidence for the broad distribution of anaerobic ammonium-oxidizing bacteria in freshwater and marine sediments. *Applied and Environmental Microbiology*, 72, 6829–6832.

Pernthaler, A. and Pernthaler, J. 2007. Fluorescence in situ hybridization for the identification of environmental microbes. *In:* Hilario, E. and Mackay, J. (eds.) *Protocols for Nucleic Acid Analysis by Nonradioactive Probes.* Totowa, NJ: Humana Press.

Pernthaler, A., Pernthaler, J. and Amann, R. 2002. Fluorescence in situ hybridization and catalyzed reporter deposition for the identification of marine bacteria. *Applied and Environmental Microbiology*, 68, 3094–3101.

Pernthaler, J. 2005. Predation on prokaryotes in the water column and its ecological implications. *Nature Reviews Microbiology*, 3, 537–546.

Perry-O'Keefe, H., Rigby, S., Oliveira, K., Sørensen, D., Stender, H., Coull, J. and Hyldig-Nielsen, J. 2001. Identification of indicator microorganisms using a standardized PNA FISH method. *Journal of Microbiological Methods*, 47, 281–292.

Pester, M., Schleper, C. and Wagner, M. 2011. The Thaumarchaeota: An emerging view of their phylogeny and ecophysiology. *Current Opinion in Microbiology*, 14, 300–306.

Pfaffl, M. W., Tichopad, A., Prgomet, C. and Neuvians, T. P. 2004. Determination of stable housekeeping genes, differentially regulated target genes and sample integrity: BestKeeper—Excel-based tool using pair-wise correlations. *Biotechnology Letters*, 26, 509–515.

Pommerening-Rösera, A., Rath, G. and Koops, H. P. 1996. Phylogenetic diversity within the genus *Nitrosomonas*. *Systematic and Applied Microbiology*, 19, 344–351.

Poulsen, L. K., Ballard, G. and Stahl, D. A. 1993. Use of rRNA fluorescence *in situ* hybridization for measuring the activity of single cells in young and established biofilms. *Applied and Environmental Microbiology*, 59, 1354–1360.

Quail, M. A., Smith, M., Coupland, P., Otto, T. D., Harris, S. R., Connor, T. R., Bertoni, A., Swerdlow, H. P. and Gu, Y. 2012. A tale of three next generation sequencing platforms: Comparison of Ion Torrent, Pacific Biosciences and Illumina MiSeq sequencers. *BMC Genomics*, 13, 341.

Rački, N., Morisset, D., Gutierrez-Aguirre, I. and Ravnikar, M. 2014. One-step RT-droplet digital PCR: A breakthrough in the quantification of waterborne RNA viruses. *Analytical and Bioanalytical Chemistry*, 406, 661–667.

Radax, R., Hoffmann, F., Rapp, H.T., Leininger, S., and Schleper, C. 2012. Ammonia-oxidizing archaea as main drivers of nitrification in cold-water sponges. *Environmental Microbiology* 14, 909–923.

Ramdhani, N. 2012. *Detection and Quantification of Nitrifying Bacteria from South African Biological Nutrient Removal Plants*. South Africa: Durban University of Technology.

Ramond, J. B., Lako, J. D. W., Stafford, W. H. L., Tuffin, M. I. and Cowan, D. A. 2015. Evidence of novel plant-species specific ammonia oxidizing bacterial clades in acidic South African fynbos soils. *Journal of Basic Microbiology*, 55, 1040–1047.

Rattray, J. E., Van De Vossenberg, J., Hopmans, E. C., Kartal, B., Van Niftrik, L., Rijpstra, W. I. C., Strous, M., Jetten, M. S., Schouten, S. and Damsté, J. S. S. 2008. Ladderane lipid distribution in four genera of anammox bacteria. *Archives of Microbiology*, 190, 51–66.

Ronaghi, M. 2001. Pyrosequencing sheds light on DNA sequencing. *Genome Research*, 11, 3–11.

Rotthauwe, J.-H., Witzel, K.-P. and Liesack, W. 1997. The ammonia mono-oxygenase structural gene amoA as a functional marker: Molecular fine-scale analysis of natural ammonia-oxidizing populations. *Applied and Environmental Microbiology*, 63, 4704–4712.

Sanapareddy, N., Hamp, T. J., Gonzalez, L. C., Hilger, H. A., Fodor, A. A. and Clinton, S. M. 2009. Molecular diversity of a North Carolina wastewater treatment plant as revealed by pyrosequencing. *Applied and Environmental Microbiology*, 75, 1688–1696.

Sanz, J. L. and Kochling, T. 2007. Molecular biology techniques used in wastewater treatment: An overview *Process Biochemistry* 42, 119–133.

Schmid, M., Schmitz-Esser, S., Jetten, M. and Wagner, M. 2001. 16S-23S rDNA intergenic spacer and 23S rDNA of anaerobic ammonium-oxidizing bacteria: Implications for phylogeny and *in situ* detection. *Environmental Microbiology*, 3, 450–459.

Schmid, M., Twachtmann, U., Klein, M., Strous, M., Juretschko, S., Jetten, M., Metzger, J. W., Schleifer, K.-H. and Wagner, M. 2000. Molecular evidence for genus level diversity of bacteria capable of catalyzing anaerobic ammonium oxidation. *Systematic and Applied Microbiology*, 23, 93–106.

Schmidt, F. R. 2005. Optimization and scale up of industrial fermentation processes. *Applied Microbiology and Biotechnology*, 68, 425–435.

Schmidt, I., Sliekers, O., Schmid, M., Bock, E., Fuerst, J., Kuenen, J. G., Jetten, M. S. and Strous, M. 2003. New concepts of microbial treatment processes for the nitrogen removal in wastewater. *FEMS Microbiology Reviews*, 27, 481–492.

Schramm, A., De Beer, D., Wagner, M. and Amann, R. 1998. Identification and activities *in situ* of Nitrosospira and Nitrospira spp. as dominant populations in a nitrifying fluidized bed reactor. *Applied and Environmental Microbiology*, 64, 3480–3485.

Seviour, E., Williams, C., Degrey, B., Soddell, J., Seviour, R. and Lindrea, K. 1994. Studies on filamentous bacteria from Australian activated sludge plants. *Water Research*, 28, 2335–2342.

Seviour, R., Lindrea, K., Griffiths, P. and Blackall, L. 1998. The activated sludge process. *In*: Seviour R. J. and Blackall L. L. (eds.), *The Microbiology of Activated Sludge*. The Netherlands: Springer.

Schmid, M.C., Maas, B., Dapena, A., Van De Pas-Schoonen, K., Van De Vossenberg, J., Kartal, B. et al. 2005. Biomarkers for in situ detection of anaerobic ammonium-oxidizing (Anammox) bacteria. *Applied and Environmental Microbiology*, 71, 1677–1684.

Shokralla, S., Spall, J. L., Gibson, J. F. and Hajibabaei, M. 2012. Next-generation sequencing technologies for environmental DNA research. *Molecular Ecology*, 21, 1794–1805.

Shu, Q., Jiao, N., Xu, G. and Shen, Z. 2011. Variation of abundance of Planctomycetes in typical aquatic environments of the China seas. *African Journal of Microbiology Research*, 5, 5208–5214.

Siripong, S., Kelly, J. J., Stahl, D. A. and Rittmann, B. E. 2006. Impact of prehybridization PCR amplification on microarray detection of nitrifying bacteria in wastewater treatment plant samples. *Environmental Microbiology*, 8, 1564–1574.

Sivaganesan, M., Siefring, S., Varma, M. and Haugland, R. A. 2011. MPN estimation of qPCR target sequence recoveries from whole cell calibrator samples. *Journal of Microbiological Methods*, 87, 343–349.

Sorokin, D. Y., Lucker, S., Vejmelkova, D., Kostrikina, N. A., Kleerebezem, R., Rijpstra, W. I., Damste, J. S. et al. 2012. Nitrification expanded: Discovery, physiology and genomics of a nitrite-oxidizing bacterium from the phylum Chloroflexi. *ISME Journal*, 6, 2245–2256.

Stahl, D. A. and De La Torre, J. R. 2012. Physiology and diversity of ammonia-oxidizing archaea. *Annual Review of Microbiology*, 66, 83–101.

Stams, A. J., Sousa, D. Z., Kleerebezem, R. and Plugge, C. M. 2012. Role of syntrophic microbial communities in high-rate methanogenic bioreactors. *Water Science and Technology*, 66(2), 352–362.

Strous, M., Kuenen, J. G. and Jetten, M. S. 1999. Key physiology of anaerobic ammonium oxidation. *Applied and Environmental Microbiology*, 65, 3248–3250.

Strous, M., Pelletier, E., Mangenot, S., Rattei, T., Lehner, A., Taylor, M. W., Horn, M. et al. 2006. Deciphering the evolution and metabolism of an anammox bacterium from a community genome. *Nature*, 440, 790–794.

Tandoi, V., Jenkins, D. and Wanner, J. 2006. *Activated Sludge Separation Problems: Theory, Control Measures, Practical Experiences*. UK, IWA Publishing.

Teira, E., Reinthaler, T., Pernthaler, A., Pernthaler, J. and Herndl, G. J. 2004. Combining catalyzed reporter deposition-fluorescence *in situ* hybridization and microautoradiography to detect substrate utilization by Bacteria and Archaea in the deep ocean. *Applied and Environmental Microbiology*, 70, 4411–4414.

Tenover, F. C. and Moellering, R. C. 2007. The rationale for revising the Clinical and Laboratory Standards Institute vancomycin minimal inhibitory concentration interpretive criteria for Staphylococcus aureus. *Clinical Infectious Diseases*, 44, 1208–1215.

Terada, A., Zhou, S. and Hosomi, M. 2011. Presence and detection of anaerobic ammonium-oxidizing (anammox) bacteria and appraisal of anammox process for high-strength nitrogenous wastewater treatment: A review. *Clean Technologies and Environmental Policy*, 13, 759–781.

Tsushima, I., Kindaichi, T. and Okabe, S. 2007. Quantification of anaerobic ammonium-oxidizing bacteria in enrichment cultures by real-time PCR. *Water Research*, 41, 785–794.

Uyom, U. U., Ama, O. K. and Ephraim, N. I. 2014. Some physical and chemical characteristics of Akpa Yafe River, Bakassi, cross river state, Nigeria. *Journal of Academia and Industrial Research (JAIR)*, 2, 631–637.

Vandenberg, L. N., Colborn, T., Hayes, T. B., Heindel, J. J., Jacobs JR, D. R., Lee, D.-H., Shioda, T., Soto, A. M., Vom Saal, F. S. and Welshons, W. V. 2012. Hormones and endocrine-disrupting chemicals: Low-dose effects and non-monotonic dose responses. *Endocrine Reviews*, 33(3), 378–455.

Van Der Star, W. R., Abma, W. R., Blommers, D., Mulder, J.-W., Tokutomi, T., Strous, M., Picioreanu, C. and Van Loosdrecht, M. C. 2007. Startup of reactors for anoxic ammonium oxidation: Experiences from the first full-scale anammox reactor in Rotterdam. *Water Research*, 41, 4149–4163.

Vanysacker, L., Denis, C., Roels, J., Verhaeghe, K. and Vankelecom, I. F. 2014. Development and evaluation of a TaqMan duplex real-time PCR quantification method for reliable enumeration of Candidatus Microthrix. *Journal of Microbiological Methods*, 97, 6–14.

Varshney, R. K., Nayak, S. N., May, G. D. and Jackson, S. A. 2009. Next-generation sequencing technologies and their implications for crop genetics and breeding. *Trends in Biotechnology*, 27, 522–530.

Vieno, N., Tuhkanen, T. and Kronberg, L. 2007. Elimination of pharmaceuticals in sewage treatment plants in Finland. *Water Research*, 41, 1001–1012.

Wagner, M., Erhart, R., Manz, W., Amann, R., Lemmer, H., Wedi, D. and Schleifer, K. 1994. Development of an rRNA-targeted oligonucleotide probe specific for the genus *Acinetobacter* and its application for *in situ* monitoring in activated sludge. *Applied and Environmental Microbiology*, 60, 792–800.

Wagner, M., Rath, G., Koops, H. P., Flood, J. and Amann, R. 1996. In situ analysis of nitrifying bacteria in sewage treatment plants. *Water Science and Technology*, 34, 237–244.

Wang, J., Li, Q., Qi, R., Tandoi, V. and Yang, M. 2014a. Sludge bulking impact on relevant bacterial populations in a full-scale municipal wastewater treatment plant. *Process Biochemistry*, 49, 2258–2265.

Wang, Z., Zhang, X.-X., Lu, X., Liu, B., Yan, L., Long, C. and Li, A. 2014b. Abundance and diversity of bacterial nitrifiers and denitrifiers and their functional genes in tannery wastewater treatment plants revealed by high throughput sequencing. *PLoS One*, 9(11), e113603.

Witzig, R., Manz, W., Rosenberger, S., Kruger, U., Kraume, M. and Szewzyk, U. 2002. Microbiological aspects of a bioreactor with submerged membranes for aerobic treatment of municipal wastewater. *Water Research*, 36, 394–402.

Woebken, D., Lam, P., Kuypers, M. M., Naqvi, S., Kartal, B., Strous, M., Jetten, M. S., Fuchs, B. M. and Amann, R. 2008. A microdiversity study of anammox bacteria reveals a novel Candidatus Scalindua phylotype in marine oxygen minimum zones. *Environmental Microbiology*, 10, 3106–3119.

Wong, M. L. and Medrano, J. F. 2005. Real-time PCR for mRNA quantitation. *Biotechniques*, 39, 75.

Xia, S., Li, J. and Wang, R. 2008. Nitrogen removal performance and microbial community structure dynamics response to carbon nitrogen ratio in a compact suspended carrier biofilm reactor. *Ecological Engineering*, 32, 256–262.

Yang, C., Zhang, W., Liu, R., Li, Q., Li, B., Wang, S., Song, C., Qiao, C. and Mulchandani, A. 2011. Phylogenetic diversity and metabolic potential of activated sludge microbial communities in full-scale wastewater treatment plants. *Environmental Science and Technology*, 45, 7408–7415.

Ye, L., Shao, M. F., Zhang, T., Tong, A. H. and Lok, S. 2011. Analysis of the bacterial community in a laboratory-scale nitrification reactor and a wastewater treatment plant by 454-pyrosequencing. *Water Research*, 45, 4390–4398.

Yilmaz, S., Haroon, M. F., Rabkin, B. A., Tyson, G. W. and Hugenholtz, P. 2010. Fixation-free fluorescence *in situ* hybridization for targeted enrichment of microbial populations. *The ISME Journal*, 4, 1352–1356.

Yilmaz, L.S., Okten, H.E., Noguera, D.R. 2006. Making all parts of the 16S rRNA of Escherichia coli accessible in situ to single DNA oligonucleotides. Appl. Environ. Microbiol. 72(1), 733–744.

You, J., Das, A., Dolan, E. M. and Hu, Z. 2009. Ammonia-oxidizing archaea involved in nitrogen removal. *Water Research*, 43, 1801–1809.

Yu, L., Wensel, P. C., Ma, J. and Chen, S. 2013. Mathematical modeling in anaerobic digestion (AD). *Journal of Bioremediation and Biodegradation S*, 4, 2.

Yu, Y., Kim, J. and Hwang, S. 2006. Use of real-time PCR for group-specific quantification of aceticlastic methanogens in anaerobic processes: Population dynamics and community structures. *Biotechnology and Bioengineering*, 93, 424–433.

Yu, Y., Lee, C., Kim, J. and Hwang, S. 2005. Group-specific primer and probe sets to detect methanogenic communities using quantitative realtime polymerase chain reaction. *Biotechnology and Bioengineering*, 89, 670–679.

Yu, Z., Yang, J. and Liu, L. 2014. Denitrifier community in the oxygen minimum zone of a subtropical deep reservoir. *PloS one*, 9(3), e92055.

Zeng, Y., De Guardia, A., Ziebal, C., De Macedo, F. J. and Dabert, P. 2012. Nitrification and microbiological evolution during aerobic treatment of municipal solid wastes. *Bioresource Technology*, 110, 144–152.

Zhang, T., Ye, L., Tong, A. H., Shao, M. F. and Lok, S. 2011. Ammonia-oxidizing archaea and ammonia-oxidizing bacteria in six full-scale wastewater treatment bioreactors. *Applied Microbiology and Biotechnology*, 91, 1215–1225.

Zhu, X., Tian, J., Liu, C. and Chen, L. 2013. Composition and dynamics of microbial community in a zeolite biofilter-membrane bioreactor treating coking wastewater. *Applied Microbiology and Biotechnology*, 97, 8767–8775.

Zielińska, M., Cydzik-Kwiatkowska, A. and Wojnowska-Baryła, I. 2012. Changes in microbial communities of nitrifying immobilized biomass: The role of operational conditions. *Polish Journal of Environmental Studies*, 21.

Ziganshin, A., Schmidt, T., Scholwin, F., Il'inskaya, O., Harms, H. and Kleinsteuber, S. 2011. Bacteria and archaea involved in anaerobic digestion of distillers grains with solubles. *Applied Microbiology and Biotechnology*, 89, 2039–2052.

Zwirglmaier, K. 2005. Fluorescence *in situ* hybridisation (FISH)–the next generation. *FEMS Microbiology Letters*, 246, 151–158.

CHAPTER 5

Thermostable Enzymes and Their Industrial Applications

Santhosh Kumar, Nanthakumar Arumugam, Kugenthiren Permaul, and Suren Singh

CONTENTS

ABSTRACT

Thermophilic organisms provide a reservoir for a plethora of enzymes that are generally thermostable and many of them can also withstand denaturants of extremely acidic or alkaline nature, which is ideal for many industrial applications. Furthermore, thermostable enzymes have better substrate solubility, high mass transfer rate, and lowered risk of contamination. Enzyme production from thermophilic microorganisms can be achieved either by using optimized fermentation conditions or cloning and expressing the functional genes in fast-growing bacteria or yeast by recombinant DNA technology. Latest developments in the field of genetic and protein engineering provide a platform for the advancement of enzymes with superior properties. In this chapter, the commonly used industrial enzymes, their source microorganisms, properties, and applications are discussed.

INTRODUCTION

Environmental and economic concerns over the application of conventional chemical processes in industries have led scientists to explore alternative technologies that are safer and that do not negatively impact the environment. Biocatalysis that exploits the catalytic potential of enzymes through enantioselectivity and regioselectivity under appropriate conditions has emerged as a promising approach to the chemical synthesis of novel and industrially significant compounds (Schmid et al. 2001; Shoemaker et al. 2003). Enzymes have been used in several manufacturing processes since ancient times, in the production of food products, such as cheese, sourdough, beer, wine and vinegar, and in the manufacture of commodities such as leather, indigo, and linen. The development of fermentation processes during the latter part of the last century has resulted in the production of enzymes on a large scale. The continuously expanding applications of enzymes for the chemical, pharmaceutical, and food industries are creating a growing demand for biocatalysts that exhibit improved or new properties. Advancements in biotechnology, especially in the area of genetics and protein engineering have made it possible to provide tailor-made enzymes displaying new activities and adapted to new process conditions, enabling a further expansion of their industrial use. Based on such favorable properties, enzymes are widely used as catalysts and processing aids in many industrial processes.

Despite the fact that more than 3000 diverse enzymes have been identified and many of these have found their way into biotechnological and industrial applications, the current enzyme pool is still insufficient to meet all demands. A major reason for this is the fact that many available enzymes are incapable of withstanding harsh industrial reaction conditions. This has resulted in the screening and characterization of thermophilic microorganisms that are able to thrive in extreme environments as a potential enzyme source. Such enzymes generally exhibit good temperature, pH and enzymatic stability, and faster reaction rates, which are important parameters for the industrial application of any enzyme. Some commonly used industrial enzymes and their application are described in this chapter.

XYLANASES

Among the various hydrolases, hemicellulases are a diverse group of enzymes that are widely used in industry. Xylanases are glycosidases (O-glycoside hydrolases) that catalyzes the endohydrolysis of 1, 4-β-D-xylosidic linkages in the xylan backbone resulting in its conversion into xylooligosaccharides and xylose (Verma and Satyanarayana 2012; Rakotoarivonina et al. 2015). However, the extend of xylanase action depends on the type, solubility, degree of polymerism, and degree of substitution of the polysaccharide. Endoxylanases have also been reported to catalyze intermolecular transglycosylation in the presence of high concentrations of xylooligomers (Masui et al. 2012). Furthermore, there are also reports regarding the existence of multiple forms of xylanases produced by microorganisms (Turner et al. 2007). Elegir et al. (1994) reported that *Streptomyces* sp. B-12-2 produced five endoxylanases when grown on oat spelt xylan. Around 15 xylanases have been reported from the culture filtrates of *Aspergillus niger* and 13 xylanases from *Trichoderma viride* (Biely 1985). From *Phanerochaete chrysosporium*, more than 30 different xylanases have been reported when grown on Avicel (Dobozi et al. 1992). Heteroxylans having a complex structure require the action of multiple xylanases with overlapping but different specificities as all of the xylosidic linkages in the substrates are not equally accessible to xylan-degrading enzymes (Elleuche et al. 2015).

Enzyme Sources

Xylanases are produced by a plethora of organisms including bacteria, algae, fungi, protozoa, gastropods, and arthropods (Knob and Carmona 2010) and some members of higher animals, including freshwater mollusks

(Yamura et al. 1997). Most of the microbial xylanases have been reported from bacteria and fungi (Kulkarni et al. 1999; Chavez et al. 2006); however, many of them are mesophilic in nature. A number of thermophilic (optimal growth at 50–80°C) and hyperthermophilic (optimal growth at >80°C) xylanase-producing microorganisms have been isolated from a variety of sources, including terrestrial and marine solfataric fields, thermal springs, hot pools, volcanic islands, composts, and self-heating decaying organic debris (Haki and Rakshit 2003; Singh et al. 2003; Cannio et al. 2004; Bouacem et al. 2014; Elleuche et al. 2015; Palavesam 2015).

Xylanases with high thermostability are better candidates for industrial applications, particularly for enzymatic hydrolysis at elevated temperatures (e.g., biopulping), where mesophilic xylanases fail to meet the desired results. Several thermophilic strains have been screened for the production of thermostable enzymes and they are found more appropriate for industrial applications, as compared with their mesophilic counterparts (Uday et al. 2016). Among the reported fungal xylanase producers, strains of *Thermomyces* have been found to produce high titers of thermostable xylanolytic enzymes (Kumar et al. 2009) with *Thermomyces lanuginosus* SSBP being the highest (Singh et al. 2000a,b).

Enzyme Production Level

Enzyme production level varies with different microorganisms depending on the class, genus and species as well as the type of media/substrate used for enzyme production. *Thermobacillus xylanilyticus* has been reported to produce xylanase with a specific activity of 480.66 U/mg (Rakotoarivonina et al. 2015) and *Stenotrophomonas maltophilia* strain X6 with 313.38 U/mg (Raj et al. 2013). Xylanase produced by *Halomonas meridiana* APCMST-KS4 has 26.13 U/mg specific activity (Palavesam 2015). *Clostridium thermocellum* has been reported to produce a family 10 xylanase with a specific activity of 93 U/mg and *Streptomyces lividans* produced family 11 xylanase with 119.5 U/mg specific activity (Gonçalves et al. 2015). Xylanase produced by *Streptomyces* sp. CS428 has 926,103 U/mg of specific activity (Pradeep et al. 2013). *Scytalidium thermophilum* has been observed to produce xylanase with a specific activity of 841.1 U/mg (Kocabaş et al. 2015). Among the different thermophilic fungal strains, *T. lanuginosus* SSBP has recorded the highest endoxylanase (family 11) activity of 3575.28 U/mg (Singh et al. 2000a,b) when grown on corn cob media. *Penicillium janczewskii* also produced xylanase with a specific activity of 179.1 U/mg (Terrasan et al. 2013).

There are several reports on the cloning and expression of xylanase on prokaryotic and eukaryotic hosts. A xylanase gene from the extremely thermophilic bacterium *Geobacillus thermoleovorans* was cloned and expressed in *Escherichia coli* BL21 (DE3) with a specific activity of 270 U/mg (Verma and Satyanarayana 2012). Similarly family 10 endoxylanase gene from *Geobacillus* sp. WSUCF1 was cloned and expressed in *E. coli* with a specific activity of 461 U/mg (Bhalla et al. 2014). The genes encoding for endoxylanase of *B. subtilis* M015 was also expressed in *E. coli* JE5505 (Banka et al. 2014). A highly thermostable xylanase from *Thermotoga thermarum* was cloned and expressed in *E. coli* BL21 (DE3) with a specific activity of 145.8 U/mg (Shi et al. 2013). Recombinant xylanases have also been expressed in *E. coli* and *Pichia pastoris* from *Neocallimastix patriciarum* with a specific activity of 5778.3 U/mg and 7995.3 U/mg, respectively (Cheng et al. 2014). An overexpression of xylanase from *Penicillium occitanis* Pol6 in *P. pastoris* was also observed with a specific activity of 8549.85 U/mg (Uday et al. 2016).

Applications

Among the different hemicellulase enzymes, xylanases represent the major commercial proportion; however, they only constitute a small percentage of the total enzyme market. Microbial xylanases have attracted a great deal of attention owing to their biotechnological potential in various industrial processes such as food, feed, and pulp and paper industries (Chavez et al. 2006; Patel and Savanth 2015). Other potential applications include the conversion of xylan in biomass from food and agricultural industry into xylose and xylooligosaccharides (Bhalla et al. 2013), and in the bioconversion of lignocellulosic materials to fuels and chemical feedstocks (Banka et al. 2014; Thomas et al. 2014; Palavesam 2015). Other less well documented putative applications include: brewing, to increase wort filterability and reduce haze in the final product (Tikhomirov et al. 2003; Raj et al. 2013; Elleuche et al. 2015); in coffee extraction and in the preparation of soluble coffee (Wong et al. 1988); in detergents (Kamal Kumar et al. 2004); in the protoplastation of plant cells (Kulkarni et al. 1999); in the production of pharmacologically active polysaccharides for use as antimicrobial agents (Christakopoulos et al. 2003) or antioxidants (Katapodis et al. 2003); in the production of alkyl glycosides for use as surfactants (Matsumura et al. 1999); and in the washing of precision devices and semiconductors (Imanaka and Sakurai 1992).

Xylanases are of significant importance to the pulp and paper industries because hydrolysis of xylan by xylanase facilitates the release of lignin from pulp and aid in reducing the level of chlorine usage as a bleaching agent (Moraïs et al. 2011; Ellis and Magnuson 2012). In the food industry, xylanases are used in a number of applications including baking, juice preparation, and starch processing (MacCabe et al. 2002). Xylanases can also be used in bread-making, together with α-amylase, malting amylase, glucose oxidase, and proteases. Synergistic action of xylanase and cellulase mixtures could result in the efficient release of sugars from lignocelluloses (Harris et al. 2014; Gonçalves et al. 2015). Xylanases break down the hemicellulose in wheat-flour thereby helping in the redistribution of water and leaving the dough softer and easier to knead (Polizeli et al. 2005; Javier et al. 2007; Butt et al. 2008).

In juice making processes, xylanases, in conjunction with cellulases, amylases, and pectinases, help to improve the yield of juice by means of liquefaction of fruit; stabilization of the fruit pulp; increased recovery of aromas, essential oils, vitamins, mineral salts, edible dyes, pigments, etc. It also helps in the reduction of viscosity, hydrolysis of substances that hinder the physical or chemical clearing of the juice, or that may cause cloudiness in the concentrate. Xylanases are also used in animal feed along with glucanases, pectinases, cellulases, proteases, amylases, phytases, galactosidases, and lipases. These enzymes break down arabinoxylans in the ingredients of the feed, reducing the viscosity of the raw material (Polizeli et al. 2005; Knob et al. 2014).

CELLULASES

Cellulases, a general term used for cellulolytic enzymes, are composed of three classes of enzymes and are recognized on the basis of their mode of action and substrate specificities: endoglucanases (EC 3.2.1.4), exoglucanases (EC 3.2.1.74 and EC 3.2.1.91), and β-glucosidases (EC 3.2.1.21) (Várnai et al. 2014; Haq et al. 2015). Endo-β-glucanase acts randomly on the cellulose polysaccharide and produces cello-oligosaccharides, while exo-β-glucanase acts on the exposed chain ends by splitting off cellobiose (Raghuwanshi et al. 2014). The release of glucose is as a result of the synergetic action of these enzymes. Action of cellulases results in either exo- or endo-cleavage of the substrate and mostly all the cellulases target specifically on the β 1,4-glycosidic bonds (Juturu and Wu 2014), whereas β-glucosidases or cellobiases cleave the products of exocellulase into monosaccharides (Jabbour et al. 2012; Lee et al. 2015). Most of the reported cellulases are from mesophilic organisms that cannot fulfill the requirements for industrial application, where the physical factors, such as pH, temperature, ionic strength, acidity, and alkalinity are at their extremes (Gunny et al. 2014; Khelila and Cheba 2014; Bhalla et al. 2015). The performance of cellulase mixtures in biomass conversion processes relies on several of their properties including stability, product inhibition, specificity, synergism between the different enzymes, productive binding to the cellulose, physical characteristics as well as the composition of cellulosic biomass (Heinzelman et al. 2009).

Enzyme Sources

Many microorganisms have been reported to produce cellulases, which include bacteria, and fungi, aerobes and anaerobes, mesophiles and thermophiles. Organisms generally adopt two strategies for utilizing their cellulases: (i) distinct noncomplexed cellulases that are typically secreted by aerobic bacteria and fungi and (ii) complexed cellulases (cellulosome) that are typically expressed on the surface of anaerobic bacteria and fungi. Cellulose degradation mechanism by aerobic bacteria and fungi are similar; however, it differs with anaerobic fungi and bacteria (Kuhad et al. 2016).

A plethora of bacteria and archea belonging to diverse genera such as *Bacillus, Clostridia, Fervidobacterium, Rhodothermus, Thermoplasma, Thermotoga, Pyrococcus, Sulfolobus, Thermococcus,* and *Desulfurococcus* have been reported to produce thermostable cellulases with different properties (Kuhad et al. 2016). Most of the fungal cellulases are produced by the genus *Trichoderma*, however, of mesophilic nature. On the other hand, a thermo-alkalistable cellulase has been produced by extremophilic fungus *Penicillium citrinum* with multiple pH optima (Dutta et al. 2008).

Enzyme Production Level

Clostridium thermocellum has been reported to produce cellulase with a specific activity of 2.4 U/mg (Thomas et al. 2014). A recombinant cellulase from *Thermococcus* sp. AM4 was expressed in *E. coli* BL21 (DE3) which has a specific activity of 700.4 U/mg (Leis et al. 2015). A marine *Bacillus* VITRKHB has been reported to produce cellulase with a specific activity of 1.92 U/mg (Singh et al. 2013) and cellulase from *Bacillus* sp. BCCS A3 has been reported with a high level of enzyme production of 50.3 U/mL (Kazemi et al. 2014).

Xanthomonas sp. EC102 has shown to produce endoglucanase with a specific activity of 1.97 U/mg (Woo et al. 2014) and a recombinant endoglucanase from *C. thermocellum* ATCC 27405 has been reported with a specific activity of 30 U/mg (Haq et al. 2015).

Endocellulase gene from *Ciboria shiraiana* has been cloned into pPIC9K and expressed in *P. pastoris* has an activity of 17.44 U/mL (Lu et al. 2015). A gene encoding for cellobiohydrolase from *Chaetomium thermophilum* was cloned into pMD 18 T and expressed in *P. pastoris* (Li et al. 2009). Similarly, cellobiohydrolase gene from *Penicillium funiculosum* NCL1 was cloned into pPICZαA4 and expressed in *P. pastoris* with a specific activity of 0.8 U/mg (Chinnathambi et al. 2015).

Carboxy methyl cellulase (CMCase) gene from *C. thermocellum* was cloned into pET21a which was expressed in *E. coli* with a specific activity of 3.5 U/mg (Mutreja et al. 2011). Similarly, CMCase gene from *Neocallimastix* sp was cloned into pCT and expressed in *E. coli* EC100 with a specific activity of 2.06 U/mg (Comlekcioglu et al. 2010). *Bacillus* sp. BSS3 has produced CMCase which showed a maximum specific activity of 104.06 U/mL (Sreedevi et al. 2013). The mutant strain *T. asperellum* SR1-7 has produced CMCase (13.2 U/g) and β-glucosidase (9.2 U/g) simultaneously under controlled conditions (Raghuwanshi et al. 2014).

Applications

Cellulases have been commercially available for more than 3 decades and have also demonstrated their potential in industries such as food, animal feed, brewing and wine, agriculture, pulp and paper, textile, and laundry (Kuhad et al. 2011; Ferreira et al. 2014). However, the type of cellulases required is completely different with respect to different industries. The predominant application of cellulolytic enzymes is in biomass processing specifically for biofuel and bioenergy generation. The biofuel industry prefers thermostable cellulases that are resistant to acidic conditions, while the detergent industry prefers enzymes operating at higher pH with good thermostability (Kumar et al. 2011). Cellulolytic enzymes are also used in detergent industries for softening and color brightening, stoning of jeans, and in the pre-treatment of industrial wastes (Salahuddin et al. 2012).

AMYLASE

Amylases are mostly a group of extracellular enzymes that break down the complex polysaccharide starch to yield assorted products such as dextrins, maltose, glucose, and maltooligosaccharides (Jyoti et al. 2011; Janeček et al. 2014). The complete hydrolysis of starch requires a combination of enzymes which include α-amylases (EC 3.2.1.1), glucoamylases or β-amylases (EC 3.2.1.3), and pullulanases (Oziengbe and Onilude 2012). α-Amylase (endo-1,4-alpha-D-glucan glucohydrolase) randomly cleaves the 1,4-α-D-glucosidic linkages between the adjacent glucose units in the linear amylose chain. These endoacting enzymes cleave the substrate in the interior of the molecules and are categorized based on their properties and mode of action. Amylases that liberate free sugars are termed as "saccharogenic" and those that liquefy starch without generating free sugars are known as "starch-liquefying." Amylolytic enzymes are categorized by the similarities in their amino acid sequences and three-dimensional structures, reaction mechanisms, and catalytic machineries which reflect their evolutionary relatedness than specificity (Janeček et al. 2014).

Enzyme Sources

Amylases can be obtained from several sources such as plants, animals, and microorganisms. However, the enzymes from microbial sources particularly obtained from extreme environments, proved to be useful for industrial processes (Joshi 2011; Jyoti et al. 2011; Ibrahim et al. 2013; Zafar et al. 2015). Furthermore, microorganisms offer easy manipulation for obtaining α-amylases of desired characteristics with good expression levels (Abdel-fattah et al. 2013). Some of the bacteria that produce amylases are *Bacillus licheniformis* (Oziengbe and Onilude 2012), *B. licheniformis* BT5.9 (Ibrahim et al. 2013), *B. circulans* (Joshi 2011), *B. subtilis* JS-2004, *B. megaterium, Bacillus* sp. Strain PM1, *Pyrococcus furiosus* (Cuong et al. 2015), *Amphibacillus* sp. NM-Ra2 (Mesbah and Wiegel 2014), and *Halobacillus* sp. LY9 (Sharma et al. 2014). Some of the fungal species that produce amylases are *Rhizomucor pusillus* (He et al. 2014), *Penicillium fellutanum*, (Sharma et al. 2014), *Penicillium camemberti, Pestalotiopsis microspore, Aspergillus oryzae*, and *Acremonium sporosulcatum* (Rana et al. 2013).

Enzyme Production Level

Bacillus licheniformis JAR-26 was reported to produce α-amylase with a specific activity of 317.9 U/mg (Jyoti et al. 2011) and *B. licheniformis* AI20 produced amylase with a specific activity of 748.98 U/mg (Abdel-fattah et al. 2013). α-Amylase gene from *B. licheniformis* was cloned into a Gateway shuttle vector pMMC and expressed in *E. coli* and *B. megaterium* (Atanassov et al. 2013). Specific activity of amylase produced from *Bacillus* sp. strain EF_TYK1-5 was 132.44 U/mg (Pathak and Rekadwad 2013). *B. circulans* PN5 produced amylase which had 2625 U/mg of specific activity

(Joshi 2011). α-Amylase from *Amphibacillus* sp. NM-Ra2 has 250 U/mg of specific activity (Mesbah and Wiegel 2014). An α-amylase gene from the thermophilic bacterium *B. subtilis* was cloned and expressed in *E. coli* DH5α which has 15950 U/mg specific activity (Park et al. 2013). *B. subtilis* A28 α-amylase gene was also cloned and expressed *in E. coli* with a specific activity of 2814 U/mg (Ozturk et al. 2013). A gene encoded for α-amylase from acidophilic bacterium *B. acidicola* was cloned into pET28a(+) vector and expressed in *E. coli* BL21 (DE3) with a specific activity of 1166 U/mg (Sharma and Satyanarayana 2012). Similarly *Thermotoga petrophila* was cloned into pET-21a(+) and expressed in *E. coli* BL21 (DE3) and has a specific activity of 126.31 U/mg (Zafar et al. 2015). Amylase produced from *Thermoactinomyces thalpophilus* KSV 17 has a specific activity of 145.8 U/mg (Rao et al. 2012). α-Amylase gene of a fungal glucoamylase and the α-amylase genes from *Rhizomucor pusillus* were cloned, and expressed in *P. pastoris* which have a specific activity of 1953 U/mg and 20732 U/mg, respectively (He et al. 2014).

Applications

Amylases are the important digestive enzymes of starch which have various applications in several industries such as food, clinical, medical, and agriculture (Pathak and Rekadwad 2013; Sharma et al. 2014). In food industry, amylases aid in starch processing which includes starch liquefaction and saccharification and also in brewing and sugar production (Nigam 2013). Amylases are also used for baking (to delay the staling of bread), in textile industries for sizing of textile fibers and in detergent manufacturing processes (Joshi 2011; Jyoti et al. 2011; Mesbah and Wiegel 2014). Amylases are also used for the production of fructose and glucose by the enzymatic conversion of starch (Van Der Maarel et al. 2002). Amylases also play an important role in removing stains from fabrics when added to detergents (Sundarram and Murthy 2014).

PULLULANASE

Pullulanase (3.2.1.41) belongs to the α-amylase family and hydrolyses the glycosidic linkages in pullulan, amylopectin, starch, and glycogen (Li et al. 2015). Pullulanase is generally used as a debranching enzyme during starch saccharification. Pullulanases are grouped into four categories: pullulan hydrolase type I (neopullulanase), pullulan hydrolase type II (isopullulanase), pullulanase type I, and pullulanase type II (amylopullulanase)

(Ramanathan 2011). Microbial pullulanases are gaining more interest due to their specific action on α-1,6 linkages in pullulan, a linear α-glucan made of maltotriosyl units (Hii et al. 2012).

Enzyme Sources

Pullulanases are mainly produced by bacteria compared with fungi or other organisms. Pullulanase has been reported from *Anaerobranca gottschalkii, Fervidobacterium pennavorans, Thermotoga neapolitana, Bacillus acidopullulyticus,* and *Bacillus* sp. CICIM 263 (Kang et al. 2011; Li et al. 2012, 2015). Pullulanase has also been produced by *Bacillus flavocaldarius, B. acidopullulyticus, B. deramifican* (Duan and Wu 2015), *B. thermoleovorans* US 105, *Aspergillus niger* (Hii et al. 2012), *Pyrococcus furiosus, Pyrococcus woesei, Thermococcus aggregans, T. hydrothermalis, T. celer, T. hydrothermalis, Sulfolobus solfataricus, Thermus caldophilus, Thermoanaerobacter ethanolicus, Clostridium thermosulfurogenes,* and *Desulfurococcus mucosus* (Bertoldo and Antranikian 2002; Kang et al. 2004; Chiang et al. 2005; Mrudula et al. 2011; Ramanathan 2011). Pullulanase is also reported from *Staphylothermus marinus* (Li et al. 2013b), *Streptomyces* sp. No. 27, and *Geobacillus* sp. LM14-3 (Sun et al. 2011).

Enzyme Production Level

Bacillus halodurans has been reported to produce pullulanase with a specific activity of 87.64 U/mg (Asha et al. 2013). A high level of amylopullulanase was noticed with *Geobacillus thermoleovorans* NP33 with a specific activity of 1260 U/mg (Nisha and Satyanarayana 2013). *Staphylothermus marinus* produced amylopullulanase with a specific activity of 42.1 U/mg (Li et al. 2013b) and *S. erumpens* had a specific activity of 98.84 U/mg (Kar et al. 2012). *Thermus thermophilus* HB27 has been reported with 280 U/mg specific activity (Wu et al. 2014). A recombinant pullulanase from *Thermococcus kodakarensis* KOD1 was reported with a specific activity of 118 U/mg (Han et al. 2013) and pullulanase from *Geobacillus* sp. was expressed in *E. coli* BL21 (DE3) with a specific activity of 134.3 U/mg (Jasilionis et al. 2014). There are reports on the cloning of pullulanase from *Anaerobranca gottaschalkii* and expression in *E. coli* BL21 (DE3) with a specific activity of 56 U/mg.

Applications

Pullulanase is used as a principal enzyme for the industrial production of high-glucose and high-maltose syrups and has major applications in starch processing (Malakar et al. 2010; Li et al. 2015). Pullulanases also have applications in other industries. This includes production of cyclo-dextrins, liquefaction and saccharification of starch, making of low-calorie beer, and as an antistaling agent to improve texture, volume, and flavor of bakery products and also as a dental plaque control agent (Sun et al. 2011; Hii et al. 2012; Asha et al. 2013; Wu et al. 2014).

XYLOSIDASE

β-xylosidases (EC 3.2.1.37) are exo-type glycosidases that catalyze the hydrolysis of 1,4-β-D-xylooligosaccharides by removing successive xylose residues from the nonreducing termini. It also releases xylose from branched or substituted xylo-oligosaccharides produced by the action of endo-1,4-β-xylanases (Subramaniyan and Prema 2002; Biely 2003; Terrasan et al. 2013). The systematic name is 1,4-β-D-xylan xylohydrolase; however, the commonly used name is β-xylosidase and is found in families 3, 39, 43, 52, and 54 (Shallom and Shoham 2003). For many industrial applications such as improving bread dough, production of xylitol, and deinking of recycled paper, β-xylosidases are used in combination with xylanases (Jordan and Wagschal 2010).

Enzyme Sources

Various microorganisms including bacteria and fungi are reported to produce β-xylosidase; however, very few yeast strains are known to produce the enzyme (Basaran and Ozcan 2008). *Bacillus subtilis* M015 was reported to produce intracellular β-xylosidase (Banka et al. 2014). β-Xylosidase production has been documented from fungal species such as *Thermomyces lanuginosus* (Singh et al. 2000a), *Aspergillus awamori* (Paredes et al. 2015), *A. terricola* and *A. ochraceus* (Michelin et al. 2012a,b), *Neocallimastix frontalis* (Hebraud and Fevre 1990), *Neocallimastix* sp. M2 (Comlekcioglu et al. 2011), *Aspergillus japonicas* (Wakiyama et al. 2008), *Penicillium janthinellum* (Kundu and Ray 2013), *Fusarium verticillioides* (Saha 2001), *F. proliferatum* (Saha 2003), *Trichoderma reesei* RUT C-30 (Herrmann et al. 1997), and *Penicillium janczewskii* (Terrasan et al. 2013).

Enzyme Production Level

Bacillus thermantarcticus has been reported to produce β-xylosidase with a specific activity of 160 U/mg (Lama et al. 2004). β-Xylosidase with a specific activity of 261.1 U/mg was produced by *Alicyclobacillus* sp. A4 (Zhang et al. 2014) and 41.43 U/mg by *A. ochraceus* (Michelin, Peixoto-Nogueira, et al. 2012). *Paecilomyces thermophila* has been reported to produce β-xylosidase with 45.4 U/mg specific activity (Teng et al. 2011). A thermo-tolerant β-xylosidase of *Aspergillus* sp. BCC125 was cloned and expressed as a secreted protein using the *P. pastoris* KM71 expression system with a specific activity of 156 U/mg (Wongwisansri et al. 2013).

The genes (Xyn A, GH Family 11 and Xyn B, GH Family 43) encoding for β-xylosidase from *B. subtilis* M015 was isolated and expressed in *E. coli* JE5505 which has an activity of 2.75 ± 0.30 U/mL and 0.41 ± 0.02 U/mL, respectively (Banka et al. 2014). A β-xylosidase gene (Tlxyn1) from the thermophilic fungus *T. lanuginosus* SSBP was cloned and expressed in *P. pastoris* GS115 with a specific activity of 2.29 U/mg (Gramany et al. 2015). A gene (designated TlXyl43) encoding β-xylosidase was cloned from *T. lanuginosus* CAU44 and expressed in *E. coli* with a specific activity of 45.4 U/mg (Chen et al. 2012). Kirikyali et al. (2014) implemented the heterologous expression of *Aspergillus oryzae* β-xylosidase (XylA) in *P. pastoris* under the control of the glyceraldehyde-3-phosphate dehydrogenase promoter which has a specific activity of 150 U/mg.

Applications

β-xylosidase has immense biotechnological potential especially in food, pharmaceutical animal feed, paper, and pulp industries. It is also used in the bioconversion of lignocellulosic wastes into value-added chemicals (Beg et al. 2001; Chapla et al. 2010). β-Xylosidase in combination with xylanase cocktails is also used in the bleaching of pulp liquor (Marques et al. 2003; Kumar et al. 2009) and processing of wood pulp (Tsujibo et al. 2001). In food industry, it is used in the extraction of juice by hydrolyzing the bitter xylosylated compounds and liberates aroma from grapes during wine making (Jordan and Wagschal 2010). β-Xylosidase also plays a role in improving bread dough baking and nutritional quality (Dornez et al. 2007) and in the release of D-xylose residues from xylan for subsequent reduction to xylitol, a sweetener used in food industry (Polizeli et al. 2005; Jordan and Wagschal 2010).

LIPASE

Lipases (triacylglycerol acylhydrolases, EC 3.1.1.3) are lipolytic enzymes that catalyze the hydrolysis of long-chain triglycerides by forming diacylglycerides, monoglycerides, glycerol, and free fatty acids at the interface between the insoluble substrate and water (Masomian et al. 2013). Thermostable microbial lipases can be used for a variety of applications including many bioconversion processes. Hydrolysis, interesterification, alcoholysis, aminolysis, esterification, and acidolysis are some of the bioconversion processes that have been performed effectively by microbial lipases (Deive et al. 2012; Borrelli and Trono 2015). Their industrial significance is due to their distinctive characteristics such as substrate specificity, stereospecificity, regioselectivity, and the ability to catalyze heterogeneous reactions at the interface of water soluble and water insoluble systems (Haki and Rakshit 2003; Saxena et al. 2003; Li and Zhang 2005; Ebrahimpour et al. 2011; Puchart et al. 2015).

Enzyme Sources

Lipases are available in most of the flora and fauna and also in microbial sources such as bacteria, fungi, and yeasts. One of the major sources for lipases are *Bacillus* sp. such as *B. acidocaldarius, B. thermocatenletus, B. thermoleovorans,* and *Bacillus* sp. RSJ-1 (Nawani and Kaur 2000; Haki and Rakshit 2003; Dror et al. 2014; Espinosa-Luna et al. 2015). *Bacillus stearothermophilus* MC 7 has expressed lipase activity even at an elevated temperature of 75°C to 80°C (Kambourova et al. 2003). Lipase QL, an extracellular enzyme produced by *Alcaligenes* sp., has optimum activity at pH and temperature of 7.0 and 50°C, respectively (Wilson et al. 2006). An organic solvent-tolerant lipase was reported from *Aneurinibacillus thermoaerophilus* HZ with an optimal temperature and pH of 65°C and 7.0, respectively (Masomian et al. 2013). Lipase from *Aspergillus carneus* is reported to be active at alkaline pH and has extreme temperature tolerance (Saxena et al. 2003). Similarly, a lipase produced by *Thermomyces lanuginosus* is optimally active at 80°C and at a pH of 10 (Ávila-Cisneros et al. 2014). Some other fungi that are reported to produce lipases are *Rhizopus homothallicus, Candida rugosa, Penicillium simplicissimum, Humicola lanuginosa, A. niger,* etc.

Enzyme Production Level

Bacillus coagulans BTS-3 produced lipase with a specific activity of 4.8 U/mg (Kumar et al. 2005). *Bacillus thermoamylovorans* CH6B produced significant levels (0.45 U/mL) of extracellular lipase (Deive et al. 2012). Lipase gene from *Geobacillus* strain T1 has been cloned and expressed in *E. coli* with a specific activity of 30.19 U/mg (Leow et al. 2004). *Alcaligenes* sp. has produced a thermostable lipase with 0.049 U/mg specific activity (Wilson et al. 2006). Gutarra et al. (2009) have reported a lipase from *Penicillium simplicissimum* with a specific activity of 4.5 U/mg. A high level production of lipase was reported from *Aspergillus carneus* with a specific activity of 502 U/mg (Saxena et al. 2003). *Thermomyces lanuginosus* produced lipase with a specific activity of 0.12 U/mg (Ávila-Cisneros et al. 2014).

Applications

Thermostable lipases have a wide range of applications in various sectors including bakery food dressing, beverages, oleochemical, agrochemical, polymer synthesis, pulp and paper, leather, synthesis of surfactants, and pharmaceutical industries (Haki and Rakshit 2003; Sharma et al. 2011; Espinosa-Luna et al. 2015). In detergents, lipase is used to remove oil stains from fabric and in cleaning products, it aids in fat removal. Lipases are also used to treat oily wastewaters and for biodiesel production from vegetable oil (Cammarota and Freire 2006; Gutarra et al. 2009). Lipases are also used to enhance the flavor and aroma in dairy products (milk, cheese, and butter) and beverages. Lipases help the interesterification of fats and oils and they catalyze the hydrolysis of lipids in butter, fats, and cream. Lipases also serve as an emulsifier in food, cosmetics, and pharmaceuticals and they prolong the shelf life of bakery foods and improve their flavor (Sharma et al. 2011).

PHOSPHOLIPASE

Phospholipases, are a group of lipolytic enzymes that cleave the ester bonds of phospholipids, are classified into A, B, C, and D classes, based on the type of reaction they catalyze. Phospholipases A1 (PLA1, 3.1.1.32) and A2 (PLA2, 3.1.1.4) catalyze the hydrolysis of the ester bond at *sn*-1 and *sn*-2 positions, respectively, of the phospholipids, thus producing a free fatty acid and 2-acyl lysophospholipid or 1-acyl lysophospholipid, respectively (Borrelli and Trono 2015). Phospholipases A1, in general, constitute a large group of 1-acyl hydrolases, some of which also degrade neutral lipids (Istivan and Coloe 2006). Phospholipases B (PLB, 3.1.1.5) can hydrolyze fatty acids esterified at both the *sn*-1 or *sn*-2 position of the phospholipid. Phospholipase C (PLC, 3.1.4.3) breaks down the glycerophosphate bond, thus releasing diacylglycerol and the phosphorylated head group, while phospholipase D (PLD, 3.1.4.4) cleaves the terminal phosphodiesteric bond, thus releasing phosphatidic acid (PA) along with the head group (Borrelli and Trono 2015). Phospholipase D is abundant in nature and is secreted by a plethora of organisms ranging from viruses to bacteria, yeast, plants, and animals (Simkhada et al. 2009). Phospholipases have various functions, ranging from nutrient digestion to bioactive molecule formation, making them a vital enzyme in life (Istivan and Coloe

2006). Substrate specificity, stability to organic solvents, tolerance to high and low temperatures, and tolerance to acidic and alkaline pHs, tolerance to proteases are some of the outstanding properties of phospholipases (Wei et al. 2015). The amphipathic character of phospholipids restricts the enzymes by forming bilayers or micelles making them very rare to have as a single soluble substrate. Since all phospholipases target phospholipid as the substrate, variations are observed with their specific active site, mode of action and regulation (Istivan and Coloe 2006).

Enzyme Sources

Commercial phospholipases are mostly produced by yeasts and fungi, followed by bacteria. The most important genera of yeasts and fungi that are exploited for the production of phospholipases include *Saccharomyces cerevisiae, Schizosaccharomyces pombe, Candida albicans, Thermomyces lanuginosus, Tuber borchii, Gibberella zeae, Magnaporthe grisea, Aspergillus oryzae, A. fumigatus, A. nidulans,* and *Neurospora crassa* (Istivan and Coloe 2006; Borrelli and Trono 2015). The most important bacteria that have been investigated for the production of phospholipases are *Serratia liquefaciens, Yersinia enterocolitica, E. coli, Streptomyces alboflavus, S. coelicolor, S. olivochromogenes, S. violaceoruber, Ochrobactrum* sp., *Bacillus subtilis, B. cereus, Clostridium perfringens, Listeria monocytogenes, L. monocytogenes, Pseudomonas fluorescens, P. aeruginosa, P. cepacia, Thermotoga lettingae, Burkholderia pseudomallei,* and *Legionella pneumophila* (Hu et al. 2013; Borrelli and Trono 2015).

Enzyme Production Level

PLA_1 has been produced from *A. oryzae* with a specific activity of 2000 U/mg which was the higher level reported from filamentous fungi (Shiba et al. 2001). PLA_2 gene from a hyperthermophilic archaeon *Aeropyrum pernix* K1, which comprised 474 bases was cloned and expressed in *E. coli* BL21 (DE3) which has a specific activity of 120 U/mg (Wang et al. 2004). Phospholipase B from *Thermotoga lettingae* TMO has been cloned, and functionally overexpressed in *E. coli* with a specific activity of 158 U/mg (Wei et al. 2015). A lysophospholipase/PLB gene from the hyperthermophilic archaeon *Thermococcus kodakarensis* KOD1 (LysoPL-tk) was cloned and expressed in *E. coli* which has a specific activity of 95.5 U/mg (Cui et al. 2012). Phospholipase D was produced from *Streptomyces* sp. CS684 with a specific activity of 37.5 U/mg (Simkhada et al. 2009) and from *Ochrobactrum* sp. ASAG-PL1 with 83.5 U/mg specific activity (Hu et al. 2013).

Applications

Phospholipases have various functions, ranging from the breakdown of nutrients to the formation of bioactive molecules (Borrelli and Trono 2015). Among the fungal phospholipases, PLA1s and PLA2s from *Fusarium oxysporum, T. lanuginosus, A. niger,* and *Trichoderma reesei* have been commercialized and are used for the degumming of vegetable oils, while PLA1s, PLA2s, and PLBs from *A. oryzae* and *A. niger* have been used primarily in food industry (Maria et al. 2007; Casado et al. 2012). Furthermore, PLA2 proteins are of great interest to the pharmaceutical industry. They are responsible for the release of arachidonic acid from membranes, and the successive conversion of fatty acids to leukotrienes and prostaglandins (Istivan and Coloe 2006). PLDs from *Actinomycetes* strains are also commercially available and are used in many industrial processes, owing to their high transphosphatidylation and hydrolytic activities (Casado et al. 2012; Borrelli and Trono 2015). Phospholipases play essential roles in a number of different physiological processes, including phospholipid metabolism, signal transduction, cell cycle progression, cytoskeletal organization, and inflammatory responses (Cockcroft 2001; Cherif et al. 2010; Wei et al. 2015).

CHITINASE

Chitinases (EC 3.2.1.14) are essential enzymes that hydrolyze the β (1, 4) linkages of chitin and convert the polysaccharide to its monomeric or oligomeric components (low-molecular-weight products). The breakdown of chitin happens in two steps. First chitinase cleaves the chitin polymer into chitin oligosaccharides and further release N-acetylglucosamine, and monosaccharides catalyzed by chitobiases (Suginta et al. 2000; Hamid et al. 2013). Chitinases have been found in a wide range of organisms, including bacteria, plants, viruses, fungi, animals, insects, and crustaceans (Dahiya et al. 2006). The chitin-binding domain of bacterial chitinases can either be located in the amino terminal or in the carboxyl terminal domains of the enzyme. Fungal chitinases play an important role in the nutrition, morphogenesis, and fungal development processes (Hamid et al. 2013).

Enzyme Sources

The thermophilic organisms such as *Bacillus licheniformis* X-7u (Takayanagi et al. 1991), *Bacillus* sp. BG-11 (Bharat and Hoondal 1998), and *Streptomyces thermoviolaceus* OPC-520 (Tsujibo et al. 1995) were reported to be the chief sources of chitinases. Thermostable exochitinases were also isolated from *B. stearothermophilus* CH-4, isolated from a compost of organic solid wastes (Haki and Rakshit 2003). *Bacillus thuringiensis* subsp. kurstaki strain HBK-51 (Kuzu et al. 2012) and *B. cereus* (Liang et al. 2014) also produced thermostable and alkaline chitinase. *Brevibacillus laterosporus* also can produce chitinase which is stable at 70°C and 6–8 pH (Prasanna et al. 2013). Chitinase is also produced from the extreme thermophilic anaerobic arachaeon *Thermococcus chitinophagus* (Huber et al. 1995).

Enzyme Production Level

Chitinase has been reported from *B. cereus* with a specific activity of 16598 U/mg (Liang et al. 2014). *Bacillus* sp. Hu1 that was isolated from hot springs produced a chitinase with a specific activity of 11.1 U/mg (Dai et al. 2011). Another chitinase enzyme with a specific activity of 494.5 U/mg has been reported from *B. licheniformis* strain LHH100 and was cloned and expressed in *E. coli* (Laribi-Habchi et al. 2015). Chitinase gene from *Aeromonas veronii* was cloned into *P. pastoris* GS115 using pPIC9 vector with a specific activity of 553.8 U/mg (Y. Zhang et al. 2014). *Aeromonas hydrophila* SBK1 has been known to produce chitinase with 71.6 U/mg of specific activity (Halder et al. 2012). *Paenibacillus barengoltzii* has secreted chitinase (30.1 U/mg) after recombination into *E. coli* (Yang et al. 2016). Chitinase II was produced from *T. lanuginosus* with 150 ± 3.48 U/mg specific activity (Zhang et al. 2015).

Applications

Chitinolytic enzymes have a wide range of applications in diverse fields. This includes preparation of pharmaceutically important chitooligosaccharides with antimicrobial, anticholesterol, and antitumor activities. Chitinases are also used for the production of *N*-acetyl ᴅ-glucosamine, preparation of single-cell protein, isolation of protoplasts from fungi and yeast, treatment of chitin wastes, etc. (Haki and Rakshit 2003; Dahiya et al. 2006; Hamid et al. 2013). Chitinases have also been implicated in plant resistance against fungal pathogens and has shown considerable antifungal activities *in vitro* (Cho et al. 2011). Chitinases are also used as mosquitocides (Halder et al. 2012) and as biocontrol agents in agricultural applications against worms and insects that cause crop damage (Hamid et al. 2013; Liang et al. 2014).

LACCASES

Laccases (1.10.3.2) are copper-containing enzymes that belong to the group of blue oxidase which is produced by many bacteria, fungi, and yeasts (Lu et al. 2013). These enzymes are characterized by their unusual substrate specificity and a wide range of oxidizable substrates that depend on the source organism (Madhavi and Lele 2009). Laccases mostly have the structure as monomeric, dimeric, and tetrameric glycoprotein. The presence of glycosylation is important for copper retention, thermal stability, and susceptibility to proteolytic degradation. After purification, laccase enzymes show considerable heterogeneity. Variation in laccase glycosylation and composition of glycoprotein is a result of various growth medium composition (Shraddha et al. 2011). Substrates that are oxidized by laccases include mono-, di-, and polyphenols, methoxyphenols, aminophenols, aromatic amines, and ascorbate, with the associated four-electron reduction of oxygen to water (Madhavi and Lele 2009; Giardina et al. 2010).

Enzyme Sources

Majority of the laccases are reported from fungi, whereas few reports are from bacteria. Fungi belonging to the class *Ascomycetes, Basidiomycetes,* and *Deuteromycetes* have been implicated for the production of most laccases (Kiiskinen et al. 2004; Gochev and Krastanov 2007). Fugal species that are known to produce laccases are *Stereum ostrea, S. hirsutam, Fomitella fraxinea, Lentinus tigrinus, Trametes versicolor, T. hirsuta, T. ochracea, T. villosa, T. gallica, Ganoderma* sp. MK05, *Cerrena unicolor, C. byrsiana, C. maxima, H. cylindrosporum, Pycnoporus sanguineus, Trichoderma harzianum, Coriolopsis polyzona, Pleurotus eryngii,* and *P. ostreatus.* Bacterial species such as *B. subtilis, Azospirillum lipoferum, Streptomyces lavendulae, S. coelicolor,* and *Stenotrophomonas maltophilia* AAP56 are also reported for the production of laccases (Morozova et al. 2007; Madhavi and Lele 2009; Desai and Nityanand 2011).

Enzyme Production Level

A thermostable laccase from *Pleurotus* sp. MAK-II has been reported with a specific activity of 1613 U/mg (Manavalan et al. 2015). Laccase produced from *Neurospora crassa* has a specific activity of 333 U/mg (Grotewold et al. 1998). A laccase produced by the basidiomycete *Marasmius quercophilus* C30 has a specific activity of 934 U/mg (Klonowska et al. 2002). The white-rot fungus *Trametes pubescens* MB 89 has been reported to produce laccase with an activity of 743 U/mL (Galhaup et al. 2002). The laccase gene *lac48424-1* from *Trametes* sp. 48424 was cloned and expressed in *P. pastoris* and the enzyme has a specific activity of 49.32 U/mg (Fan et al. 2011). Similarly a thermo-alkali-stable laccase gene from *Bacillus licheniformis* was cloned and expressed in *P. pastoris* with a maximum activity of 227.9 U/L (Lu et al. 2013). *Pleurotus ostreatus* strain 32 has been induced with ABTS and produced laccase with 410 U/mL activity (Hou et al. 2004). Phenol has been used as an inducer for laccase production by *T. versicolor* with an activity level of 2.575 U/mL (Pazarlioğlu et al. 2005).

Applications

Laccases are used for a variety of applications including textile, food, wood processing, chemical, and pharmaceutical industries (Kunamneni et al. 2007). Laccases are also used for coupling reactions during organic synthesis (Kudanga and Le Roes-Hill 2014). Laccases also catalyze some of the processes such as dye decolorization, degradation of xenobiotics, and effluent treatment. They also play a vital role in the oxidation of toxins and contaminants from industries. Laccases also have potential for biological delignification of pulp. In food industry they are used to eliminate the phenolics, haze formation, and turbidity development in fruit juice, wine, and beer. Laccases are also used in hair dyes, which are less irritant and are easier to handle (Kunamneni et al. 2007; Roriz et al. 2009; Desai and Nityanand 2011).

PROTEASE

Proteases are lytic enzymes that cleave other proteins by hydrolyzing the peptide linkages (Li et al. 2013a). Cleavage of peptide bonds result in the degradation of protein substrates into their principal amino acids, or it can be specific, leading to selective protein chopping for post-translational modification and processing. Proteases are classified as peptide hydrolases or peptidases (EC 3.4) and are commonly grouped into two categories (exopeptidases—that slice off amino acids from the ends of the protein chain and endopeptidases—which cleave peptide bonds within the protein) (de Souza et al. 2015). Proteases have wide applications in the food, pharmaceutical, leather, and textile industries (Fan et al. 2001; Mozersky et al. 2002).

Proteases play a critical role in many physiological and pathological functions such as protein catabolism, blood coagulation, cell growth and migration, tissue arrangement, morphogenesis in cell line development, tumor growth and metastasis, activation of zymogens, release of hormones and pharmacologically active peptides from precursor proteins, and transport of secretory proteins across membranes (Li et al. 2013a; de Souza et al. 2015). Extracellular proteases convert the proteins into smaller peptides and amino acids for subsequent absorption into cells, thereby playing a vital role in nitrogen metabolism (Sabotic and Kos 2012).

Enzyme Sources

The dominant producers of proteases are microorganisms of the genera *Pyrococcus*, *Thermococcus*, and *Staphylothermus*. Few examples are *Pyrococcus* sp. KODI, *S. marinus*, *T. aggreganes*, *T. celer*, *T. litoralis*, *Thermoacidophiles* (archeal and bacterial origin), and *Thermotoga maritima*. Few thermophilic protease-producing *Bacillus* spp. have been identified, such as *B. brevis*, *B. licheniformis*, *B. stearothermophilus*, *B. stearothermophilus* TP26, *Bacillus* sp. JB-99, and *B. thermoruber* (Haki and Rakshit 2003). A huge number of fungal strains have also been known to produce proteases such as *Aspergillus*, *Penicillium*, *Rhizopus*, *Mucor*, *Humicola*, *Thermoascus*, *Thermomyces* sp. (de Souza et al. 2015), *Trichoderma asperellum* (Yang et al. 2013), and *Aureobasidium pullulans* (Banani et al. 2014).

Enzyme Production Level

Bacillus megaterium has been reported to produce protease with 41.09 U/mg specific activity (Asker et al. 2013) and *Anoxybacillus* sp. KP1 with 16.39 U/mg (Bekler et al. 2015) activity. *Bacillus subtilis* has shown a good protease production level of 205.87 U/mg (Pant et al. 2015). Protease with 1052 U/mg of specific activity was produced by *B. pumilus* (Jayakumar et al. 2012). A nattokinase/subtilisin (serine proteases family) from *B. subtilis* VTCC-DVN-12-01 was expressed in *B. subtilis* WB800 with a specific activity of 12.7 U/mg (Nguyen et al. 2013). Haloalkaliphilic bacteria isolated from saline habitats have been reported to produce protease with 6765.76 U/mg specific activity (Purohit and Singh 2013). A protease gene APL5 of *Aureobasidium pullulans* strain PL5 was cloned and expressed in *E. coli* BL21 with a specific activity of 129 U/mg (Zhang et al. 2012).

Applications

A vast array of commercial proteases is available which has potential applications in various industrial processes such as textile, food, dairy, and pharmaceutical preparations (Nguyen et al. 2013; de Souza et al. 2015). Proteases such as alkaline protease, pancreatic protease, rennin, and papain have major applications in detergent formulations (to enhance the ability to remove tough stains and making the detergent environmentally safe), baking (altering the viscoelastic properties of dough), meat tenderization, and leather industries (dehairing of animal hides and skin) (Zambare et al. 2011; Kumari et al. 2012; de Souza et al. 2015).

PEROXIDASES

Peroxidases (1.11.1.x) are pervasive enzymes that oxidize a wide range of reducing substrates with the help of H_2O_2 or other peroxides (Fodil et al. 2012). This extensively utilized group of heme-containing peroxidases is produced from fungal or plant source and has advantages due to its wide range of substrates (Loncar and Fraaije 2015). Catalase has been transformed into either peroxidase or oxidase and is resistant to inactivation by hydrogen peroxide, this being the drawback of many peroxidases. Bacterial enzymes are preferred for industrial application because of their thermostability as fungal/plant is normally quite labile at higher temperatures. Thermostable peroxidases may be used *in situ* for the treatment of process water enabling fast recycling and with low energy usage (Loncar and Fraaije 2015).

Enzyme Sources

A catalase which has peroxidase activity has been reported from *Thermobifida fusca* (Loncar and Fraaije 2015). An extracellular thermostable humic acid peroxidase (HaP3) was isolated from a *Streptomyces* sp. strain AH4 (Fodil et al. 2012). Dye-decolorizing peroxidases were isolated from *B. subtilis*, *P. putida* MET94 (Santos et al. 2014), and *Kocuria rosea* MTCC 1532 (Parshetti et al. 2012). Microorganisms belonging to the genera *Agaricales*, *Corticiales*, *Polyporales*, and *Hymenochaetales* are reported for manganese peroxidase production (Tello et al. 2000; Hilden et al. 2008; Morgenstern et al. 2010; Janusz et al. 2013). Fungal species such as *Phlebia* sp. MG60, *P. radiata* 79, *Dichomitus squalens* , *Lentinula edodes*, *Phanerochaete chrysosporium*, *P. sordida*, *Ganoderma lucidum*, *Ceriporiopsis subvermispora*, and *Trametes versicolor* 9522-1 have genes encoded for Mn peroxidase (Janusz et al. 2013). A novel peroxidase (SviDyP) was isolated, purified, and characterized from *Saccharomonospora viridis* DSM 43017, a pentachlorophenol-degrading thermophilic *actinomycete* (Webb et al. 2001).

Enzyme Production Level

Kocuria rosea MTCC 1532 produced lignin peroxidase which was having 168.33 U/mg specific activity (Parshetti et al. 2012). Humic acid peroxidase (HaP3) was produced from a *Streptomyces* sp. strain AH4 which has a specific activity of 9.45 U/mg (Fodil et al. 2012). Peroxidase from *Saccharomonospora viridis* DSM 43017 was reported with 17.8 U/mg specific activity (Yu et al. 2014). *Thermobifida fusca* also produced peroxidase with a specific activity of 67 U/mg (Loncar and Fraaije 2015). Peroxidase genes from *B. subtilis* and *P. putida* MET94 were cloned into plasmid pET-21a(+) to yield plasmids pRC-1 and pRC-2, respectively. These were introduced into the host expression strains *E. coli* BL21 and *E. coli* BL21 star, respectively, in which the target genes were expressed under the control of the T7lac promoter with a production level of 40 U/mg and 15 U/mg, respectively (Santos et al. 2014).

Applications

Peroxidases have applications in the selective delignification of lignocellulosic materials for the production of cellulose or conversion into feed and biofuels. They also have applications in the treatment of toxic industrial effluents, such as those containing synthetic dyes, generated in numerous industrial practices as they have also potential to be used as biological decolorizing agents (Santos et al. 2014). Peroxidase has potential applications in biopulping, biobleaching (Yu et al. 2014), biodegradation, and bioremediation (Fujii et al. 2013; Janusz et al. 2013). Peroxidase is also used in analytical chemistry, immune chemistry, biosensor construction, food processing, and food storage (Mall et al. 2013).

LIGASES

Ligases are enzymes that join the DNA fragments together by catalyzing bond formation between neighboring nucleotides. DNA and RNA ligases are ubiquitous enzymes that catalyze the formation of phosphodiester bonds between opposing 5′-phosphate and 3′-hydroxyl termini in nucleic acids (Wang et al. 2013). They belong to the nucleotidyl transferase superfamily together with the RNA capping enzymes and tRNA ligases. All of the enzymes in this superfamily catalyze phosphodiester bond formation in a conserved, three-step mechanism that utilizes ATP, GTP, or NAD+ as a high-energy cofactor (Wang and Shuman 2005; Shuman 2009; Chambers and Patrick 2015). Ligases are the most important catalysts in the central biological processes, including DNA replication, recombination, and rearrangement of immunoglobulin genes. Their activities *in vitro* have also been exploited in numerous molecular biology protocols, making them crucial tools for modern biotechnology (Chambers and Patrick 2015). For decades, DNA ligases have been used to create recombinant DNA molecules (i.e., cloning) and for genetic disease detection using the ligation chain reactions (Gibson et al. 2009).

In vivo, DNA ligases catalyze the formation of phosphodiester bonds at single-stranded nicks in double-stranded DNA. This activity is critical for maintaining genomic integrity during DNA replication, DNA recombination, and DNA excision repair (Wang et al. 2013). They are essential in all organisms and they are conventionally categorized into two families according to their cofactor specificity (Doherty and Suh 2000; Wang et al. 2013). ATP-dependent ligases (EC 6.5.1.1) are typically found in Eukarya, Archaea, and viruses (including bacteriophages), while the NAD+-dependent DNA ligases (EC 6.5.1.2) are typically found in bacteria and some eukaryotic viruses. Most notably, the archaeal species *Haloferax volcanii* holds two active DNA ligases: one ATP-dependent (LigA) and the other NAD+-dependent (LigN) (Zhao et al. 2006). However, there is some protein sequence uniformity between bacterial and eukaryotic DNA ligases. Further, bacterial DNA ligase requires NAD+ as a cofactor, while eukaryotic and most viral DNA ligases utilize ATP (Stokes et al. 2011).

Enzyme Sources

Several bacteria are capable of producing the ligase enzymes. This include *Thermus aquaticus, Aeropyrum pernix, Desulfurolobus ambivalens, Staphylothermus marinus, Sulfolobus acidocaldarius, S. shibatae, S. solfataricus, Sulfophobococcus zilligii, Archaeoglobus fulgidus, Methanothermobacter thermautotrophicus, Pyrococcus horikoshii, Thermococcus* sp. 1519, *T. fumicolans, T. kodakaraensis, T. onnurineus, T. sibiricus, Methanocaldococcus jannaschii, Thermotoga maritima*, and *Schizosaccharomyces pombe* (Lai et al. 2002; Lohman et al. 2011; Stewart et al. 2011; Le et al. 2013; Wang et al. 2013; Chambers and Patrick 2015).

Applications

DNA ligases are most importantly having a role in DNA repair and most archeal DNA ligases have the ability to seal single-stranded nicks in double-stranded DNA (Kotani et al. 2012; Le et al. 2013). Their ability to ligate the double-stranded, cohesive-, or blunt-end fragments made these enzymes receive great attention in biotechnological applications. Ligations of cohesive-ended fragments have been performed by the enzymes from *Aeropyrum pernix, Staphylothermus marinus, Thermococcus* sp. 1519, and *T. fumicolans*. In addition, the *S. marinus* (Seo et al. 2007) and *T. fumicolans*

DNA ligases could also ligate blunt-ended fragments (Rolland et al. 2004; Chambers and Patrick 2015). A number of next-generation sequencing methods also depend on DNA ligases (Quail et al. 2008; Lohman et al. 2011) either for adapter ligation during sample preparation (e.g., Illumina and 454 sequencing) or for the sequencing reaction itself (SOLiD sequencing) (Chambers and Patrick 2015).

DNA POLYMERASE

The polymerase chain reaction (PCR) process has directed a huge advancement in genetic engineering due to its capacity to amplify DNA. The three sequential steps in this process include denaturation or melting of the DNA strand (separation) carried out at 90–95°C, renaturation, or primer annealing at 55°C followed by synthesis or primer extension at around 75°C. Development in this process has been to a large extent facilitated by the availability of thermostable DNA polymerases, which catalyze the elongation of primer DNA strand (Haki and Rakshit 2003). Many microbial sources have been used for the production of DNA polymerases.

Enzyme Sources

DNA polymerase has been reported from *E. coli,* however, did not retain the activity at higher temperatures. *Taq* polymerase from the bacterium *Thermus aquaticus* was the first thermostable DNA polymerase identified and characterized biochemically (Chien et al. 1976; Kaledin et al. 1980). Some commercially used DNA polymerases are produced from bacteria which includes *Thermus caldophilus, T. filiformis, T. flavis, T. thermophiles, Pyrococcus species* GB-D, *P. abyssi, P. furiosus, Thermococcus kodakaraensis, T. brokianus, T. fumiculans, T. gorgonarius, T. litoralis, T. peptonophilus, T. zilligii, Thermotoga maritima,* and *T. neopolitana* (Terpe 2013). The DNA polymerase I (Taq Pol I) gene from *T. aquaticus* was cloned into a plasmid expression vector that utilizes the strong bacteriophage PL promoter. It was transferred to *E. coli* for expression with a specific activity of 292,000 U/mg (Lawyer et al. 1993). *Escherichia coli* BL21 was transformed with p*Taq* gene and expressed the DNA polymerase which has a specific activity of 5263.16 U/mg (Engelke et al. 1990).

Applications

DNA polymerases have applications in forensic science, nucleic acid sequencing industries, molecular characterization of plants, animals, and microorganisms. The enzyme is involved in molecular processes such as the construction of gene cloning (Ikehara et al. 2004; Herrin et al. 2005), genomic DNA cloning (Nisole et al. 2004), synthesis of second-strand cDNA (Sasaki et al. 2004), knockout targeting vector (Kim et al. 2005), and synthetic gene manufacture (Wu et al. 2006).

REVERSE TRANSCRIPTASE

Reverse transcriptase (RTase) is the enzyme that catalyzes DNA polymerization using RNA as a template (RNA-dependent DNA polymerase) (Baranauskas et al. 2012). The enzyme is used in various genetic experiments, such as in microarray analysis by synthesizing cDNA or the start-site mapping of transcript mRNA. RTases play central roles in these genetic experiments, and the enzymes used are derived from retroviruses, such as the Moloney murine leukemia virus (MMLV) or the avian myeloblastosis virus which are not thermostable (AMV) (Arezi and Hogrefe 2007; Sano et al. 2012). To develop thermostable RTase, several strategies have been attempted (Yasukawa et al. 2008; Arezi and Hogrefe 2009; Kranaster et al. 2010; Mizuno et al. 2010; Jozwiakowski and Connolly 2011), and some genetically engineered enzymes are commercially available. RTase possesses three enzymatic activities: the RNA-dependent DNA polymerase, the DNA-dependent DNA polymerase, and RNase H, which degrade RNA strand in the RNA–DNA hybrid (Sambrook and Russell 2001). The synthesis of cDNA is probably the second most important technique in present molecular biology after the PCR and its modifications (Baranauskas et al. 2012).

Enzyme Sources

Thermus thermophiles has been reported for RTase activity in the presence of Mn^{2+} (Mohr et al. 2013). A DNA polymerase from *T. aquaticus* has been mutated to have RTase activity (Kranaster et al. 2010). RTase can also be produced from Moloney murine leukemia virus RTase (M-MuLV RTase) variants (Baranauskas et al. 2012). A mutant RTase has been produced from *Thermotoga petrophila* K4 (Sano et al. 2012). Some other organisms that produce RTase are avian myeloblastosis virus (Arezi and Hogrefe 2009), human immunodeficiency virus type 1 (HIV-1) (Sarafianos et al. 2009), and *Carboxydothermus hydrogenoformans* (Vieille and Zeikus 2001). Thermostable DNA-dependent DNA polymerase of *Bacillus stearothermophilus* has been reported which have RTase activity (Jestin et al. 2015).

Applications

RTases are extensively used to generate cDNA libraries for cloning, end-point and quantitative RT-polymerase chain reaction (RT-PCR), RACE technique, microarray analysis, RNA amplification (Sambrook and Russell 2001; Baranauskas et al. 2012), transcriptome and miRNA profiling, next-generation RNA sequencing (RNA-seq), RNA structure mapping, and the analysis of protein- or ribosome-bound RNA fragments (Wang et al. 2009; Mayer et al. 2011; Ozsolak and Milos 2011).

CONCLUSIONS

The application of thermostable enzymes as effective catalysts in industry would result in substantial savings of resources, mainly energy and water. With the fast increasing global population and shrinking natural resources, enzyme technology offers a great perspective for many industries to help meet the future challenges. Furthermore, with a paradigm shift in industry moving from natural to renewable resource utilization, the need for thermostable microbial catalysts is predicted to increase in the future. This warrants for further research in identifying novel thermostable enzymes with superior properties that address specific industrial needs.

REFERENCES

Abdel-fattah, Y.R. et al., 2013. Production, purification, and characterization of thermostable-amylase produced by *Bacillus licheniformis* isolate AI20. *Journal of Chemistry*, 2013, pp.1–11. http://dx.doi.org/10.1155/2013/673173.

Arezi, B. and Hogrefe, H., 2009. Novel mutations in Moloney Murine Leukemia Virus reverse transcriptase increase thermostability through tighter binding to template-primer. *Nucleic Acids Research*, 37, pp. 473–481.

Arezi, B. and Hogrefe, H.H., 2007. *Escherichia coli* DNA polymerase III epsilon subunit increases Moloney murine leukemia virus reverse transcriptase fidelity and accuracy of RT-PCR procedures. *Analytical Biochemistry*, 360, pp. 84–91.

Asha, R., Niyonzima, F.N., and Sunil, S.M., 2013. Purification and properties of pullulanase from *Bacillus halodurans*. *International Research Journal of Biological Sciences*, 2, pp. 35–43.

Asker, M.M.S. et al., 2013. Purification and characterization of two thermostable protease fractions from *Bacillus megaterium*. *Journal of Genetic Engineering and Biotechnology*, 11, pp. 103–109. Available at: http://dx.doi.org/10.1016/j.jgeb.2013.08.001.

Atanassov, I. et al., 2013. Seamless GFP and GFP-Amylase cloning in gateway shuttle vector, expression of the recombinant proteins in *E. coli* and *Bacillus megaterium* and assessment of the GFP-amylase thermostability. *Biotechnology and Biotechnological Equipment*, 27, pp. 4172–4180. Available at: http://www.tandfonline.com/doi/abs/10.5504/BBEQ.2013.0079.

Ávila-Cisneros, N. et al., 2014. Production of thermostable lipase by Thermomyces lanuginosus on solid-state fermentation: Selective hydrolysis of sardine oil. *Applied Biochemistry and Biotechnology*, 174, pp. 1859–1872. Available at: http://dx.doi.org/10.1007/s12010-014-1159-9.

Banani, H. et al., 2014. Biocontrol activity of an alkaline serine protease from *Aureobasidium pullulans* expressed in Pichia pastoris against four postharvest pathogens on apple. *International Journal of Food Microbiology*, 182–183, pp. 1–8. Available at: http://dx.doi.org/10.1016/j.ijfoodmicro.2014.05.001.

Banka, A.L., Albayrak Guralp, S., and Gulari, E., 2014. Secretory expression and characterization of two hemicellulases, xylanase, and β-xylosidase, isolated from Bacillus subtilis M015. *Applied Biochemistry and Biotechnology*, 174, pp. 2702–2710. Available at: http://www.pubmedcentral.nih.gov/articlerender.fcgi?artid=4237932&tool=pmcentrez&rendertype=abstract.

Baranauskas, A. et al., 2012. Generation and characterization of new highly thermostable and processive M-MuLV reverse transcriptase variants. *Protein Engineering Design and Selection*, 25, pp. 657–668. Available at: http://peds.oxfordjournals.org/cgi/doi/10.1093/protein/gzs034.

Basaran, P. and Ozcan, M., 2008. Characterization of b-xylosidase enzyme from a *Pichia stipitis* mutant. *Bioresource Technology*, 99, pp. 38–43.

Beg, Q.K. et al., 2001. Microbial xylanases and their industrial applications: A review. *Applied Microbiology and Biotechnology*, 56(3–4), pp. 326–338.

Bekler, F.M., Acer, O., and Guven, K., 2015. Production and purification of novel thermostable alkaline protease from *Anoxybacillus* sp. KP1. *Cellular and Molecular Biology*, 61, pp. 113–120.

Bertoldo, C. and Antranikian, G., 2002. Starch-hydrolyzing enzymes from thermophilic archaea and bacteria. *Current Opinion in Chemical Biology*, 6, pp. 151–160.

Bhalla, A. et al., 2013. Improved lignocellulose conversion to biofuels with thermophilic bacteria and thermostable enzymes. *Bioresource Technology*, 128, pp. 751–759. Available at: http://www.ncbi.nlm.nih.gov/pubmed/23246299 (Accessed September 29, 2014).

Bhalla, A. et al., 2014. Novel thermostable endo-xylanase cloned and expressed from bacterium *Geobacillus* sp. WSUCF1. *Bioresource Technology*, 165, pp. 314–318. Available at: http://linkinghub.elsevier.com/retrieve/pii/S0960852414004179.

Bhalla, A. et al., 2015. Improved lignocellulose conversion to biofuels with thermophilic bacteria and thermostable enzymes. *Bioresource Technology*, 128, pp. 751–759.

Bharat, B. and Hoondal, G., 1998. Isolation, purification and properties of thermostable chitinase from an alkalophilic *Bacillus* sp. BG-11. *Biotechnology Letters*, 20, pp. 157–159.

Biely, P., 1985. Microbial xylanolytic systems. *Trends in Biotechnology*, 3, pp. 286–290.

Biely, P., 2003. Xylanolytic enzymes. In J.R. Whitaker, A.G.J. Voragen, and D.W.S. Wong, eds. *Handbook of Food Enzymology*. Marcel Dekker, Inc., New York, NY, pp. 879–915.

Borrelli, G. and Trono, D., 2015. Recombinant lipases and phospholipases and their use as biocatalysts for industrial applications. *International Journal of Molecular Sciences*, 16, pp. 20774–20840. Available at: http://www.mdpi.com/1422-0067/16/9/20774/.

Bouacem, K. et al., 2014. Partial characterization of xylanase produced by *Caldicoprobacter algeriensis*, a new thermophilic anaerobic bacterium isolated from an Algerian hot spring. *Applied Biochemistry and Biotechnology*, 174, pp. 1969–1981. doi:10.1007/s12010-014-1153-2.

Butt, M.S. et al., 2008. Xylanases and their application in baking industry. *Food Technology and Biotechnology*, 46, pp. 22–31.

Cammarota, M.C. and Freire, D.M.G., 2006. A review on hydrolytic enzymes in the treatment of wastewater with high oil and grease content. *Bioresource Technology*, 97, pp. 2195–2210.

Cannio, R., Di Prizito, N., and Rossi Alessandra Morana, M., 2004. A xylan-degrading strain of *Sulfolobus solfataricus*: Isolation and characterization of the xylanase activity. *Extremophiles*, 8, pp. 117–124.

Casado, V. et al., 2012. Phospholipases in food industry: A review. In G. Sandoval, ed. *Lipases and Phospholipases: Methods and Protocols*. Springer, New York, NY, pp. 495–523.

Chambers, C.R. and Patrick, W.M., 2015. *Archaeal* nucleic acid ligases and their potential in biotechnology. Archaea, 2015, pp.1–10. http://dx.doi.org/10.1155/2015/170571.

Chapla, D. et al., 2010. Utilization of agro-industrial waste for xylanase production by *Aspergillus foetidus* MTCC 4898 under solid state fermentation and its application in saccharification. *Biochemical Engineering Journal*, 49, pp. 361–369.

Chavez, R., Bull, P., and Eyzaguirre, J., 2006. The xylanolytic enzyme system from the genus Penicillium. *Journal of Biotechnology*, 123, pp. 413–433.

Chen, Z. et al., 2012. Secretory expression of a β-xylosidase gene from *Thermomyces lanuginosus* in *Escherichia coli* and characterization of its recombinant enzyme. *Letters in Applied Microbiology*, 55, pp. 330–337.

Cheng, Y.-S. et al., 2014. Structural analysis of a glycoside hydrolase family 11 xylanase from neocallimastix patriciarum: Insight into the molecular basis of a thermophilic enzyme. *Journal of Biological Chemistry*, 289, pp. 11020–11028. Available at: http://www.jbc.org/cgi/doi/10.1074/jbc.M114.550905.

Cherif, S. et al., 2010. Crab digestive phospholipase: A new invertebrate member. *Bioresource Technology*, 101, pp. 366–371.

Chiang, C.M. et al., 2005. Expression of a bi-functional and thermostable amylopullulanase in transgenic rice seeds leads to autohydrolysis and altered composition of starch. *Molecular Breeding*, 15, pp. 125–143.

Chien, A., Edgar, D.B., and Trela, J.M., 1976. Deoxyribonucleic acid polymerase from the extreme thermophile Thermus aquaticus. *Journal of Bacteriology*, 127, pp. 1550–1557.

Chinnathambi, V. et al., 2015. Molecular cloning and expression of a family 6 Cellobiohydrolase gene cbhII from *Penicillium funiculosum* NCL1. *Advances in Bioscience and Biotechnology*, 6, pp. 213–222.

Cho, E.K., Choi, I.S., and Choi, Y.J., 2011. Overexpression and characterization of thermostable chitinase from *Bacillus atrophaeus* SC081 in *Escherichia coli*. *BMB Reports*, 44(3), pp. 193–198.

Christakopoulos, P. et al., 2003. Antimicrobial activity of acidic xylo-oligosaccharides produced by family 10 and 11 endoxylanases. *International Journal of Biological Macromolecules*, 31, pp. 171–175.

Cockcroft, S., 2001. Signalling roles of mammalian phospholipase D1 and D2. *Cellular and Molecular Life Science*, 58, pp. 1674–1687.

Comlekcioglu, U. et al., 2010. Cloning and characterization of cellulase and xylanase coding genes from anaerobic fungus *Neocallimastix* sp. GMLF1. *International Journal of Agriculture and Biology*, 12(5), pp. 691–696.

Comlekcioglu, U. et al., 2011. Effects of various agro-wastes on xylanase and b-xylosidase production of anaerobic ruminal fungi. *Journal of Scientific and Industrial Research*, 70(4), pp. 293–299.

Cui, Z. et al., 2012. High level expression and characterization of a thermostable lysophospholipase from *Thermococcus kodakarensis* KOD1. *Extremophiles*, 16, pp. 619–625.

Cuong, N.P. et al., 2015. Continuous production of pure maltodextrin from cyclodextrin using immobilized Pyrococcus furiosus thermostable amylase. *Process Biochemistry*. Available at: http://linkinghub.elsevier.com/retrieve/pii/S1359511315301318.

Dahiya, N. et al., 2006. Production of an antifungal chitinase from *Enterobacter* sp. NRG4 and its application in protoplast production. *World Journal of Microbiology and Biotechnology*, 21, pp. 1611–1616.

Dai, D. et al., 2011. Purification and characterization of a novel extracellular chitinase from thermophilic *Bacillus* sp. Hu1. *African Journal of Biotechnology*, 10, pp. 2476–2484.

Deive, F. et al., 2012. A process for extracellular thermostable lipase production by a novel Bacillus thermoamylovorans strain. *Bioprocess and Biosystems Engineering*, 35, pp. 931–941. Available at: http://dx.doi.org/10.1007/s00449-011-0678-9.

Desai, S.S. and Nityanand, C., 2011. Microbial laccases and their applications: A review. *Asian Journal of Biotechnology*, 3, pp. 98–124.

de Souza, P.M. et al., 2015. A biotechnology perspective of fungal proteases. *Brazilian Journal of Microbiology*, 46, pp. 337–346. Available at: http://www.scielo.br/scielo.php?script=sci_arttext&pid=S1517-83822015000200337&lng=en&nrm=iso&tlng=en.

Dobozi, M.S., Szakács, G., and Bruschi, C.V., 1992. Xylanase activity of *Phanerochaete chrysosporium*. *Applied and Environmental Microbiology*, 58, pp. 3466–3471.

Doherty, A.J. and Suh, S.W., 2000. Structural and mechanistic conservation in DNA ligases. *Nucleic Acids Research*, 28, pp. 4051–4058.

Dornez, E. et al., 2007. Impact of wheat flour-associated endoxylanases on arabinoxylan in dough after mixing and resting. *Journal of Agricultural and Food Chemistry*, 55, pp. 7149–7155.

Dror, A. et al., 2014. Protein engineering by random mutagenesis and structure-guided consensus of *Geobacillus stearothermophilus* Lipase T6 for enhanced stability in methanol. *Applied and Environmental Microbiology*, 80, pp. 1515–1527. Available at: http://aem.asm.org/cgi/doi/10.1128/AEM.03371-13.

Duan, X. and Wu, J., 2015. Enhancing the secretion efficiency and thermostability of a *Bacillus deramificans* pullulanase mutant (D437H/D503Y) by N-terminal domain truncation. *Applied and Environmental Microbiology*, 81, pp. 1926–1931. Available at: http://www.ncbi.nlm.nih.gov/pubmed/25556190.

Dutta, T. et al., 2008. Novel cellulases from an extremophilic filamentous fungi *Penicillium citrinum*: Production and characterization. *Journal of Industrial Microbiology and Biotechnology*, 35, pp. 275–282.

Ebrahimpour, A. et al., 2011. High level expression and characterization of a novel thermostable, organic solvent tolerant, 1,3-regioselective lipase from *Geobacillus* sp. strain ARM. *Bioresource Technology*, 102, pp. 6972–6981. Available at: http://www.sciencedirect.com/science/article/pii/S0960852411004408 (Accessed October 29, 2015).

Elegir, G., Szakács, G., and Jeffries, T.W., 1994. Purification, characterization, and substrate specificities of multiple xylanases from *Streptomyces sp.* Strain B-12-2. *Applied and Environmental Microbiology*, 60, pp. 2609–2615. Available at: http://www.pubmedcentral.nih.gov/articlerender.fcgi?artid=2 01691&tool=pmcentrez&rendertype=abstract.

Elleuche, S. et al., 2015. Exploration of extremophiles for high temperature biotechnological processes. *Current Opinion in Microbiology*, 25, pp. 113–119. Available at: http://www.sciencedirect.com/science/article/pii/S1369527415000624.

Ellis, J.T. and Magnuson, T.S., 2012. Thermostable and alkalistable xylanases produced by the thermophilic bacterium *Anoxybacillus flavithermus* TWXYL3. ISRN Microbiol, 2012, pp.1–8. Available at: http://www.ncbi. nlm.nih.gov/pubmed/23762752.

Engelke, D.R. et al., 1990. Purification of *Thermus aquaticus* DNA polymerase expressed in *Escherichia coli*. *Analytical Biochemistry*, 191, pp. 396–400. Available at: http://www.ncbi.nlm.nih.gov/pubmed/2085185\n, http://www.sciencedirect.com/science/article/pii/0003269790902385.

Espinosa-Luna, G. et al., 2015. Gene cloning and characterization of the *Geobacillus thermoleovorans* CCR11 Carboxylesterase CaesCCR11, a New Member of Family XV. *Molecular Biotechnology*. Available at: http://link. springer.com/10.1007/s12033-015-9901-2.

Fan, F. et al., 2011. Cloning and functional analysis of a new laccase gene from Trametes sp. 48424 which had the high yield of laccase and strong ability for decolorizing different dyes. *Bioresource Technology*, 102, pp. 3126–3137. Available at: http://dx.doi.org/10.1016/j.biortech.2010.10.079.

Fan, Z., Zhu, Q., and Dai, J., 2001. Enzymatic treatment of wool. *Journal of Dong Hua University*, 18 (2), pp. 112–115. https://www.scopus.com/record/display.uri?eid=2-s2.0-0035382286&origin=inward&txGid=0

Ferreira, N.L. et al., 2014. Use of cellulases from *Trichoderma reesei* in the twenty-first century—Part I: Current industrial uses and future applications in the production of second ethanol generation. In: Gupta, V. K. et al. (eds.), *Biotechnology and Biology of Trichoderma*. The Netherlands: Elsevier Ltd, pp. 245–261. http://dx.doi.org/10.1016/B978-0-444-59576-8.00017-5.

Fodil, D. et al., 2012. A thermostable humic acid peroxidase from *Streptomyces* sp. strain AH4: Purification and biochemical characterization. *Bioresource Technology*, 111, pp. 383–390. Available at: http://dx.doi.org/10.1016/j.biortech.2012.01.153.

Fujii, K. et al., 2013. Environmental control of lignin peroxidase, manganese peroxidase, and laccase activities in forest floor layers in humid Asia. *Soil Biology and Biochemistry*, 57, pp. 109–115. Available at: http://dx.doi.org/10.1016/j.soilbio.2012.07.007.

Galhaup, C. et al., 2002. Increased production of laccase by the wood-degrading basidiomycete *Trametes pubescens*. *Enzyme and Microbial Technology*, 30, pp. 529–536.

Giardina, P. et al., 2010. Laccases: A never-ending story. *Cellular and Molecular Life Sciences*, 67, pp. 369–385.

Gibson, D.G. et al., 2009. Enzymatic assembly of DNA molecules up to several hundred kilobases. *Nature Methods*, 6, pp. 343–345.

Gochev, V.K. and Krastanov, A.I., 2007. Isolation of laccases producing *Trichoderma* pp. Bulgarina. *Journal of Agricultural Science*, 13, pp. 171–176.

Gonçalves, G.A.L. et al., 2015. Synergistic effect and application of xylanases as accessory enzymes to enhance the hydrolysis of pretreated bagasse. *Enzyme and Microbial Technology*, 72, pp. 16–24.

Gramany, V. et al., 2015. Cloning, expression, and molecular dynamics simulations of a xylosidase obtained from *Thermomyces lanuginosus*. *Journal of Biomolecular Structure and Dynamics*, 1102(11), pp. 1–12. Available at: http://www.tandfonline.com/doi/full/10.1080/07391102.2015.1089186.

Grotewold, E. et al., 1998. Purification of an extracellular fungal laccase. *Mircen Journal of Applied Microbiology and Biotechnology*, 4, pp. 357–363.

Gunny, A.A.S. et al., 2014. Potential halophilic cellulases for *in situ* enzymatic saccharification of ionic liquids pretreated lignocelluloses. *Bioresource Technology*, 155, pp. 177–181.

Gutarra, M.L.E. et al., 2009. Production of an acidic and thermostable lipase of the mesophilic fungus *Penicillium simplicissimum* by solid-state fermentation. *Bioresource Technology*, 100, pp. 5249–5254. Available at: http://dx.doi.org/10.1016/j.biortech.2008.08.050.

Haki, G.D. and Rakshit, S.K., 2003. Developments in industrially important thermostable enzymes: A review. *Bioresource Technology*, 89, pp. 17–34.

Halder, S.K. et al., 2012. Chitinolytic enzymes from the newly isolated *Aeromonas hydrophila* SBK1: Study of the mosquitocidal activity. *BioControl*, 57, pp. 441–449.

Hamid, R. et al., 2013. Chitinases: An update. *Journal of Pharmacy and Bioallied Sciences*, 5, pp. 21–29. Available at: http://www.ncbi.nlm.nih.gov/pmc/articles/PMC3612335/.

Han, T. et al., 2013. Biochemical characterization of a recombinant pullulanase from *Thermococcus kodakarensis* KOD1. *Letters in Applied Microbiology*, 57, pp. 336–343.

Haq, I. ul et al., 2015. CenC, a multidomain thermostable GH9 processive endoglucanase from *Clostridium thermocellum*: Cloning, characterization and saccharification studies. *World Journal of Microbiology and Biotechnology*, 31, pp. 1699–1710. Available at: http://link.springer.com/10.1007/s11274-015-1920-4.

Harris, P. V. et al., 2014. New enzyme insights drive advances in commercial ethanol production. *Current Opinion in Chemical Biology*, 19, pp. 162–170. Available at: http://linkinghub.elsevier.com/retrieve/pii/S1367593114000271.

He, Z. et al., 2014. Cloning of a novel thermostable glucoamylase from thermophilic fungus *Rhizomucor pusillus* and high-level co-expression with α-amylase in *Pichia pastoris*. *BMC Biotechnology*, 14, pp. 1–10. Available at: http://www.biomedcentral.com/1472-6750/14/114.

Hebraud, M. and Fevre, M., 1990. Purification and characterization of an extracellular beta-xylosidase from the rumen anaerobic fungus *Neocallimastix frontalis*. *FEMS Microbiology Letters*, 60(1–2), pp. 11–16. Available at: http://search.ebscohost.com/login.aspx?direct=true&db=mnh&AN=2126511&site=ehost-live.

Heinzelman, P. et al., 2009. A family of thermostable fungal cellulases created by structure-guided recombination. *Proceedings of the National Academy of Sciences of the United States of America*, 106, pp. 5610–5615. Available at: http://www.ncbi.nlm.nih.gov/pmc/articles/PMC2667002/.

Herrin, B.R., Groeger, A.L., and Justement, L.B., 2005. The adaptor proteinHSH2 attenuates apoptosis in response to ligation of the B cell antigen receptor complex on the B lymphoma cell line, WEHI-231. *Journal of Biological Chemistry*, 280, pp. 3507–3515.

Herrmann, M.C. et al., 1997. The b-D-xylosidase of *Trichoderma ressei* is a multifunctional b-D-xylan xylohydrolase. *Biochemistry Journal*, 321, pp. 375–381.

Hii, S.L. et al., 2012. Pullulanase: Role in starch hydrolysis and potential industrial applications. *Enzyme Research*, 2012, pp. 1–14. Available at: http://www.hindawi.com/journals/er/2012/921362/.

Hilden, K.S. et al., 2008. Molecular characterization of the basidiomycete isolate *Nematoloma frowardii* b19 and its manganese peroxidase places the fungus in the corticioid genus Phlebia. *Microbiology*, 154, pp. 2371–9.

Hou, H. et al., 2004. Enhancement of laccase production by *Pleurotus ostreatus* and its use for the decolorization of anthraquinone dye. *Process Biochemistry*, 39, pp. 1415–1419.

Hu, F. et al., 2013. A novel phospholipase D constitutively secreted *by Ochrobactrum* sp. ASAG-PL1 capable of enzymatic synthesis of phosphatidylserine. *Biotechnology Letters*, 35, pp. 1317–1321. Available at: http://dx.doi.org/10.1007/s10529-013-1207-5.

Huber, R. et al., 1995. *Thermococcus chitonophagus* sp. Nov., a novel chitin degrading, hyperthermophilic archeum from the deep sea hydrothermal vent environment. *Archives Microbiology*, 164, pp. 255–264.

Ibrahim, D. et al., 2013. *Bacillius licheniformis* BT5.9 isolated from Changar hop spring, Malang, Indonesia as a potential producer of Thermostable a-amylase. *Tropical Life Sciences*, 24, pp. 72–84.

Ikehara, Y., Ikehara, S.K., and Paulson, J.C., 2004. Negative regulation of T cell receptor signaling by Siglec-7 (p70/AIRM) and Siglec-9. *Journal of Biological Chemistry*, 279, pp. 43117–43125.

Imanaka, T. and Sakurai, S., 1992. Method of washing super precision devices,semiconductors, with enzymes. https://www.google.com/patents/US5078802.

Istivan, T.S. and Coloe, P.J., 2006. Phospholipase A in Gram-negative bacteria and its role in pathogenesis. *Microbiology*, 152, pp. 1263–1274.

Jabbour, D., Klippel, B., and Antranikian, G., 2012. A novel thermostable and glucose-tolerant β-glucosidase from *Fervidobacterium islandicum*. *Applied Microbiology and Biotechnology*, 93, pp. 1947–1956. Available at: http://link.springer.com/10.1007/s00253-011-3406-0.

Janeček, Š., Svensson, B., and MacGregor, E.A., 2014. α-Amylase: An enzyme specificity found in various families of glycoside hydrolases. *Cellular and Molecular Life Sciences*, 71, pp. 1149–1170.

Janusz, G. et al., 2013. Fungal laccase, manganese peroxidase and lignin peroxidase: Gene expression and regulation. *Enzyme and Microbial Technology*, 52, pp. 1–12. Available at: http://dx.doi.org/10.1016/j.enzmictec.2012.10.003.

Jasilionis, A., Petkauskaite, R., and Kuisiene, N., 2014. A novel type I thermostable pullulanase isolated from a thermophilic starch enrichment culture. *Microbiology*, 83, pp. 227–234. Available at: http://link.springer.com/10.1134/S0026261714030084.

Javier, P.I. et al., 2007. Xylanases: Molecular properties and applications. In J. Polaina and A. MacCabe, eds. *Industrial Enzymes*. Springer, Dordrecht, the Netherlands, pp. 65–82.

Jayakumar, R. et al., 2012. Characterization of thermostable serine alkaline protease from an alkaliphilic strain *Bacillus pumilus* MCAS8 and its applications. *Applied Biochemistry and Biotechnology*, 168, pp. 1849–1866.

Jestin, J.-L., Vichier-Guerre, S., and Ferries, S., 2015. Methods for obtaining thermostable enzymes, DNA polymerase I variants from Thermus aquaticus having new catalytic activities, methods for obtaining the same, and applications to the same. https://www.google.com/patents/US8927699

Jordan, D.B. and Wagschal, K., 2010. Properties and applications of microbial β-D-xylosidases featuring the catalytically efficient enzyme from *Selenomonas ruminantium*. *Applied Microbiology and Biotechnology*, 86, pp. 1647–1658.

Joshi, B.H., 2011. A novel thermostable alkaline α-amylase from Bacillus circulans PN5: Biochemical characterization and production. *Asian Journal of Biotechnology*, 3, pp. 58–67.

Jozwiakowski, S.K. and Connolly, B.A., 2011. A modified family-B archaeal DNA polymerase with reverse transcriptase activity. *Chem Bio Chem*, 12, pp. 35–37.

Juturu, V. and Wu, J.C., 2014. Microbial cellulases: Engineering, production and applications. *Renewable and Sustainable Energy Reviews*, 33, pp. 188–203.

Jyoti, J. et al., 2011. Partial purification and characterization of an acidophilic extracellular A—Amylase from *Bacillus Licheniformis* Jar-26 abstract: *International Journal of Advanced Biotechnology and Research*, 2, pp. 315–320.

Kaledin, A.S., Sliusarenko, A.G., and Gorodetskii, S.I., 1980. Isolation and properties of DNA polymerase from extremely thermophilic bacterium *Thermus aquaticus* YT1. *Biokhimiia*, 45, pp. 644–651.

Kamal Kumar, B., Balakrishnan, H., and Rele, M.V., 2004. Compatibility of alkaline xylanases from an alkaliphilic *Bacillus* NCL (87-6-10) with commercial detergents and proteases. *Journal of Industrial Microbiology and Biotechnology*, 31, pp. 83–87.

Kambourova, M. et al., 2003. Purification and properties of thermostable lipase from a thermophilic *Bacillus stearothermophilus* MC 7. *Journal of Molecular Catalysis B: Enzymatic*, 22(5–6), pp. 307–313.

Kang, J. et al., 2011. Molecular cloning and biochemical characterization of a heat-stable type I pullulanase from *Thermotoga neapolitana*. *Enzyme and Microbial Technology*, 48, pp. 260–266.

Kang, S., Vieille, C., and Zeikus, J.G., 2004. Identification of *Pyrococcus furiosus* amylopullulanase catalytic residues. *Applied Microbiology and Biotechnology*, 66, pp. 408–413.

Kar, S., Ray, R.C., and Mohapatra, U.B., 2012. Purification, characterization and application of thermostable amylopullulanase from *Streptomyces erumpens* MTCC 7317 under submerged fermentation. *Annals of Microbiology*, 62, pp. 931–937. Available at: http://www.scopus.com/inward/record.url?eid=2-s2.0-84871613270&partnerID=40&md5=d95672b70ab4fa6dbcd309cea02 2589e.

Katapodis, P. et al., 2003. Enzymatic production of a feruloylated oligosaccharide with antioxidant activity from wheat flour arabinoxylan. *European Journal of Nutrition*, 42, pp. 55–60.

Kazemi, A. et al., 2014. Isolation, identification and media optimization of high level cellulase production by Bacillus sp. BCCS A3, in a fermentation system using response surface methodology. *Preparative Biochemistry and Biotechnology*, 44, pp. 107–118.

Khelila, O. and Cheba, B., 2014. Thermophilic cellulolytic microorganisms from western Algerian sources: Promising isolates for cellulosic biomass recycling. *Procedia Technology*, 12, pp. 519–528.

Kiiskinen, L.L. et al., 2004. Expression of *Melanospora albamyces* laccase in Trichoderma reesei and characterization of the purified enzyme. *Microbiology*, 150, pp. 3065–3074.

Kim, T.S. et al., 2005. Delayed dark adaptation in 11-cis-retinol dehydrogenase-deficient mice: A role of RDH11 in visual processes *in vivo. Journal of Biological Chemistry*, 280, pp. 8694–8704.

Kirikyali, N., Wood, J., and Connerton, I.F., 2014. Characterisation of a recombinant β-xylosidase (xylA) from *Aspergillus oryzae* expressed in *Pichia pastoris. AMB Express*, 4, p.68. Available at: http://www.amb-express.com/content/4/1/68.

Klonowska, A. et al., 2002. Characterization of a low redox potential laccase from the basidiomycete c30. *European Journal of Biochemistry*, 269, pp. 6119–6125.

Knob, A. and Carmona, E.C., 2010. Purification and characterization of two extracellular xylanases from *Penicillium sclerotiorum*: A novel acidophilic xylanase. *Applied Biochemistry and Biotechnology*, 162, pp. 429–443. Available at: http://www.ncbi.nlm.nih.gov/pubmed/19680819.

Knob, A. et al., 2014. Agro-residues as alternative for xylanase production by filamentous fungi. *BioResources*, 9, pp. 5738–5773.

Kocabaş, D.S., Güder, S., and Özben, N., 2015. Purification strategies and properties of a low-molecular weight xylanase and its application in agricultural waste biomass hydrolysis. *Journal of Molecular Catalysis B: Enzymatic*, 115, pp. 66–75. Available at: http://linkinghub.elsevier.com/retrieve/pii/S1381117715000296.

Kotani, A. et al., 2012. EndoV/DNA ligase mutation scanning assay using microchip capillary electrophoresis and dual-color laser-induced fluorescence detection. *Analytical Methods*, 4, p.58.

Kranaster, R. et al., 2010. One-step RNA pathogen detection with reverse transcriptase activity of a mutated thermostable *Thermus aquaticus* DNA polymerase. *Biotechnology Journal*, 5, pp. 224–231.

Kudanga, T. and Le Roes-Hill, M., 2014. Laccase applications in biofuels production: Current status and future prospects. *Applied Microbiology and Biotechnology*, 98, pp. 6525–6542.

Kuhad, R.C., Gupta, R., and Singh, A., 2011. Microbial cellulases and their industrial applications. *Enzyme Research*, 2011, pp. 1–10. Available at: http://www.hindawi.com/journals/er/2011/280696/.

Kuhad, R.C. et al., 2016. Revisiting cellulase production and redefining current strategies based on major challenges. *Renewable and Sustainable Energy Reviews*, 55, pp. 249–272. Available at: http://linkinghub.elsevier.com/retrieve/pii/S1364032115012113.

Kulkarni, N., Shendye, A., and Rao, M., 1999. Molecular and biotechnological aspects of xylanases. *FEMS Microbiology Reviews*, 23, pp. 411–456.

Kumar, K.S. et al., 2009. Production of beta-xylanase by a *Thermomyces lanuginosus* MC 134 mutant on corn cobs and its application in bio bleaching of bagasse pulp. *Journal of Bioscience and Bioengineering*, 107, pp. 494–498. Available at: http://www.ncbi.nlm.nih.gov/pubmed/19393546 (Accessed November 14, 2014).

Kumar, L., Awasthi, G., and Singh, B., 2011. Extremophiles: A novel source of industrially important enzymes. *Biotechnology*, 10, pp. 121–135.

Kumar, S. et al., 2005. Production, purification, and characterization of lipase from thermophilic and alkaliphilic *Bacillus coagulans* BTS-3. *Protein Expression and Purification*, 41, pp. 38–44.

Kumari, M., Sharma, A., and Jagannadham, M.V., 2012. Religiosin B, a milk-clotting serine protease from *Ficus religiosa*. *Food Chem*, 131, pp. 1295–1303.

Kunamneni, A. et al., 2007. Fungal laccase—a versatile enzyme for biotechnological applications. In A. Mendez-Vias, ed. *Communication Current Research and Educational Topics and Trends in Applied Microbiology*. Formex, Badajoz, pp. 233–245.

Kundu, A. and Ray, R.R., 2013. Production of intracellular b-xylosidase from the submerged fermentation of citrus wastes by *Penicillium janthinellum* MTCC. *3 Biotech*, 3, pp. 241–246.

Kuzu, S.B., Güvenmez, H.K., and Denizci, A.A., 2012. Production of a thermo-stable and alkaline chitinase by *Bacillus thuringiensis* subsp. kurstaki Strain HBK-51. *Biotechnology Research International*, 2012, p. 135498. Available at: http://www.pubmedcentral.nih.gov/articlerender.fcgi?artid=3532916& tool=pmcentrez&rendertype=abstract.

Lai, X. et al., 2002. Biochemical characterization of an ATP-dependent DNA ligase from the hyperthermophilic crenarchaeon *Sulfolobus shibatae*. *Extremophiles*, 6, pp. 469–477.

Lama, L. et al., 2004. Purification and characterization of thermostable xylanase and beta-xylosidase by the thermophilic bacterium *Bacillus thermantarcti-cus*. *Research in Microbiology*, 155, pp. 283–289. Available at: http://www. ncbi.nlm.nih.gov/pubmed/15142626.

Laribi-Habchi, H. et al., 2015. Purification, characterization, and molecular clon-ing of an extracellular chitinase from *Bacillus licheniformis* stain LHH100 isolated from wastewater samples in Algeria. *International Journal of Biological Macromolecules*, 72, pp. 1117–1128. Available at: http://dx.doi. org/10.1016/j.ijbiomac.2014.10.035.

Lawyer, F.C. et al., 1993. High-level expression, purification, and enzymatic characterization of full-length *Thermus aquaticus* DNA polymerase and a truncated form deficient in 5′ to 3′ Exonuclease activity. *Research—PCR Methods and Applications*, 2, pp. 275–287.

Le, Y. et al., 2013. Thermostable DNA ligase-mediated PCR production of cir-cular plasmid (PPCP) and its application in directed evolution via *in situ* error-prone PCR. *DNA Research*, 20, pp. 375–382.

Lee, I., Evans, B.R., and Woodward, J., 2015. The mechanism of cellulase action on cotton fibres: Evidence from atomic force microscopy. *Ultramicroscopy*, 82, pp. 213–221.

Leis, B. et al., 2015. Functional screening of hydrolytic activities reveals an extremely thermostable cellulase from a deep-sea archaeon. *Frontiers in Bioengineering and Biotechnology*, 3(7), p. 95. Available at: http://www.ncbi. nlm.nih.gov/pubmed/26191525.

Leow, T.C. et al., 2004. High level expression of thermostable lipase from *Geobacillus* sp. strain T1. *Bioscience, Biotechnology, and Biochemistry*, 68, pp. 96–103.

Li, H. and Zhang, X., 2005. Characterization of thermostable lipase from ther-mophilic *Geobacillus* sp. TW1. *Protein Expression and Purification*, 42, pp. 153–159.

Li, Q. et al., 2013a. Commercial proteases: Present and future. *FEBS Letters*, 587, pp. 1155–1163.

Li, S. et al., 2015. Structure and sequence analysis-based engineering of pullula-nase from *Anoxybacillus* sp. LM18-11 for improved thermostability. *Journal of Biotechnology*, 210, pp. 8–14. Available at: http://linkinghub.elsevier.com/ retrieve/pii/S0168165615300328.

Li, X., Li, D., and Park, K.-H., 2013b. An extremely thermostable amylopullulanase from *Staphylothermus marinus* displays both pullulan- and cyclodex-
trin-degrading activities. *Applied Microbiology and Biotechnology*, 97, pp.
5359–5369. Available at: http://www.ncbi.nlm.nih.gov/pubmed/23001056.

Li, Y. et al., 2012. Cloning, expression, characterization, and biocatalytic investigation of a novel *Bacilli thermostable* type i pullulanase from *Bacillus* sp.
CICIM 263. *Journal of Agricultural and Food Chemistry*, 60, pp. 11164–11172.

Li, Y.-T. et al., 2009. Preparation of homogenous oligosaccharide chains from gly-
cosphingolipids. *Glycoconjugate Journal*, 26, pp. 929–33. Available at: http://
www.ncbi.nlm.nih.gov/pubmed/18415015.

Liang, T.-W., Hsieh, T.-Y., and Wang, S.-L., 2014. Purification of a thermostable chitinase from *Bacillus cereus* by chitin affinity and its application in microbial
community changes in soil. *Bioprocess and Biosystems Engineering*, 37, pp.
1201–9. Available at: http://www.ncbi.nlm.nih.gov/pubmed/24342954.

Lohman, G.J., Tabor, S., and Nichols, N.M., 2011. DNA ligases. In: Ausubel, F.M.
et al. (eds.), *Current Protocols in Molecular Biology*. 94:III:3.14:3.14.1–3.14.7.
DOI:10.1002/0471142727.mb0314s94.

Loncar, N. and Fraaije, M.W., 2015. Not so monofunctional—A case of thermostable *Thermobifida fusca* catalase with peroxidase activity. *Applied
Microbiology and Biotechnology*, 99, pp. 2225–2232. Available at: http://
www.ncbi.nlm.nih.gov/pubmed/25227535\n, http://download.springer.
com/static/pdf/175/art%253A10.1007%252Fs00253-014-6060-5.pdf?auth66
=1427795261_56f36dc55730aafb5ce2f7d5f3cf4665&ext=.pdf.

Lu, L. et al., 2013. Cloning and expression of thermo-alkali-stable laccase of
Bacillus licheniformis in *Pichia pastoris* and its characterization. *Bioresource
Technology*, 134, pp. 81–86. Available at: http://dx.doi.org/10.1016/j.
biortech.2013.02.015.

Lu, R. et al., 2015. Screening cloning and expression analysis of a cellulase derived
from the causative agent of hypertrophy sorosis scleroteniosis *Ciboria shiraiana*. *Gene*, 565, pp. 221–227.

MacCabe, A.P. et al., 2002. Improving extracellular production of food-use
enzymes from *Aspergillus nidulans*. *Journal of Biotechnology*, 96, pp. 43–54.

Madhavi, V. and Lele, S.S., 2009. Laccase: Properties and applications. *Bio
Resources*, 4, pp. 1694–717.

Malakar, R., Tiwari, D.A., and Malviya, S.N., 2010. Pullulanase: A potential
enzyme for industrial application. *International Journal of Biomedical
Research*, 1, pp. 10–20.

Mall, R. et al., 2013. Purification and characterization of a thermostable soluble peroxidase from *Citrus medica* leaf. *Preparative Biochemistry and
Biotechnology*, 43, pp. 137–151. Available at: http://www.tandfonline.com/
doi/abs/10.1080/10826068.2012.711793.

Manavalan, A. et al., 2015. Characterization of a solvent, surfactant and tem-
perature-tolerant laccase from *Pleurotus* sp. MAK-II and its dye decolorizing property. *Biotechnology Letters*, 37, pp. 2403–2409. Available at: http://
dx.doi.org/10.1007/s10529-015-1937-7.

Maria, L. De et al., 2007. Phospholipases and their industrial applications. *Applied Microbiology and Biotechnology*, 74, pp. 290–300.

Marques, S. et al., 2003. Characterisation and application of glycanases secreted by *Aspergillus terreus* CCMI 498 and *Trichoderma viride* CCMI 84 for enzymatic deinking of mixed office wastepaper. *Journal of Biotechnology*, 100, pp. 209–219.

Masomian, M. et al., 2013. A new thermostable and organic solvent-tolerant lipase from *Aneurinibacillus thermoaerophilus* strain HZ. *Process Biochemistry*, 48, pp. 169–175. Available at: http://www.sciencedirect.com/science/article/pii/S135951131200387X (Accessed October 12, 2015).

Masui, D.C. et al., 2012. Production of a xylose-stimulated β-glucosidase and a cellulase-free thermostable xylanase by the thermophilic fungus *Humicola brevis* var. *thermoidea* under solid state fermentation. *World Journal of Microbiology and Biotechnology*, 28, pp. 2689–2701.

Matsumura, S., Sakiyama, K., and Toshima, K., 1999. Preparation of octyl β-D-xylobioside and xyloside by xylanase catalyzed direct transglycosylation reaction of xylan and octanol. *Biotechnology Letters*, 21, pp. 17–22.

Mayer, G., Muller, J., and Lunse, C.E., 2011. RNA diagnostics: Real-time RT-PCR strategies and promising novel target RNAs. *Wiley Interdisciplinary Reviews RNA*, 2, pp. 32–41.

Mesbah, N.M. and Wiegel, J., 2014. Halophilic alkali- and thermostable amylase from a novel polyextremophilic *Amphibacillus sp*. NM-Ra2. *International Journal of Biological Macromolecules*, 70C, pp. 222–229. Available at: http://www.sciencedirect.com/science/article/pii/S014181301400436X.

Michelin, M. et al., 2012a. A novel xylan degrading β-d-xylosidase: Purification and biochemical characterization. *World Journal of Microbiology and Biotechnology*, 28, pp. 3179–3186.

Michelin, M. et al., 2012b. Production of xylanase and β-xylosidase from auto-hydrolysis liquor of corncob using two fungal strains. *Bioprocess and Biosystems Engineering*, 35, pp. 1185–1192. Available at: http://link.springer.com/10.1007/s00449-012-0705-5.

Mizuno, M., Yasukawa, K., and Inouye, K., 2010. Insight into the mechanism of the stabilization of moloney murine leukaemia virus reverse transcriptase by eliminating RNase H activity. *Bioscience, Biotechnology and Biochemistry*, 74, pp. 440–442.

Mohr, S. et al., 2013. Thermostable group II intron reverse transcriptase fusion proteins and their use in cDNA synthesis and next-generation RNA sequencing. *RNA*, 19, pp. 958–970. Available at: http://www.pubmedcentral.nih.gov/articlerender.fcgi?artid=3683930&tool=pmcentrez&rendertype=abstract.

Moraïs, S. et al., 2011. Assembly of xylanases into designer cellulosomes promotes efficient hydrolysis of the xylan component of a natural recalcitrant cellulosic substrate. *MBio*, 2(6), pp.1–11. Available at: http://www.ncbi.nlm.nih.gov/pubmed/22086489.

Morgenstern, I., Robertson, D.L., and Hibbett, D.S., 2010. Characterization of three mnp genes of *Fomitiporia mediterranea* and report of additional class II peroxidases in the order Hymenochaetales. *Applied and Environment Microbiology*, 76, pp. 6431–6440.

Morozova, O. V et al., 2007. "Blue" laccases. *Biochemistry (Mosc)*, 72, pp. 1136–1150.

Mozersky, S., Marmer, W., and Dale, A.O., 2002. Vigorous proteolysis: Relining in the presence of an alkaline protease and bating (Post-Liming) with an extremophile protease. *JALCA*, 97, pp. 150–155.

Mrudula, S., Gopal, R., and Seenayya, G., 2011. Effect of substrate and culture conditions on the production of amylase and pullulanase by thermophilic *Clostridium thermosulforegenes* SVM17 in solid state fermentation. *Malaysian Journal of Microbiology*, 7, pp. 19–25.

Mutreja, R. et al., 2011. Bioconversion of agricultural waste to ethanol by SSF using recombinant cellulase from *Clostridium thermocellum*. *Enzyme Research*, 2011, pp. 1–6. Available at: http://www.hindawi.com/journals/er/2011/340279/.

Nawani, N. and Kaur, J., 2000. Purification, characterization and thermostability of lipase from a thermophilic *Bacillus* sp. J33. *Molecular and Cellular Biochemistry*, 206, pp. 91–96.

Nguyen, T.T., Quyen, T.D., and Le, H.T., 2013. Cloning and enhancing production of a nattokinase from *Bacillus subtilis* VTCC-DVN-12-01 by using an eight-protease-gene-deficient *Bacillus subtilis* WB800. *Microbial Cell Factories*, 12, pp. 1–11. Available at: Microbial Cell Factories.

Nigam, P.S., 2013. Microbial enzymes with special characteristics for biotechnological applications. *Biomolecules*, 3, pp. 597–611.

Nisha, M. and Satyanarayana, T., 2013. Characterization of recombinant amylopullulanase (gt-apu) and truncated amylopullulanase (gt-apuT) of the extreme thermophile *Geobacillus thermoleovorans* NP33 and their action in starch saccharification. *Applied Microbiology and Biotechnology*, 97, pp. 6279–6292.

Nisole, S. et al., 2004. A Trim5-cyclophilin A fusion protein found in owl monkey kidney cells can restrict HIV-1. *Proceedings of National Academy of Sciences*, 101, pp. 13324–13328.

Oziengbe and Onilude, 2012. Production of a thermostable a-amylase and its assay using *Bacillus licheniformis* isolated from excavated land sites in Ibadan, Nigeria. *Bayero Journal of Pure Applied Sciences*, 5, pp. 132–138.

Ozsolak, F. and Milos, P.M., 2011. RNA sequencing: Advances, challenges and opportunities. *Nature Review Genetics*, 12, pp. 87–98.

Ozturk, M.T. et al., 2013. Ligase-independent cloning of amylase gene from a local *Bacillus subtilis* isolate and biochemical characterization of the purified enzyme. *Applied Biochemistry and Biotechnology*, 171, pp. 263–78. Available at: http://www.ncbi.nlm.nih.gov/pubmed/23832859.

Palavesam, A., 2015. Investigation on lignocellulosic saccharification and characterization of haloalkaline solvent tolerant endo-1,4 β-d-xylanase from *Halomonas meridiana* APCMST-KS4. *Biocatalysis and Agricultural Biotechnology*, 4, pp.761–766. Available at: http://linkinghub.elsevier.com/retrieve/pii/S1878818115001279.

Pant, G. et al., 2015. Production, optimization and partial purification of protease from *Bacillus subtilis*. *Journal of Taibah University for Sciences*, 9, pp. 50–55. Available at: http://dx.doi.org/10.1016/j.jtusci.2014.04.010.

Paredes, R. de S. et al., 2015. Production of xylanase, α-l-arabinofuranosidase, β-xylosidase, and β-glucosidase by *Aspergillus awamori* using the liquid stream from hot-compressed water treatment of sugarcane bagasse. *Biomass Conversion and Biorefinery*, pp. 3–11. Available at: http://link.springer.com/10.1007/s13399-015-0159-5.

Park, J.-T. et al., 2013. Molecular cloning and characterization of a thermostable α-amylase exhibiting an unusually high activity. *Food Science and Biotechnology*, 23, pp. 125–132. Available at: http://www.scopus.com/inward/record.url?eid=2-s2.0-84894434916&partnerID=tZOtx3y1.

Parshetti, G.K. et al., 2012. Industrial dye decolorizing lignin peroxidase from *Kocuria rosea* MTCC 1532. *Annals of Microbiology*, 62(1), pp.217–223. doi:10.1007/s13213-011-0249-y.

Patel, S.J. and Savanth, V.D., 2015. Review on fungal xylanases and their applications xylan xylanase fungal xylanases xylanase production. *International Journal of Advanced Research*, 3, pp. 311–315.

Pathak, A.P. and Rekadwad, B.N., 2013. Isolation of thermophilic *Bacillus* sp. strain EF _TYK1-5 and production of industrially important thermostable a-amylase using suspended solids for fermentation. *Journal of Scientific and Industrial Research*, 72, pp. 685–689.

Pazarlioğlu, N.K., Sariişik, M., and Telefoncu, A., 2005. Laccase: Production by *Trametes versicolor* and application to denim washing. *Process Biochemistry*, 40, pp. 1673–1678.

Polizeli, M.L.T.M. et al., 2005. Xylanases from fungi: Properties and industrial applications. *Applied Microbiology and Biotechnology*, 67, pp.577–591.

Pradeep, G.C. et al., 2013. A novel thermostable cellulase free xylanase stable in broad range of pH from *Streptomyces* sp. CS428. *Process Biochemistry*, 48, pp. 1188–1196. Available at: http://linkinghub.elsevier.com/retrieve/pii/S1359511313002766.

Prasanna, L. et al., 2013. A novel strain of *Brevibacillus laterosporus* produces chitinases that contribute to its biocontrol potential. *Applied Microbiology and Biotechnology*, 97, pp. 1601–1611.

Puchart, V. et al., 2015. A unique CE16 acetyl esterase from *Podospora anserina* active on polymeric xylan. *Applied Microbiology and Biotechnology*, pp. 10515–10526. Available at: http://link.springer.com/10.1007/s00253-015-6934-1.

Purohit, M.K. and Singh, S.P., 2013. A metagenomic alkaline protease from saline habitat: Cloning, over-expression and functional attributes. *International Journal of Biological Macromolecules*, 53, pp. 138–143. Available at: http://dx.doi.org/10.1016/j.ijbiomac.2012.10.032.

Quail, M.A. et al., 2008. A large genome center's improvements to the Illumina sequencing system. *Nature Methods*, 12, pp. 1005–1010.

Raghuwanshi, S. et al., 2014. Bioprocessing of enhanced cellulase production from a mutant of *Trichoderma asperellum* RCK2011 and its application in hydrolysis of cellulose. *Fuel*, 124(2014), pp. 183–189. Available at: http://dx.doi.org/10.1016/j.fuel.2014.01.107.

Raj, A., Kumar, S., and Singh, S.K., 2013. A highly thermostable xylanase from *Stenotrophomonas maltophilia*: Purification and partial characterization. *Enzyme Research*, 2013, p. 429305. Available at: http://www.ncbi.nlm.nih. gov/pubmed/24416589.

Rakotoarivonina, H. et al., 2015. Engineering the hydrophobic residues of a GH11 xylanase impacts its adsorption onto lignin and its thermostability. *Enzyme and Microbial Technology*, 81, pp. 47–55. Available at: http://linkinghub. elsevier.com/retrieve/pii/S0141022915300351.

Ramanathan, S., 2011. *Production of Thermostable Pullulanase from Bacillus flavothermus KWF-1 in Fed-Batch Culture*. Universiti Teknologi Malaysia.

Rana, N., Walia, A., and Gaur, A., 2013. α-Amylases from microbial sources and its potential applications in various industries. *National Academy Science Letters*, 36, pp. 9–17.

Rao, S., Ellaiah, P., and Biradar, K.V., 2012. Purification and characterization of thermostable amylase from a strain of thermoactinomyces thalpophilus KSV 17. *RGUHS Journal of Pharmaceutical Sciences*, 2, pp. 83–89. Available at: http://www.rjps.in/article/65.

Rolland, J.-L. et al., 2004. Characterization of a thermophilic DNA ligase from the archaeon *Thermococcus fumicolans*. *FEMS Microbiology Letters*, 236, pp. 267–273.

Roriz, M.S. et al., 2009. Application of response surface methodological approach to optimise Reactive black 5 decolouration by crude laccase from *Trametes pubescens*. *Journal of Hazardous Materials*, 169, pp. 691–696.

Sabotic, J. and Kos, J., 2012. Microbial and fungal protease inhibitors—Current and potential applications. *Applied Microbiology and Biotechnology*, 93, pp. 1351–1375.

Saha, B.C., 2001. Purification and characterization of an extracellular Fusarium verticillioides. *Journal of Industrial Microbial Biotechnology*, 27, pp. 241–245.

Saha, B.C., 2003. Purification and properties of an extracellular beta-xylosidase from a newly isolated *Fusarium proliferatum*. *Bioresource Technology*, 90, pp. 33–38.

Salahuddin, K. et al., 2012. Biochemical characterization of thermostable cellulase enzyme from mesophilic strains of actinomycete. *African Journal of Biotechnology*, 11, pp. 10125–10134. Available at: http://www.academicjournals. org/ajb/abstracts/abs2012/29May/Salahuddin et al.htm.

Sambrook, J. and Russell, D.W., 2001. *Molecular Cloning: A Laboratory Manual*. Cold Spring Harbor Laboratory, Cold Spring Harbor, NY.

Sano, S. et al., 2012. Mutations to create thermostable reverse transcriptase with bacterial family A DNA polymerase from *Thermotoga petrophila* K4. *Journal of Bioscience and Bioengineering*, 113, pp. 315–21. Available at: http://dx.doi.org/10.1016/j.jbiosc.2011.11.001.

Santos, A. et al., 2014. New dye-decolorizing peroxidases from *Bacillus subtilis* and *Pseudomonas putida* MET94: Towards biotechnological applications. *Applied Microbiology and Biotechnology*, 98, pp. 2053–2065.

Sarafianos, S.G. et al., 2009. Structure and function of HIV-1 reverse transcriptase: Molecular mechanisms of polymerization and inhibition. *Journal of Molecular Biology*, 385, pp. 693–713. Available at: http://dx.doi.org/10.1016/j.jmb.2008.10.071.

Sasaki, Y. et al., 2004. TNF family member B cell-activating factor (BAFF) receptor dependent and -independent roles for BAFF in B cell physiology. *Journal of Immunology*, 173, pp. 2245–2252.

Saxena, R.K. et al., 2003. Purification and characterization of an alkaline thermostable lipase from *Aspergillus carneus*. *Process Biochemistry*, 39, pp. 239–247.

Schmid, A. et al., 2001. Industrial biocatalysis today and tomorrow. *Nature*, 409, pp. 258–268.

Seo, M.S. et al., 2007. Cloning and expression of a DNA ligase from the hyperthermophilic archaeon *Staphylothermus marinus* and properties of the enzyme. *Journal of Biotechnology*, 128, pp. 519–530.

Shallom, D. and Shoham, Y., 2003. Microbial hemicellulases. *Current Opinion in Microbiology*, 6, pp. 219–228.

Sharma, A. and Satyanarayana, T., 2012. Cloning and expression of acid-stable, high maltose-forming, Ca2+-independent-amylase from an acidophile *Bacillus acidicola* and its applicability in starch hydrolysis. *Extremophiles*, 16, pp. 515–522.

Sharma, D., Sharma, B., and Shukla, A.K., 2011. Biotechnological approach of microbial lipase: A review. *Biotechnology*, 10, pp. 23–40.

Sharma, K. et al., 2014. Isolation, identification and optimization of culture conditions of *Bacillus* sp. strain PM1 for alkalo-thermostable amylase production. *British Microbiology Research Journal*, 4, pp. 369–380.

Shi, H. et al., 2013. A novel highly thermostable xylanase stimulated by Ca2+ from *Thermotoga thermarum*: Cloning, expression and characterization. *Biotechnology for Biofuels*, 6, p. 26. Available at: http://www.ncbi.nlm.nih.gov/pubmed/23418789\n, http://www.biotechnologyforbiofuels.com/content/6/1/26.

Shiba, Y. et al., 2001. High-level secretory production of phospholipase A1 by *Saccharomyces cerevisiae* and *Aspergillus oryzae*. *Bioscience, Biotechnology, and Biochemistry*, 65(2), pp. 94–101.

Shoemaker, G.K., Juers, D.H., and Coombs, J.M.L., 2003. Crystallization of β-galactosidase does not reduce the range of activity of individual molecules. *Biochemistry*, 42, pp. 1707–1710.

Shraddha et al., 2011. Laccase: Microbial sources, production, purification, and potential biotechnological applications. *Enzyme Research*, p. 217861.

Shuman, S., 2009. DNA ligases: Progress and prospects. *The Journal of Biological Chemistry*, 284, pp. 17365–17369.

Simkhada, J.R. et al., 2009. A novel low molecular weight phospholipase D from *Streptomyces* sp. CS684. *Bioresource Technology*, 100, pp. 1388–93. Available at: http://www.sciencedirect.com/science/article/pii/S0960852408007748.

Singh, K. et al., 2013. Statistical media optimization and cellulase production from marine *Bacillus* VITRKHB. *3 Biotech*, 4, pp. 591–598. Available at: http://link.springer.com/10.1007/s13205-013-0173-x (Accessed November 17, 2014).

Singh, S. et al., 2000a. Production and properties of hemicellulases by a *Thermomyces lanuginosus* strain. *Journal of Applied Microbiology*, 88, pp. 975–982.

Singh, S. et al., 2000b. Relatedness of *Thermomyces lanuginosus* strains producing a thermostable xylanase. *Journal of Biotechnology*, 81(2–3), pp. 119–128.

Singh, S., Madlala, A.M., and Prior, B.A., 2003. *Thermomyces lanuginosus*: Properties of strains and their hemicellulases. *FEMS Microbiology Reviews*, 27(1), pp. 3–16.

Sreedevi, S., Sajith, S., and Benjamin, S., 2013. Cellulase producing bacteria from the wood-yards on Kallai river bank. *Advances in Microbiology*, 3(8), pp. 326–332. Available at: http://afrjournal.org/index.php/afr/article/view/80 (Accessed November 17, 2014).

Stewart, E.V. et al., 2011. Yeast SREBP cleavage activation requires the Golgi Dsc E3 ligase complex. *Molecular Cell*, 42, pp. 160–171. Available at: http://linkinghub.elsevier.com/retrieve/pii/S1097276511002528.

Stokes, S.S. et al., 2011. Discovery of bacterial NAD+-dependent DNA ligase inhibitors: Optimization of antibacterial activity. *Bioorganic and Medicinal Chemistry Letters*, 21, pp. 4556–4560. Available at: http://www.sciencedirect.com/science/article/pii/S0960894X11007694.

Subramaniyan, S. and Prema, P., 2002. Biotechnology of microbial xylanases: Enzymology, molecular biology, and application. *Critical Reviews in Biotechnology*, 22, pp. 33–64.

Suginta, W. et al., 2000. Chitinases from vibrio: Activity screening and purification of chiA from *V. carchariae*. *Journal of Applied Microbiology*, 89, pp. 76–84.

Sun, S. et al., 2011. Cloning, expression and characterization of a thermostable pullulanase from newly isolated thermophilic Geobacillus sp. LM14-3. In *4th International Conference on Biochemical Engineering and Informatics*. Shanghai, China, pp. 1567–1570. DOI: 10.1109/BMEI.2011.6098577.

Sundarram, A. and Murthy, T.P.K., 2014. α-Amylase production and applications: A review. *Journal of Applied and Environmental Microbiology*, 2, pp. 166–175.

Takayanagi, T. et al., 1991. Isolation and characterization of thermostable chitinases from *Bacillus licheniformis* X_7u. *Biochemica et Biophysica Acta*, 1078, pp. 404–410.

Tello, M. et al., 2000. Characterization of three new manganese peroxidase genes from the ligninolytic basidiomycete *Ceriporiopsis subvermispora*. *Biochimica et Biophysica Acta*, 1490, pp. 137–44.

Teng, C. et al., 2011. High-level expression of extracellular secretion of a β-xylosidase gene from *Paecilomyces thermophila* in *Escherichia coli*. *Bioresource Technology*, 102, pp. 1822–1830. Available at: http://dx.doi.org/10.1016/j.biortech.2010.09.055.

Terpe, K., 2013. Overview of thermostable DNA polymerases for classical PCR applications: From molecular and biochemical fundamentals to commercial systems. *Applied Microbial Biotechnology*, 97, pp. 10243–10254.

Terrasan, C.R.F. et al., 2013. Xylanase and β-xylosidase from *Penicillium janc-zewskii*: Production, physico-chemical properties, and application of the crude extract to pulp biobleaching. *BioResources*, 8, pp. 1292–1305.

Thomas, L., Joseph, A., and Gottumukkala, L.D., 2014. Xylanase and cellu-lase systems of *Clostridium* sp.: An insight on molecular approaches for strain improvement. *Bioresource Technology*, 158, pp. 343–350. Available at: http://www.ncbi.nlm.nih.gov/pubmed/24581864 (Accessed October 19, 2014).

Tikhomirov, D.F. et al., 2003. Non-starch polysaccharide hydrolysing microbial enzymes in grain processing. In C.M. Courtin, W.S. Veraverbeke, and J.A. Delcour, eds. *Recent Advances in Enzymes in Grain Processing*. Katholieke University, Leuven, pp. 423–418.

Tsujibo, H. et al., 1995. Expression in *Escherichia coli* of a gene encoding a thermo-stable chitinase from *Streptomyces thermoviolaceus* OPC-520. *Bioscience, Biotechnology and Biochemistry*, 59, pp. 145–146.

Tsujibo, H. et al., 2001. Cloning, sequencing, and expression of the gene encoding an intracellular β-D-xylosidase from *Streptomyces thermoviolaceus* OPC-520. *Bioscience, Biotechnology, Biochemistry*, 65, pp. 1824–1831.

Turner, P., Mamo, G., and Karlsson, E.N., 2007. Potential and utilization of thermophiles and thermostable enzymes in biorefining. *Microbial Cell Factories*, 6, p. 9.

Uday, U.S.P. et al., 2016. Classification, mode of action and production strat-egy of xylanase and its application for biofuel production from water hyacinth. *International Journal of Biological Macromolecules*, 82, pp. 1041–1054. Available at: http://linkinghub.elsevier.com/retrieve/pii/S014181301530088X.

Van Der Maarel, M.J. et al., 2002. Properties and applications of starch-con-verting enzymes of the α-amylase family. *Journal of Biotechnology*, 94, pp. 137–155.

Várnai, A. et al., 2014. Expression of endoglucanases in *Pichia pastoris* under control of the GAP promoter. *Microbial Cell Factories*, 13, p. 57. Available at: http://www.microbialcellfactories.com/content/13/1/57.

Verma, D. and Satyanarayana, T., 2012. Cloning, expression and applicability of thermo-alkali-stable xylanase of *Geobacillus thermoleovorans* in generat-ing xylooligosaccharides from agro-residues. *Bioresource Technology*, 107, pp. 333–338. Available at: http://www.ncbi.nlm.nih.gov/pubmed/22212694.

Vieille, C. and Zeikus, G.J., 2001. Hyperthermophilic enzymes: Sources, uses, and molecular mechanisms for thermostability. *Microbiology and Molecular Biology Reviews*, 65, pp. 1–43.

Wakiyama, M. et al., 2008. Purification and properties of an extracellular b-xylo-sidase from *Aspergillus japonicus* and sequence analysis of the encoding gene. *Journal of Bioscience and Bioengineering*, 106, pp. 398–404.

Wang, B. et al., 2004. A novel phospholipase A2/esterase from hyperthermo-philic archaeon *Aeropyrum pernix* K1. *Protein Expression and Purification*, 35, pp. 199–205. Available at: http://www.sciencedirect.com/science/article/pii/S1046592804000440 (Accessed October 30, 2015).

Wang, L.K. and Shuman, S., 2005. Structure-function analysis of yeast tRNA ligase. *RNA*, 11(6), pp. 966–975.

Wang, Y. et al., 2013. Expression, purification and biochemical characterization of *Methanocaldococcus jannaschii* DNA ligase. *Protein Expression and Purification*, 87(2), pp. 79–86. Available at: http://www.ncbi.nlm.nih.gov/pubmed/23147204.

Wang, Z., Gerstein, M., and Snyder, M., 2009. RNA-Seq: A revolutionary tool for transcriptomics. *Nature Review Genetics*, 10, pp. 57–63.

Webb, M.D. et al., 2001. Metabolism of pentachlorophenol by *Saccharomonospora viridis* strains isolated from mushroom compost. *Soil Biology and Biochemistry*, 33, pp. 1903–1914.

Wei, T. et al., 2015. Characterization of a novel thermophilic phospholipase B from *Thermotoga lettingae* TMO: Applicability in enzymatic degumming of vegetable oils. *Journal of Industrial Microbiology and Biotechnology*, 42, pp. 515–522. Available at: http://link.springer.com/10.1007/s10295-014-1580-7.

Wilson, L. et al., 2006. Improvement of the functional properties of a thermostable lipase from *Alcaligenes* sp. via strong adsorption on hydrophobic supports. *Enzyme and Microbial Technology*, 38, pp. 975–980.

Wong, K.K.Y., Tan, L.U.L., and Saddler, J.N., 1988. Multiplicity of β-1, 4-xylanase in microorganisms: Functions and applications. *Microbiology Reviews*, 52, pp. 305–317.

Wongwisansri, S. et al., 2013. High-level production of thermotolerant β-xylosidase of *Aspergillus* sp. BCC125 in *Pichia pastoris*: Characterization and its application in ethanol production. *Bioresource Technology*, 132, pp. 410–3. Available at: http://www.ncbi.nlm.nih.gov/pubmed/23265813.

Woo, M.H. et al., 2014. First thermostable endo-β-1,4-glucanase from newly isolated *Xanthomonas* sp. EC102. Protein Journal, 33, pp. 110–117.

Wu, G. et al., 2006. Simplified gene synthesis: A one-step approach to PCR-based gene construction. *Journal of Biotechnology*, 124, pp. 496–503.

Wu, H. et al., 2014. Cloning, overexpression and characterization of a thermostable pullulanase from *Thermus thermophilus* HB27. *Protein Expression and Purification*, 95, pp. 22–27. Available at: http://www.sciencedirect.com/science/article/pii/S1046592813002544.

Yamura, I., Koga, T., and Matsumoto, T., 1997. Purification and some properties of endo-1,4-β xylanase from a fresh water mollusk Pomacea insularus (de Oringny). *Bioscience Biotechnology and Biochemistry*, 61, pp. 615–620.

Yang, S. et al., 2016. Cloning, expression, purification and application of a novel chitinase from a thermophilic marine bacterium *Paenibacillus barengoltzii*. *Food Chemistry*, 192, pp. 1041–1048. Available at: http://dx.doi.org/10.1016/j.foodchem.2015.07.092.

Yang, X., Cong, H., and Song, J., 2013. Heterologous expression of an aspartic protease gene from biocontrol fungus Trichoderma asperellum in *Pichia pastoris*. *World Journal of Microbiology and Biotechnology*, 29, pp. 2087–2094.

Yasukawa, K., Nemoto, D., and Inouye, K., 2008. Comparison of the thermal stabilities of reverse transcriptases from avian myeloblastosis virus and Moloney murine leukaemia virus. *Journal of Biochemistry*, 143, pp. 261–268.

Yu, W. et al., 2014. Application of a novel alkali-tolerant thermostable DyP-type peroxidase from *Saccharomonospora viridis* DSM 43017 in biobleaching of Eucalyptus Kraft Pulp. *PLoS ONE*, 9, p. e110319. Available at: http://dx.plos.org/10.1371/journal.pone.0110319.

Zafar, A. et al., 2015. Cloning, purification and characterization of a highly thermostable amylase gene of *Thermotoga petrophila* into *Escherichia coli*. *Applied Biochemistry and Biotechnology*. Available at: http://link.springer.com/10.1007/s12010-015-1912-8.

Zambare, V., Nilegaonkar, S., and Kanekar, P., 2011. A novel extracellular protease from *Pseudomonas aeruginosa* MCMB-327: Enzyme production and its partial characterization. *New Biotechnology*, 28, pp. 173–181.

Zhang, D. et al., 2012. Cloning, characterization, expression and antifungal activity of an alkaline serine protease of *Aureobasidium pullulans* PL5 involved in the biological control of postharvest pathogens. *International Journal of Food Microbiology*, 153, pp. 453–464. Available at: http://dx.doi.org/10.1016/j.ijfoodmicro.2011.12.016.

Zhang, M. et al., 2015. The multi-chitinolytic enzyme system of the compost-dwelling thermophilic fungus *Thermomyces lanuginosus*. *Process Biochemistry*, 50, pp. 237–244. Available at: http://dx.doi.org/10.1016/j.procbio.2014.11.008.

Zhang, S. et al., 2014. Cloning, expression, and characterization of a thermostable β-xylosidase from thermoacidophilic *Alicyclobacillus* sp. A4. *Process Biochemistry*, 49, pp. 1422–1428. Available at: http://linkinghub.elsevier.com/retrieve/pii/S1359511314003225.

Zhang, Y. et al., 2014. High-yield production of a chitinase from *Aeromonas veronii* B565 as a potential feed supplement for warm-water aquaculture. *Applied Microbiology and Biotechnology*, 98, pp. 1651–1662.

Zhao, A., Gray, F.C., and MacNeill, S.A., 2006. ATP- and NAD+-dependent DNA ligases share an essential function in the halophilic archaeon *Haloferax volcanii*. *Molecular Microbiology*, 59, pp. 743–752.

CHAPTER 6

Microbial Enzymes for Pulp and Paper Industry

Prospects and Developments

Puneet Pathak, Prabhjot Kaur,
and Nishi K. Bhardwaj

CONTENTS

ABSTRACT

The traditional pulp- and papermaking processes involve huge quantity of water, energy, and chemicals. In today's scenario, due to increasing environmental and economic concerns associated with traditional processes, papermakers are hard pressed to introduce eco-friendly processes for the reduction in energy consumption and toxic chemicals. Therefore, microbial enzymes may augment various pulp and paper processes with less effort and better results including savings in electricity, water, and harsh chemicals, thereby reducing environmental impacts. Currently, the most promising areas for enzyme applications in pulp and paper industry are amylase for the starch modification in surface sizing; xylanase as well as mannanase for pulp bleaching and deinking of recycled pulp; cellulase for pulp refining, pulp drainage improvement, and deinking of recycled pulp; and lignin-degrading enzymes for pulp bleaching, wastewater treatment, etc. Some applications have already been commercialized; however, the majority of the potential applications are still in the pilot or lab scale. Therefore, this chapter covers the prospects and developments of different enzymes applied in various areas in pulp and paper industry including current research on laboratory, pilot, and mill scales.

INTRODUCTION

Pulp and paper industry is believed as one of the largest as well as most polluting industries in the world. Approximately 380 million tons of paper and paper board is manufactured all around the world with the most rapid growth rate in Asia. Although China is expanding the industry at good pace, the production in North America has reduced. Asia also exhibited the good amount of consumption with nearly 40% of global consumption while producing over 33% of world paper and paper board. Still the dominating countries for this industry include North America (United States and Canada), North Europe (Finland, Sweden), and East Asia along with Latin America and Australia. Region-wise consumption of paper products has shown significant variations (7–300 kg per capita per annum) whereas global consumption for European and Northern American countries accounts for nearly 25% each. The average value of paper consumption for a person is around 60 kg per year. Also, it is expected that consumption rate in Asia will keep increasing in the foreseeable future (Bajpai 2012).

   Pulp and paper industry is known as a large contributor to the economic growth of a country and also to the environmental load. Today, the main targets of pulp and paper industry are to deliver improved product quality in a cost-effective manner with eco-friendly techniques. Interest of papermakers for using target-specific enzymes was triggered shortly after the realization of natural polymeric nature of paper consisting of cellulose, hemicelluloses, and lignin. Since then, extensive investigations for employing enzymes in various processes of pulp and paper manufacture were initiated. Historically, the first enzyme to be used in pulp and paper manufacture was cellulase for facilitating fiber beating in 1959 by Bolaski and Gallatin (Kirk and Jeffries 1996). The first introduction of enzymes at mill scale took place during the 1980s followed by rapid increase in exploring the potential applications of enzymes. Moreover, recent years have registered impressive research done on the applications of enzymes mainly due to more concern for environmental issues as well as to strengthen a sustainable approach for the utilization of forest resources. Depleting nonrenewable resources, increasing pollution with increasing consumption of pulp and paper have been the main driving force for necessitating the use of environmentally benign processes to replace or facilitate chemical processes in pulp and paper industry. The attractiveness of using enzymes for this purpose lies in their substrate-specific and eco-friendly nature.

Enzymes are now known for improving a vast number of pulp and paper properties. An enzyme technology utilizing xylanases has reduced or even eliminated the requirement of chlorine in the bleaching process leading to elemental chlorine free (ECF) and total chlorine free (TCF) bleaching of pulp, hence reducing the environmental load. Besides, xylanases can be used for improving the mechanical properties of paper and the refining process. Xylanases along with cellulases can be implemented in deinking of waste paper and production of dissolving pulp from paper-grade pulp. Cellulases alone can be utilized in various processes such as to improve pulp drainage, bioethanol production, and reactivity enhancement of dissolving pulp. Furthermore, lipases have the potential to be utilized in pitch removal in pulp and stickies control. Thus, enzymes can be utilized as potential tools to obtain tailored products as per the specific needs. There is a range of commercial applications of enzymes such as xylanase-aided bleaching, improved runnability of paper machine, and starch modification for surface sizing of paper. A number of applications has reached to pilot scale including laccase-mediator system for bleaching of pulp, xylanases and cellulases for dissolving pulp production, xylanase and ligninase for biochemical pulping, etc. More recent application at laboratory scale includes cellulases and hemicellulases in nanocellulose production (Zhu et al. 2011).

ENZYMES

Enzymes are chiefly proteins (except ribozyme) that catalyze all the vital chemical reactions in living beings. These "biocatalysts" are capable of carrying out both synthetic and degradative pathways of different chemical compounds required for smooth functioning of the organism. Moreover, they have their unique substrate specificity and a defined range of environmental and physiological conditions, yet found to work under a diversified environment.

Similar to all catalysts, enzymes accelerate the rate of reaction by lowering the activation energy for a reaction. These enzymes are neither consumed by the reactions during catalysis nor alter the equilibrium (Bairoch 2000). The enzyme activity is influenced by temperature, chemical environment (like pH, temperature), substrate concentration, and presence of inhibitors or activators. Enzymes are commonly globular proteins having a three-dimensional (3D) structure, which decides the activities of the enzyme (Anfinsen 1973). Similar to all proteins, enzymes are composed of long, linear chains of amino acids, which fold to create a 3D structure. Owing to their unique amino acid sequences, enzymes produce a specific structure, which has distinctive properties. Enzymes are generally very specific toward the reactions to be catalyzed in terms of substrates that are involved in these reactions. The specificity of enzymes is mainly determined by its complementary shape, charge, hydrophilic/hydrophobic characteristics of enzymes, and substrates. Enzymes can also show impressive levels of chemo-selectivity, regio-selectivity, and stereospecificity (Jaeger and Eggert 2004).

GLOBAL SCENARIO FOR ENZYME INDUSTRY

The innovations in enzyme technology have shown a vast impact on almost every sector of industrial activity that ranges from a technical field to food, feed, and healthcare activity. This is attributed to enzymes' biodegradable nature and cost-effectiveness that the enzymatic processes have rapidly become better financial and ecological alternatives to chemical processes (http://www.bccresearch.com).

The global market for industrial enzymes has shown a 5-year compound annual growth rate (CAGR) of 8.2% and is predicted to reach to approximately $7.1 billion by 2018. Enzymes market has been dedicated to different niches including carbohydrases, nucleases and polymerases, lipases, and proteases. Among them, carbohydrase was the dominant enzyme that accounted for over 40% of the market revenue in 2013. Its major application segments included biofuels, food and beverages, animal feed, and detergent (http://www.hexaresearch.com). Region wise, North America accounted for nearly 35% of the global demand in 2013 and is estimated to have its leadership maintained till 2018. High growth rate is also predicted for Asia Pacific in the regional market at projected CAGR of over 9.0% from 2014 to 2020. This can be mainly due to the availability of raw material for biofuels production and increasing adoption, particularly in Japan and China (http://www.hexaresearch.com). The industrial enzyme market is dominated by Novozymes, DuPont, and DSM (http://www.bccresearch.com). Globally, a major proportion of these industrial enzymes are being used for detergent, starch industry, baking, and textile industries. Recently, pulp and paper industry also got entry as enzyme user industry due to their application in various pulp and paper processes.

In a study by Demuner et al. (2011), focused on technology prospecting in enzymes for their application in pulp and paper sector, development of enzymes was concentrated to only a few biotech enterprises with the domination of Novonordisk and Novozymes. Novozymes has been consistently increasing the investments in research and development (R&D) and increasing its sales in enzymes for the pulp and paper industry. Among different types of enzymes used for pulp and paper industry, the most commonly used and important enzymes are cellulase, xylanase, laccase, and lipase (Figure 6.1a). The most related objectives of enzyme among different applications in pulp and paper processes are bleaching boosting (mostly using xylanases), enhancement of delignification from pulp (using laccases), fiber modification (using cellulase), and pitch control (using lipase). Besides, great efforts have been made to increase thermostability of these four enzymes and to turn xylanase and cellulase more alkali tolerant (Figure 6.1b) (Demuner et al. 2011).

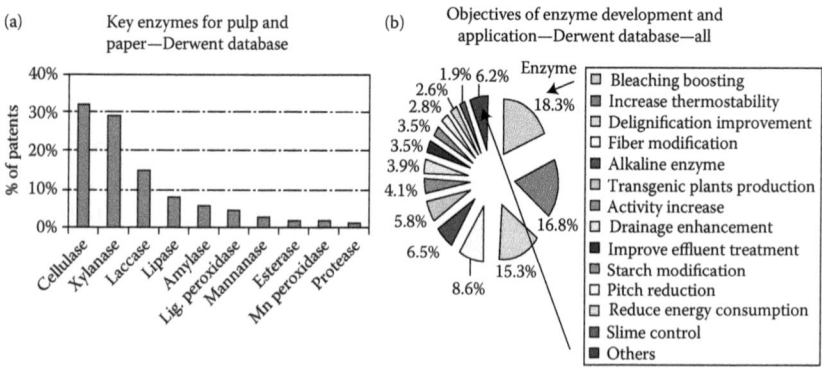

FIGURE 6.1   (a) Key enzymes extracted in the patents. (b) Major objectives of development and applications of enzymes, as extracted in the patents on enzyme for pulp and paper in the Derwent database from 1963 to July 2010. (Adapted from Demuner, B. J., Pereira Junior, N., and Antunes, A. 2011. *Journal of Technology Management and Innovation* 6:148–158.)

BASIC MECHANISMS AND FUNCTIONS OF ENZYMES

Cellulase

Cellulases are components of large systems or complexes that hydrolyze β-1, 4-glucosidic linkages in cellulose which produce water-soluble sugars. Cellulases can be divided into three major classes. These are endocellulase (endoglucanases or endo-1, 4-β-glucanase), exocellulase (exoglucanase or cellobiohydrolase (CBH)), and β-glucosidase (cellobiase). These three hydrolytic enzymes act synergistically. Endoglucanase randomly acts on the cellulose fiber, resulting in a rapid decrease in the degree of polymerization (DP), that is, chain length of CM-cellulose or H_3PO_4-swollen cellulose and yields glucose, cellobiose, cellotriose, and other higher oligomers (Eriksson et al. 1990, Karmakar and Ray 2011).

CBH acts possessively on the existing chain ends and on those created by the endoglucanases, releasing cellobiose molecules. β-Glucosidase cleaves the released cellobiose into two glucose molecules (Eriksson et al. 1990, Karmakar and Ray 2011) (Figure 6.2).

Hemicellulase

Hemicellulases are the hydrolytic enzymes required for the enzymatic breakdown or modification of hemicelluloses. Since hemicelluloses exhibit a complex chemical structure, a complement of about 24 enzymes is required to act cooperatively for the complete hydrolysis to take place (Battan et al. 2007). The two main glycosyl hydrolases that cleave the

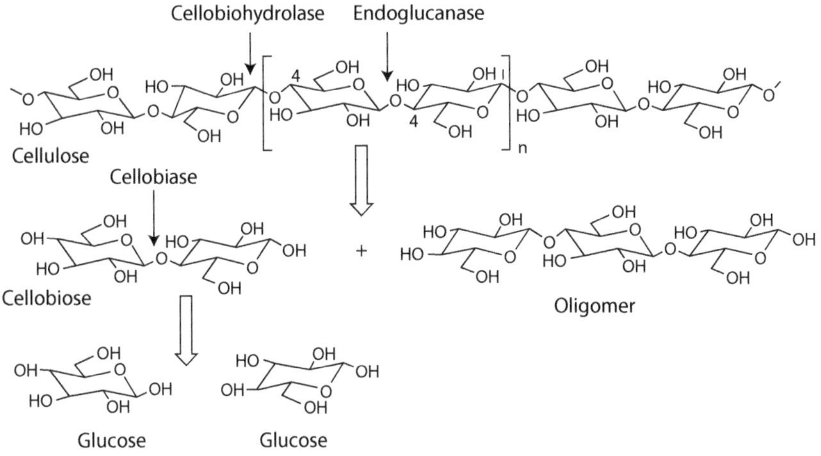

FIGURE 6.2 Mechanism of cellulase complex. (Modified from Karmakar, M., and Ray, R. R. 2011. *Research Journal of Microbiology* 6(1):41–53.)

hemicellulose backbone are endo-1, 4-β-D-xylanase and endo-1, 4-β-D-mannanase (Suurnäkki et al. 1997). Xylan is the major component of hardwood hemicelluloses. Owing to its complex chemical structure, an array of enzymes is needed for complete depolymerization of xylan. These enzymes include L-xylanase, L-xylosidase, and enzymes such as β-L-arabinofuranosidase, β-glucuronidase, acetylxylan esterase, and hydroxycinnamic acid esterases leading to cleavage of side-chain residues present within the xylan (Sunna and Antranikian 1997). The liberated products of randomly cleaved xylan backbone include both substituted and nonsubstituted shorter-chain oligomers, positional isomers of xylose and glucose, xylotrioses and xylobioses, etc. (Woodward et al. 1994, Motta et al. 2013). Xylosidases also play a significant role in complete hydrolysis of xylan as they depolymerize xylo-oligosaccharides to xylose. Moreover, synergistic action of enzymes such as arabinosidase, α-glucuronidase, and acetylxylan esterase with the xylanases and xylosidases is essential for liberating the substituents on xylan backbone to attain complete breakdown of xylan to monosaccharide (Eriksson et al. 1990) (Figure 6.3).

While xylan is the major component of hardwood hemicelluloses, mannan is chiefly found in softwood hemicelluloses. Their hydrolysis is catalyzed by mannases that randomly act to cleave D-1, 4-mannopyranosyl linkages within the mannan backbone and also different polysaccharides that consist of glucomannan, galactomannan, and galactoglucomannan

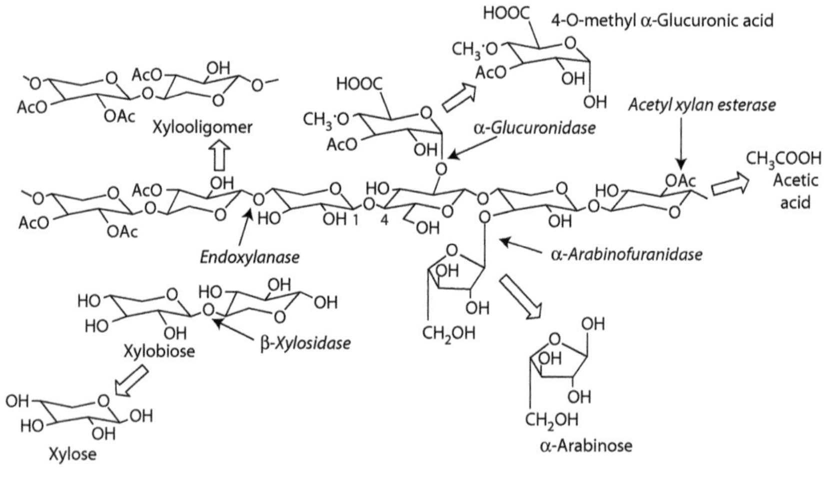

FIGURE 6.3 Mechanism of xylanase complex. (Modified from Sunna, A., and Antranikian, G. 1997. *Critical Reviews in Biotechnology* 17:39–67.)

(Suurnäkki et al. 1997). Mannosidases and galactosidases are also involved significantly in the degradation of mannan since the β-mannosidases catalyze the cleavage of terminal nonreducing β-ᴅ-mannose residues in the mannan backbone whereas α-galactosidases depolymerize terminal nonreducing α-ᴅ-galactosides from galactose oligosaccharides, galactomannan, and galactoglucomannan (Suurnäkki et al. 1997). Besides, the latter are involved in releasing the α-1, 6-bound galactosyl units from polymeric galactomannan.

Lignin-Degrading Enzymes

Laccase

Laccase belongs to oxygen oxidoreductase type (EC 1.10.3.2) that is copper-containing polyphenol oxidase; therefore, it is also named as multi-copper oxidase (Thurston 1994). With the help of four copper atoms in the catalytic core, laccase catalyzes the reaction via mono-electronic oxidation of substrate molecules, which generates reactive radicals. In this redox reaction, one molecule of oxygen is reduced to two molecules of water and simultaneously four substrate molecules are also oxidized to four radicals (Riva 2006).

However, laccase cannot oxidize all substrates directly because of high redox potential and restriction of penetration of substrate molecule into the active site of enzyme due to the large size of the substrate. The problem of high redox potential can be overcome using suitable chemical mediators as intermediate substrate for laccase. The oxidized radical form of these mediators interacts with the target substrate having a high redox potential. Syringaldazine is considered to be a matchless laccase substrate. Laccases can oxidize polyphenols, methoxysubstituted phenols, aromatic diamines, and a range of other compounds. Laccases catalyze different types of reactions such as demethylation, demethoxylation, polymerization, and depolymerization, and also cleavage of bonds (alkyl–phenyl/alkyl–alkyl bonds, phenolic lignin dimmers, and Cα–Cβ bond) (Riva 2006, Madhavi and Lele 2009, Arora and Sharma 2010) (Figure 6.4).

Lignin Peroxidase

LiPs, a heme-containing glycoprotein, catalyze the H_2O_2-dependent oxidative depolymerization of different types of non-phenolic compounds (e.g., diarylpropane), β-O-4 non-phenolic lignin model compounds, and phenolic compounds of lignin (e.g., catechol, syringic acid, vanillyl

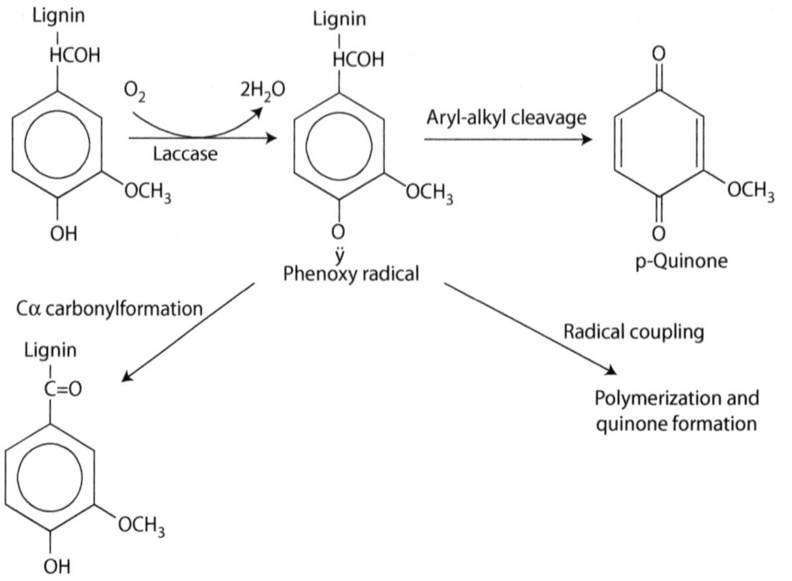

FIGURE 6.4   Oxidation of phenolic subunits of lignin by laccase. (Adapted from Madhavi, V., and Lele, S. S. 2009. *BioResources* 4:1694–1717.)

alcohol, acteosyringone, and guaiacol) having redox potentials up to 1.4 V (Piontek et al. 2001, Dashtban et al. 2010). Oxidization of these lignin compounds by LiP is an electron transfer reaction in multistep resulting into the formation of intermediate radicals such as veratryl alcohol radical cations and phenoxy radicals. Finally, these radicals undergo side-chain cleavage, intramolecular addition and rearrangement, demethylation, radical coupling, and polymerization (Figure 6.5). Owing to its high redox potential, LiP oxidizes non-phenolic aromatic substrates without any participation of mediators. For the peroxidase activity of enzyme, the presence of extracellular hydrogen peroxide is essential (Dashtban et al. 2010).

Mn-Peroxidase

Mn-peroxidases (MnPs) are extracellular enzymes, which are secreted in multiple isoforms, containing one heme molecule as iron protoporphyrin IX (Asgher et al. 2008). MnP oxidizes Mn(II) (reducing substrate) to Mn(III) in the presence of peroxide. This Mn(III) makes a chelated Mn(III) complex with oxalate or with other chelators and then gets released from the surface of enzyme acting as a diffusible, reactive, and

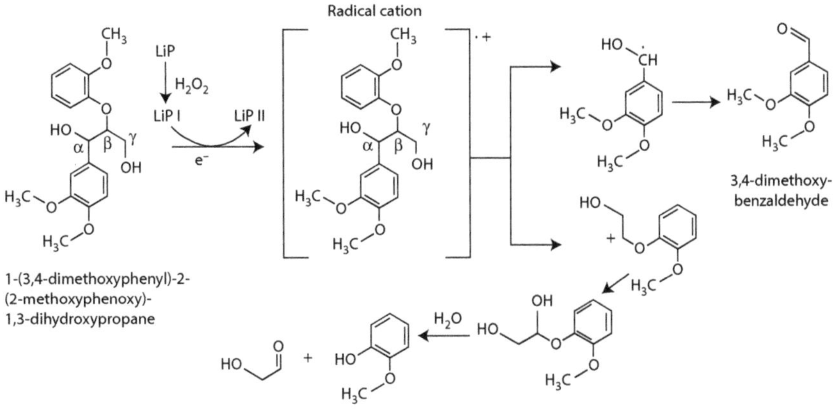

FIGURE 6.5 LiP-catalyzed oxidation of non-phenolics β-O-4-lignin model compound. (Adapted from Sánchez, O., Alméciga-Díaz, C.J., and Sierra, R. 2011. *Alternative Fuel*, Intech Open Access Publisher. 111–154. DOI: 10.5772/851.)

low-molecular-weight redox mediator of phenolic substrates including simple phenols, amines, dyes, phenolic lignin substructures, and dimers. The Mn(III) chelator can act on phenolic lignin structures only due to definite oxidation potential (Dashtban et al. 2010).

Also, non-phenolic substrates are oxidized by Mn(III) via reactive radicals formation such as acetic acid radicals, peroxyl radicals, formate radicals, and superoxide in the presence of a second mediator such as oxalate and malonate ions (Wesenberg et al. 2003, Asgher et al. 2008, Wong 2009) (Figure 6.6). On the other hand, fungi devoid of H_2O_2-generating oxidases use these radicals via MnP as an alternative source of H_2O_2 so that its lignin-degrading efficiency gets increased (Hofrichter et al. 1998, Wesenberg et al. 2003, Wong 2009).

Lipase

Lipases catalyze both the hydrolysis and the synthesis of esters. This reversible reaction includes the formation of ester from glycerol and long-chain fatty acids (Karigar and Rao 2011) (Figure 6.7).

Amylase

Amylases are the hydrolytic enzymes, which degrade starch. There are mainly two types of amylases; α- and β-amylases (debranching enzyme) according to the anomeric type of sugars produced by the enzyme

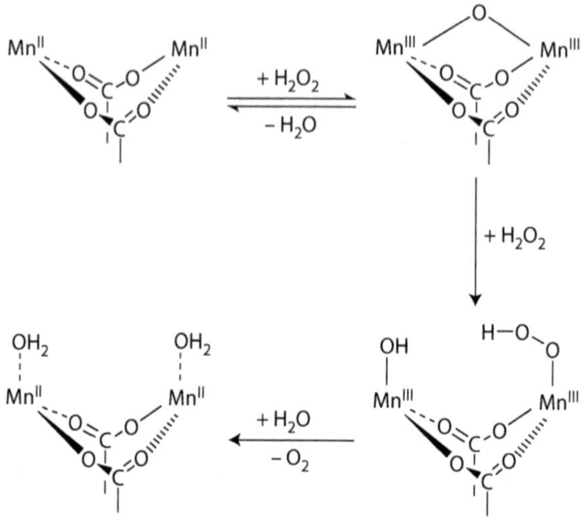

FIGURE 6.6 Mechanism of Mn-peroxidase via chelation to degrade phenolic compounds. (Adapted from Shul'pin, G. B. et al. 2002. *New Journal of Chemistry* 26:1238–1245.)

reaction. α-Amylase catalyzes the hydrolysis of α-1,4 glycosidic bond present in starch that produces glucose, dextrins, and limit dextrins. Amylases further can be classified into endoamylases and exoamylases. Endoamylases hydrolyze the interior α-1,4 glycosidic bond of the starch molecule in a random manner producing linear and branched oligosaccharides of various chain lengths. Exoamylases act on the nonreducing end

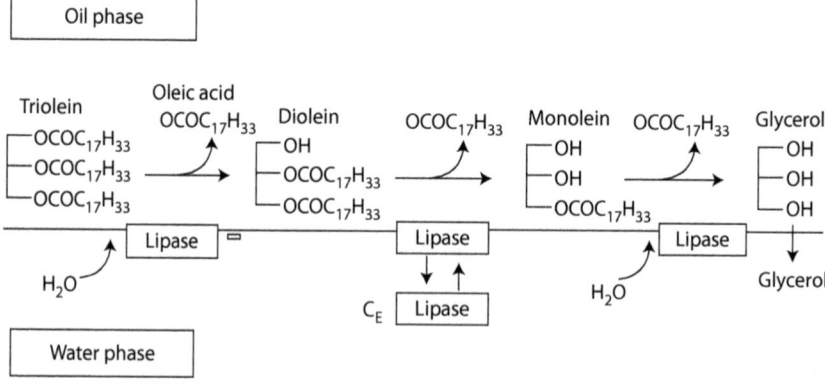

FIGURE 6.7 Action of lipases on the triacylglycerol. (Adapted from Karigar, C. S., and Rao, S. S. 2011. *Enzyme Research*. DOI: 10.4061/2011/805187.)

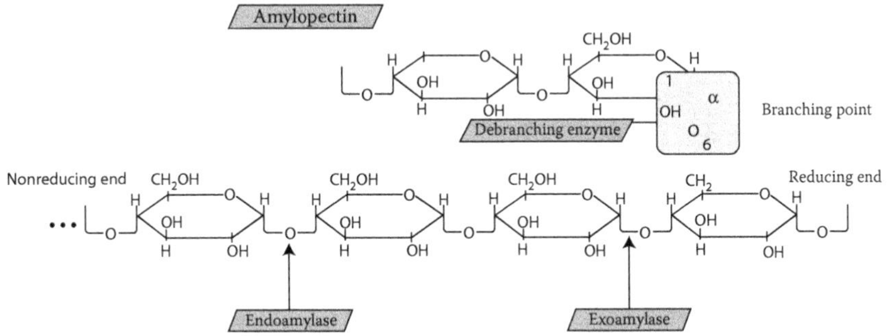

FIGURE 6.8 Action of amylase's components on the starch. (From Springer Science+Business Media: *The Prokaryotes*, Diversity and biotechnological applications of prokaryotic enzymes, 2013, 213–240, Vermelho, A. B., Noronha, E. F., Filho, E. X. et al., eds. E. Rosenber, E. F. DeeLong, S. Lory, E. Stackebrandt, and F. Thompson.)

resulting in short- chain products. β-Amylases are debranching enzymes that cleave β-1,6 glycosidic bond (Figure 6.8) (Vermelho et al. 2013).

Protease

Protease catalyzes protein hydrolysis by cleaving the peptide bonds joining different amino acids together to form a polypeptide chain; it is also called as peptidase or proteinase. These proteases are of two types: 1. Endopeptidase attacks on the internal peptide bonds of proteins. 2. Exopeptidase attacks on the terminal amino acids from the protein chain. Exopeptidases are of two types, that is, aminopeptidase (cleaves from amino ends) and carboxypeptidase (cleaves from carboxylic ends) (Mótyán et al. 2013) (Figure 6.9).

Pectinase

Pectic enzymes are classified into two main groups depending upon their action toward the galacturon part of the pectins including pectin methylesterases or pectin esterases and pectin depolymerases.

Pectinesterases (PE) transform pectin to low ester pectin and pectic acid by cleaving its methoxy groups while pectin depolymerases act on the glycosidic linkages in galacturonan backbone.

Polymethylgalacturonases (PMG) hydrolyze α-1,4-glycosidic bonds. Endo-PMG cleaves α-1,4-glycosidic bonds of pectin randomly, preferentially a highly esterified pectin, Exo-PMG hydrolyzes α-1,4-glycosidic bonds sequentially from the nonreducing end of the pectin chain.

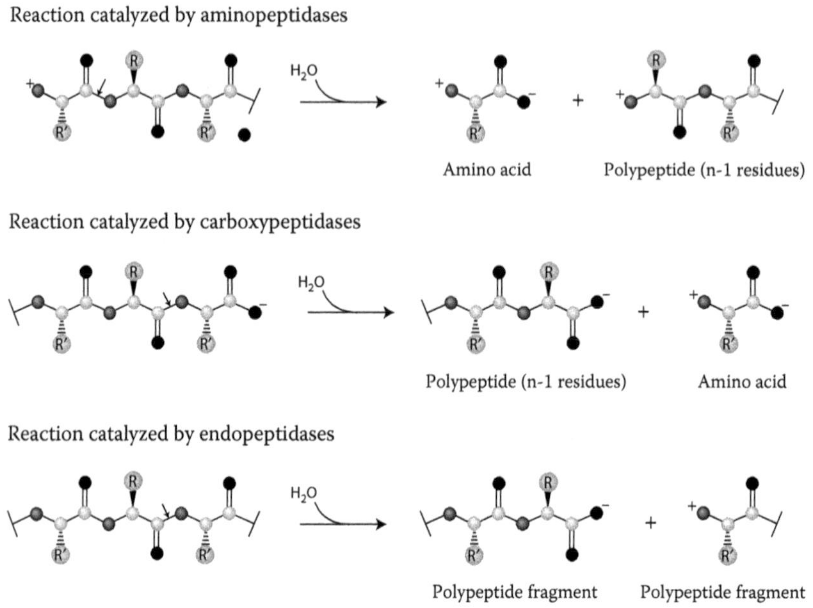

Reaction catalyzed by aminopeptidases

Amino acid   Polypeptide (n-1 residues)

Reaction catalyzed by carboxypeptidases

Polypeptide (n-1 residues)   Amino acid

Reaction catalyzed by endopeptidases

Polypeptide fragment   Polypeptide fragment

FIGURE 6.9 Mechanism of proteins hydrolysis by protease enzyme. (Adapted from Mótyán, J. A., Tóth, F., and Tőzser, J. 2013. *Biomolecules* 3:923–942.)

Polygalacturonases (PG) hydrolyze α-1,4-glycosidic linkages in pectic acid (polygalacturonic acid). Endo-PG hydrolyzes random α-1,4-glycosidic linkages in pectic acid and Exo-PG cleaves α-1,4-glycosidic linkages of pectic acid in a sequential manner.

Polymethylgalacturonate lyases (PMGL) degrade the pectin by *trans*-eliminative cleavage, which results in galacturonide having an unsaturated bond between the fourth and fifth carbon at the nonreducing end of the galacturonic acid formed. They are Endo-PMGL catalyzes random cleavage of α-1,4-glycosidic linkages in pectin and Exo-PMGL catalyzes the sequential breakdown of pectin by *trans*-eliminative cleavage.

Polygalacturonate lyases (PGL) cleave the α-1,4-glycosidic linkage in pectic acid by *trans*-elimination. Endo-PGL catalyzes random cleavage of α-1,4-glycosidic linkages in pectic acid and Exo-PGL catalyzes the sequential cleavage of α-1,4-glycosidic linkages in pectic acid. Protopectinase solubilizes protopectin producing highly polymerized soluble pectin (Kashyap et al. 2001).

PAPERMAKING PROCESS (OVERVIEW)

Papermaking from wood involves different processes as follows.

From Wood

Debarking and Chipping

Debarking is the process to remove the outer layer of the wood known as bark, which contains mainly tannins, resin acids, etc. Softwoods contain a higher amount of resin acid in comparison to hardwoods while agro-residues may not have resin acids. Chipping is the process to break the woody material in small size chips to provide a higher surface area, which facilitates the entry of cooking chemicals during the pulping process.

Pulping

Pulping is the process, which converts the wood chips into fibrous form (i.e., pulp) by the removal or solubilization of the most of lignin and partial hemicellulose present in the wood. Thus, obtained pulp is rich in cellulose having lower hemicellulose and lignin. The pulping can be carried out by different methods such as mechanical and chemical. Mechanical pulping of wood is done using a large amount of electrical energy. During chemical pulping, wood chips are cooked under high pressure and high temperature at extreme pH using chemicals (Eriksson and Cavaco-Paulo 1998, Bajpai 2012).

Bleaching

The undesirable residual lignin provides the brown color, which causes the reduction in final brightness. Therefore, bleaching processes are employed to convert the low- brightness brown pulp into high-brightness white pulp. The final brightness and color is dictated by the product standards. Depending on the raw material, pulping process, and required final target brightness, different bleaching sequences may be applied using various types of bleaching agents such as chlorine, chlorine dioxide, hydrogen peroxide, oxygen, and ozone, alone or in combination. During this process, residual lignin, phenolics, and resin acids are converted into respective chlorinated compounds that are finally transformed into highly toxic xenobiotics. These compounds are considered as mutagens, persistent, bioaccumulative, which finally disturb the aquatic ecosystem (Bajpai 2012).

Refining

A typical wood fiber is mainly composed of an outer primary layer (P) and secondary layers (S1 and S2). The inner layer S2 is a cellulose-rich layer (Figure 6.10). During the refining process, the peeling off of undesired

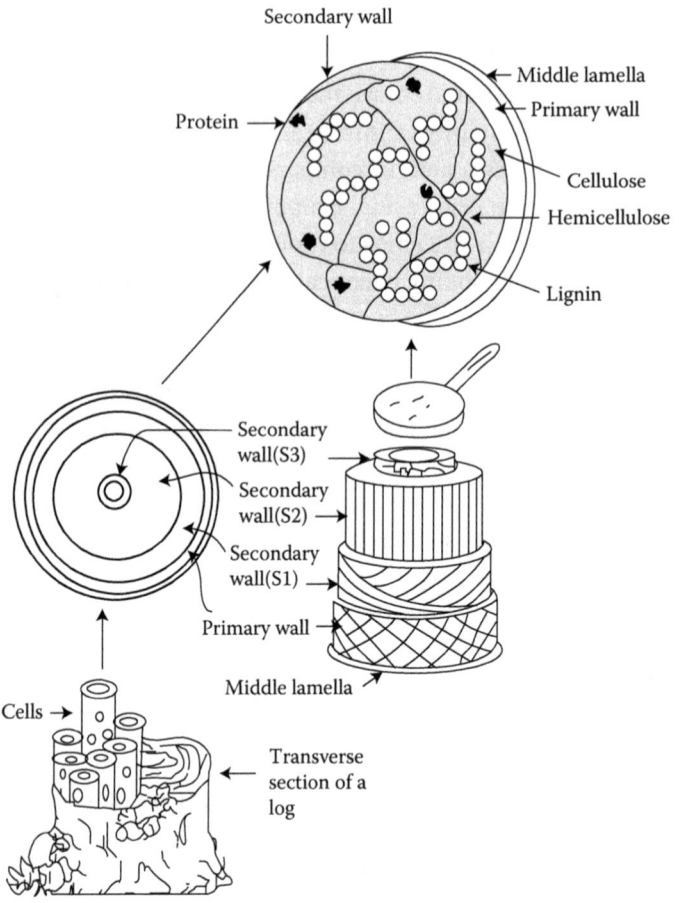

FIGURE 6.10   Fiber wall layers in a typical wood fiber. (Adapted from Sánchez, C. 2009. *Biotechnology Advances* 27:185–194.)

P and S1 layer occurs that modifies the fiber. Thereby, the inner S2 layer is exposed that provides more hydroxyl groups for hydration of fiber resulting in better swelling and higher surface area of fibers than unrefined fibers. Owing to the increase in surface area, more contact among fibers occurs by interfiber bonding such as hydrogen bonding, vander-Waals interaction, or molecular entanglement, which makes the sheet of a complex fiber network, that is, paper. During refining, fiber cutting/ shortening, fibrillation, fines development, and their partial solubilization also occur. All these effects influence different properties of pulp as well as paper (Sánchez 2009, Singh and Bhardwaj 2010).

Papermaking

During papermaking, different types of chemicals such as filler (clay, calcium carbonate, titanium di oxide, etc.), sizing agents (rosin, alkylketene dimer [AKD], alkyl succinic anhydride [ASA], starch, etc.), coloring agent, wet-strength additives, etc. are added to washed pulp to produce paper with desired specifications. The coating of base paper is also done to achieve the desired functional properties of coated paper including better printability, smoothness, etc.

From Recycled Fibers

Deinking is a process for detaching and removing printing inks from the fibers of recovered printed materials to be recycled to improve the optical characteristics of pulp and paper. Deinking involves the dislodgement of ink particles from the fiber surface (ink detachment) and separating dispersed ink from fiber suspension by flotation and/or washing (ink removal) (Pathak et al. 2010, 2011, 2014). Unsatisfactory separation results in poor-quality paper having low brightness and dirt specks. Larger ink particles >50 μm are visible to the naked eye as black or colored spots in the paper. The efficiency of this method also depends on the technique and printing conditions, the kind of ink, and the kind of printing substrate (Prasad et al. 1993, Bajpai and Bajpai 1998). The conventional deinking process requires large amounts of chemicals such as sodium hydroxide, sodium silicate, hydrogen peroxide, and surfactants (Pathak et al. 2010, 2011, 2014).

There may be variation in the methods/chemicals applied in each papermaking process from wood and recycled fiber. Figure 6.11 shows different processes in papermaking from wood and status of enzymes application at mill, pilot, and laboratory scale. Table 6.1 summarizes the functions and status of different enzymes used in pulp and paper industry.

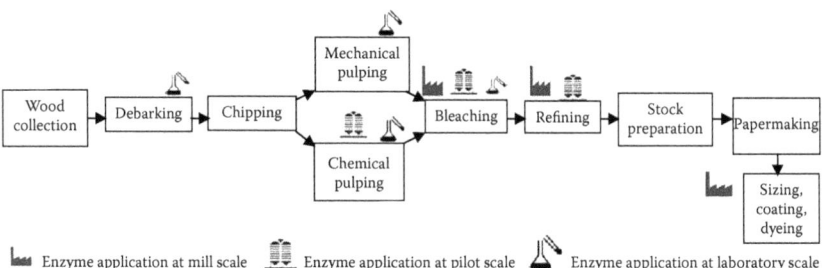

FIGURE 6.11 Schematic diagram showing different processes in papermaking and status of enzymes application at mill, pilot, and laboratory scale.

TABLE 6.1 Functions and Status of Different Enzymes Used in Pulp and Paper Industry

Enzymes	Substrate	Functions	Application in Paper Industry	Status (Mill/Pilot/Lab Scale)
Cellulase	Cellulose	Partial hydrolysis of cellulose	Refining/fiber modification, vessel picking	Mill
		The release of ink from the fiber surface	Deinking	Mill
		Hydrolysis of the colloidal material in paper mill drainage	Drainage improvement	Mill
Xylanase	Xylan	Degradation of redeposited xylan and lignin–carbohydrate complexes	Bleach boosting	Mill
			Deinking of newsprint and magazines	Mill
			Production of dissolving grade pulp	Pilot
			Biopulping	Pilot
			Removal of shives	Lab
			Drainage	Mill
			Refining	Mill
			Debarking	Lab
Mannanase	Gluco-mannan	Removal of glucomannan	Bleach boosting	Mill
			Biopulping	Pilot
Laccase	Lignin	Degradation of lignin in the presence of mediators (transition metal complexes)	Bleaching, effluent treatment	Pilot
Mn-peroxidase	Lignin	Degradation of lignin in the presence of additives	Bleaching, effluent treatment	Lab
Lipase	Fat/oil	The hydrolysis of triglyceride	Pitch control, contaminant control, and deinking of oil-based ink	Mill
Amylase	Starch	Hydrolysis of α-1,4 and/or α-1,6 bonds of starch	Surface sizing, starch coating, deinking, drainage improvement, and slime control	Mill
Esterase	Macrostickies	Breaks ester bonds	Stickies control	Mill
Protease	Protein	Hydrolysis of cell wall proteins	Biofilm removal	Mill
Pectinase	Pectin	Hydrolysis of cambial layer	Refining	Mill
			Energy saving in debarking, decreased cationic demand	Lab

EFFECTIVENESS OF ENZYMES IN DIFFERENT PROCESSES OF PULP AND PAPER INDUSTRY

Pulping

Employing white-rot fungi or related enzymes for the treatment of wood chips prior to mechanical or chemical pulping is called biopulping. In bio-mechanical pulping, the purpose of using fungi is to avoid chemicals in the pretreatment of wood for mechanical pulping, reduce energy consumption, and improve paper strength because fibers suffer less damage during refining steps. Pulp quality depends on refiner's design, wood species, and the desired refining level. For biochemical pulping, biopulping is aimed to reduce the amount of pulping chemical, to enhance the capacity of cooking, and to enable extended cooking that is resulted into lower consumption of bleaching chemicals. With improved delignification efficiency, energy and chemicals can be saved during pulping and bleaching resulting into reduction of pollution. The mechanical pulping processes (such as refining and grinding) of the wood lead to production of pulps with high amount of fines, bulk, and stiffness. On the other hand, cellulase-assisted biomechanical pulping could save energy by 20%–40% during refining with improvements in strength properties of hand sheet (Akhtar 1994, Bhat 2000, Pere et al. 2001, Singh et al. 2007). The pretreatment of wood chips with the white-rot fungi can degrade the lignin present in wood chips, which saves the energy during the mechanical pulping and chemical consumption during chemical pulping. Most of the white-rot fungi are capable of degrading all the wood components simultaneously resulting in cellulose degradation that is undesirable. On the other hand, endoglucanases, hemicellulases have limited action on wood cell walls due to their limited diffusion across the lignocellulosic matrices (Ferraz et al. 2008). To degrade the lignin without affecting the cellulose, mutant strains were also used (Eriksson et al. 1976). In general, fungal hyphae penetrate the wood chips through the lumen of the vessels and fiber cells, natural wood cell pits, and fungal boreholes. Extensive removal of the extractive material during the wood degradation by some white-rot fungi is also responsible for the reduction of alkali doses during kraft pulping because of dissolution of major obstacle resin canals; resulting into better penetration of the liquor and reduction of alkali consumption by non-phenolic compounds. According to Schwanninger et al. (2004), during biotreatment of white-rot fungi, minor changes in the hydrogen bonding between the fiber surfaces would also be responsible for the softening of pulp, which helps the disruption of lignocellulosic material by disk refiners. According to Hunt et al. (2004), water saturation point of the fibers is increased by the esterification of oxalate ions secreted by the fungus on the polysaccharide chains.

For biopulping, a commercially viable fungus should be relatively faster in growth rate, have the ability to grow on hardwood as well as softwood, ability to produce hemicellulase and lignolytic enzymes with low cellulase activity, enable to elicit allergies, minimal pigmentation to avoid the decrease in pulp brightness, and fast-rate capability to sporulate in order to provide spores for the inoculation of the wood chips.

During a typical biopulping process, decontaminated (using steam) wood chips are sprayed with a dilute inoculum of specific lignin-degrading fungi. These inoculated chips are incubated and aerated in a chip pile for 2 weeks. Under warm and moistened conditions, fungi colonize on chip surfaces and penetrate into chip's interiors with a network of hyphae. These treated chips are easy to break down during subsequent refining that produces flexible and intact fibers (Figure 6.12) (Scott et al. 1997).

Leatham et al. (1990a,b) have studied on the energy savings for biomechanical pulping after 4-weeks incubation using different fungal strains namely *Phanerochaete chrysosporium* (14%), *Hyphodontia setulosa* (26%), *Phlebia brevispora* (28%), *P. subseritalis* (32%), *P. tramellosa* (36%), and *Ceriporiopsis subvermispora* (42%). Clariant, UK has launched a product named as Cartapip containing spores of *Ophiostoma piliferum*. Treatment of hardwood chips with Cartapip®97 for 21 days improved the kraft-pulping efficiency by decreasing the Kappa number by 29% and active alkali consumption by 20% (Wall et al. 1996). *C. subvermispora* has the ability to grow on both softwoods and hardwoods. Mill trial of *Eucalyptus grandis* wood chips biotreated by *C. subvermispora* on a 50-tonne pilot plant

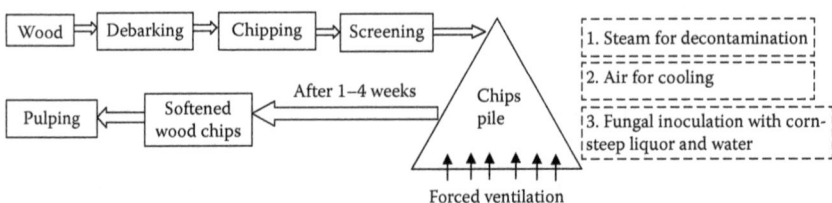

FIGURE 6.12 Overview of different processes used in biopulping. (Modified from Scott, G. M., Akhtar M., and Lentz, M. J.: *Environmentally Friendly Technologies for Pulp and Paper Industry*. 341–383. 1997. Copyright Wiley-VCH Verlag GmbH & KGaA.)

demonstrated that equivalent energy savings could be obtained in lab- and mill-scale biopulping (Ferraz et al. 2008). The inoculation of wood chips with the white-rot fungus *P. brevispora* BAFC 633 selectively removed the lignin during biotreatment, causing a reduction of 6.3% in Kappa number associated with decreased levels of lignin in the pulp and 7.1% increase in paper brightness (Fonseca et al. 2014). Pretreatment of poplar chips with *Trametes versicolor* for 1, 2, and 3 weeks resulted into Kappa number of around 20 after kraft pulping (Garmaroody et al. 2011). Pretreatment of empty fruit bunch by *Marasmius* sp. for 30 days reduced lignin by 35.94%. The Kappa number was found to be 38.63 and 31.10 for untreated and treated pulps, respectively (Risdianto and Sugesty 2015). *E. grandis* wood chips pretreated with *P. chrysosporium* RP-78 in a wood yard of chemi-thermomechanical pulp (CTMP) mill resulted in 18.5% net energy savings including lower shives and improved strength properties as compared to control (Masarin et al. 2009).

The main limitations of the biopulping are (1) Regular requirement of inoculum for the commercial-scale applications, which involves additional work and expense. (2) Large-scale production of basidiomycetes is usually difficult for the treatment of a large quantity of wood. (3) Fungal treatment is a lengthy process due to its slow growth, which requires a minimum of 2-weeks of incubation to obtain the desired effects (Scott et al. 1997).

Bleaching

Lignin is responsible for the brown color of the pulp. Bleaching is a process of lignin removal from chemical pulps and/or modification of chromophoric groups present in pulp (unbleached) to attain high-brightness pulps (bleached). The undesirable residual lignin provides the brown color, which causes the reduction in final brightness. To improve the brightness, pulp and paper industry uses different types of bleaching chemicals mainly chlorine. The effluent generated after bleaching contains toxic chlorinated organic compounds, which are considered as mutagens, persistent, bioaccumulative, which is detrimental for the aquatic ecosystem (Onysko 1993). Owing to environmental concerns related with the chlorine, papermakers have diverted their mind to find out alternative ways to reduce the amount of chlorine for the bleaching purpose. These alternatives are incorporation of less toxic or nontoxic bleaching chemicals such as chlorine dioxide/hydrogen peroxide/oxygen/ozone in place of chlorine and extended pulping. But these alternative methods require a capital investment and modifications in the process. Therefore, to avoid these expenditures, enzymes have been proven as cost-effective and an environmental-friendly option to get better results to reduce or completely avoid chlorine usage. In the beginning, lignin-degrading enzymes were the most important enzyme for bleaching purpose. However, for the first time in 1986, Viikari et al. verified the efficacy of xylanases as a pre-bleaching agent for pulps (Viikari et al. 1986).

During the biobleaching process, hemicellulase enzyme does not directly attack on residual lignin present in the pulp. It modifies pulp hemicelluloses, which enhances the removal of lignin by bleaching chemicals. According to one hypothesis, xylanases partially hydrolyze the re-precipitated xylan and remove the xylan from the lignin–carbohydrate complexes (Beg et al. 2001).

In another hypothesis, enzyme facilitates the diffusion of entrapped lignin from the fiber wall. Even a limited exclusion of pulp xylan is able to improve the leaching of residual lignin from pulp. Both of these hypotheses ultimately lead to the enhanced diffusion of entrapped lignin from the fiber wall. Even limited removal of pulp xylan results in increased leachability of residual lignin from kraft pulps and thus pulp bleachability increases during subsequent bleaching stages. Xylan containing hexenuronic acid groups are more prone toward enzymatic removal of lignin to get lower Kappa number.

Xylanases have the potential to eliminate the first chlorination stage in a $C/DE_{OP}DE_{P}D$ bleaching sequence and replace it with an enzyme stage (X) to become $XE_{OP}DE_{P}D$. The filtrates from the E_{OP} stage can be recirculated to the recovery system without risk of chloride-initiated corrosion. It may also contribute to closed water circulation system of the mill to minimize the final effluent discharge. Reduction in chlorine-containing compounds in the bleach plant effluent decreases the effluent color.

Pulp viscosity is affected due to partial removal of pulp carbohydrates (mainly hemicellulose) during enzymatic pre-bleaching process. Xylanase treatment improves the pulp viscosity by hydrolyzing pulp xylan with relatively low DP but overall viscosity of the pulp is lowered. Treatment of pulps by cellulase-free mannanase has not been observed to influence the pulp viscosity significantly. Any side activity of cellulase present in the hemicellulase preparations during enzyme treatment can cause a detrimental effect on pulp since endoglucanase and CBH activities will work in a synergistic manner resulting in rapid depolymerization of cellulose, which is undesirable for the papermakers. Among all the three components of cellulase, endoglucanases have been observed to be the most damaging component due to their action on the amorphous regions of pulp cellulose while the CBHs alone (even in high doses) did not affect pulp viscosity significantly.

Endoxylanases are the main enzymes in most of the commercial bleaching enzyme preparations needed to enhance the delignification of kraft pulp but other enzymes such as mannanase, lipase, and β-galactosidase have also the ability to improve the effect of enzymatic treatment of kraft pulp (Elegir et al. 1995, Gübitz et al. 1997, Beg et al. 2001). These enzymes are mostly active at acidic or neutral pH while they perform better under alkaline conditions. Usually, bacterial and fungal xylanases have their optimum pH between 6 to 9 and 4 to 6, respectively. The optimum temperature ranges may vary from 35°C to 60°C depending on enzymes (Beg et al. 2001, Bajpai 2012).

Implementation of bio-bleaching has been jumped directly from the laboratory scale to the large industrial scale (1000 tons per day [tpd]) without intermediate pilot stages. During mill trial, higher brightness can be achieved than those achievable in the laboratory due to more efficient mixing systems at relatively higher pulp consistencies. Generally, there is no requirement of expensive capital investments for mill-scale runs except the additional requirement of pH adjustment facilities. The enzyme solution, usually mixed with water, is allowed to react in the high-density storage tower tank for up to 2–3 h prior to the subsequent chemical-bleaching steps (Figure 6.13) (Bajpai 2004).

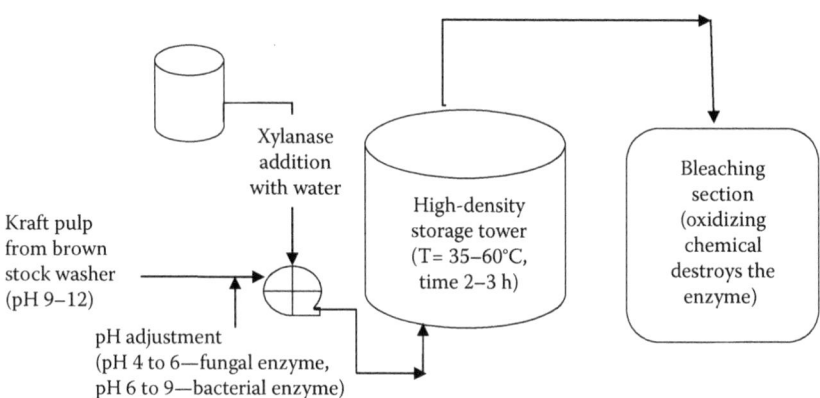

FIGURE 6.13 Process flow sheet for the xylanase addition in mill trials. (Modified from Springer Science+Business Media: *Biotechnology for Pulp and Paper Processing*, 2012, Bajpai, P.)

In an Indian paper mill trial, xylanase pretreatment reduced chlorine from 5 to 4 kg/tp, hypochlorite reduction from 45 to 38–40 kg/tp maintaining the final target brightness of 82%–83%. The absorbable organic halides (AOX) were reduced from 5.25 to 4.01 kg/tp (−23%) without affecting the chemical oxygen demand (COD) and biological oxygen demand (BOD) (Table 6.2) (Thakur et al. 2012).

Before the bleaching process, xylanase pretreatment of pulps is able to considerably reduce bleaching chemical consumption, which reduces cost with higher brightness particularly for mills using large amounts of peroxide or chlorine dioxide. A lessening in the use of chlorine-containing bleaching chemicals, primarily by decreasing chlorine gas usage, undoubtedly reduces the production and discharge of chlorinated organic compounds in the wastewater and the pulps, which also results into reduced AOX discharge. The consumption of chlorine dioxide can also be reduced that helps to eliminate the requirement of increased chlorine dioxide generation capacity that is beneficial for de-bottlenecking mills limited by chlorine dioxide generator capacity. Installation of oxygen delignification facilities may be avoided, which is a costly affair. Application of enzymes in ECF and TCF bleaching sequences increases the final pulp brightness value. Xylanase treatment increases the brightness ceiling, mainly for mills considering ECF and TCF bleaching sequences. Enzyme treatment stage before or simultaneously with the chelating stage (before hydrogen peroxide bleaching stage) is beneficial to get the maximal benefit of enzymatic treatment in pulp bleaching because the neutral pH is suitable for enzyme treatment as well as for the chelation of magnesium, iron, and manganese ions (Bajpai 1999, 2012).

Sometimes, enzyme treatment creates problems in application as well as in bleaching plant control due to unavailability of online observations or rapid testing. A decreased tear strength and pitch formation was also reported in some mills (Bajpai 2012). Table 6.3 shows different industrial suppliers of commercial xylanase enzyme used for bleaching purpose (Bajpai 1999). Table 6.4 summarizes the bleaching results of mill trials with commercial xylanase enzymes (Bajpai and Bajpai 1996, Bajpai 1999).

Laccase-Mediator System
Laccase, an oxidative enzyme, can also be used as a biobleaching agent as it degrades and decolorizes the residual lignin in pulp using atmospheric oxygen as its electron acceptor. Laccase-mediator system has also shown to possess the possibility to substitute chlorine-based bleaching

TABLE 6.2 Consolidated Results of Xylanase Prebleaching and Nature of Bleach Effluents (Combined) during Enzyme Treatment of Mill Trial

Particulars	Without Enzyme	With Enzyme	Particulars	Without Enzyme	With Enzyme
BSW pulp Kappa no.	23–25	23–25	*E_P-second stage*		
Screened pulp Kappa no.	22–24	21–22	Applied NaOH, %	0.5	0.5
Kappa no. reduction	1.0–1.5	2.0–3.0	Applied H_2O_2, %	0.5	0.5
Pulp brightness, %	27–30	29–32	Micro Kappa no.	3.6–3.8	3.0–3.5
CED viscosity, cm³/g	727–836	779–892	ISO brightness, %	53–54	54–59
Chlorination stage			ISO brightness improvement	–	3–4
Appl. Cl_2 dose, %	5.0	4.8	*Hypo stage*		
Appl. Cl_2 dose, kg/t pulp	50	48	Hypo flow, m³/h	75	62–65
Cl_2 savings, kg/t pulp	–	2.0	Applied hypo, kg/t pulp	45	38–40
Cl_2 savings, %	–	4%	Applied hypo, %	4.5	3.8–4.0
Residual Cl_2 in filtrate, mg/L	50–135	100–450	Hypo savings, %	–	11–15
Residual Cl_2 in pulp, mg/L	20–36	30–70	Residual Cl_2 in filtrate, mg/L	460–560	400–1000
Micro Kappa no.	7.0–8.0	6.0–7.0	Residual Cl_2 in pulp, mg/L	25–85	50–300
ISO brightness, %	36–38	40–43	ISO brightness, %	82–83	82–83
ISO brightness improvement	–	3–4	PC no.	3.0–4.0	2.0–2.5
E_P-first stage			CED viscosity, cm³/g	239–377	299–401
Applied NaOH, %	1.5	1.5	*Effluent characteristics*		
Applied H_2O_2, %	0.5	0.5	AOX, kg/t pulp	5.25	4.01
Micro Kappa no.	5.0–5.9	4.0–5.0	AOX reduction, %	–	23.6
ISO brightness, %	42–47	43–50	COD, kg/t pulp	700–820	672–800
ISO brightness improvement	–	2–3	BOD, kg/t pulp	230–250	230–248

Source: Data from Thakur, V. V., Jain, R. K., and Mathur, R. M. 2012. *BioResources,* 7(2):2220–2235.

TABLE 6.3   Industrial Suppliers of Commercial Xylanase Enzyme

Name of the Suppliers	Trade Name of Xylanase Enzyme
Clariant, UK	Cartazyme HS 10, Cartazyme HT, Cartazyme SR 10, Cartazyme PS 10, Cartazyme 9407 E, Cartazyme NS 10, and Cartazyme MP
Genencor, Finland/Ciba Geigy,[a] Switzerland	Irgazyme 40-4X/Albazyme 40-4X, Irgazyme-10A/Albazyme 10A
Voest Alpine, Austria	VAI Xylanase
Novo Nordisk, Denmark	Pulpzyme HA, Pulpzyme HB, and Pulpzyme HC
Biocon India, Bangalore	Bleachzyme F
Rohm Enzyme OY, Finland	Ecopulp X-100, Ecopulp X-200, Ecopulp X-200/4, Ecopulp TX-100, Ecopulp TX-200, and Ecopulp XM
Solvay Interox, USA	Optipulp L-8000
Thomas Swan Co., UK	Ecozyme
Iogen Corp., Canada	GS-35, HS-70

Source: Bajpai, P. 1999. *Biotechnology Progress* 15(2):147–157.
[a] Ciba Geigy has sold its pulp and paper enzyme techonology to Nalco Chemical Co.

reagents and therefore has the potential to reduce the pollution load due to chloroorganic compounds (Call 1994, Wesenberg et al. 2003). The laccase enzyme system of the *T. versicolor* alongwith the *N*-hydroxy compounds has been effectively used for the delignification process for the kraft pulp at pilot scale. Laccase-mediator system has made the possibility to develop alternative TCF bleaching sequences, but expensive and potentially toxic mediators are the major disadvantages (Yoon and Jung 2014). The most common mediators for the lignin are the 1-hydroxybenzotriazole (1-HBT) and 2, 2′-azinobis-3-ethylbenzthiazoline-6-sulfonate (ABTS), but these are very expensive. There is still a need of cheaper, nontoxic, effective, and easily available mediators for pulp and paper industry (Arias et al. 2003).

Sequential Xylanase and Laccase Treatment

For biobleaching, a sequential xylanase (X) and laccase- mediator treatment (LME) was found to be more efficient pretreatment that resulted in lower Kappa numbers pulp. In comparison to laccase enzyme alone (LE) and LME sequences, about 7.25% and 24.8% reduction in Kappa number was achieved for XLE and XLME sequences, respectively, which signifies the advantages of xylanase treatment in the reduction of bleaching chemicals consumption and decrease in amounts of organochlorine compounds in the bleach plant effluents (Dedhia et al. 2014). In a plant trial,

TABLE 6.4   Results of Mill Trials with Commercial Xylanase Enzymes

Name of the Mills	Bleaching Sequences	Increase in Brightness (% ISO)	Reduction in Chemicals	Reduction in AOX (%)	Pulp Properties
Bukoza Pulp Mill, Czechoslovakia	$(CD)EDE_HD$	–	30% Cl_2, 30% hypochlorite	nd	Unchanged
Tasman Pulp and Paper Co., New Zealand	$(DC)E_oDED$	–	20% ClO_2	20	Unchanged
Canfor's International Mill, BC, Canada	$DE_{op}DED$	–	15.6% ClO_2	–	Unchanged
Metsa Botnia, Finland	OPPP	2–4	–	nd	Unchanged
Enso Gutzeit, Finland	$(D_{50}C_{50})E_1D_1E_2D_1$	–	22% active chlorine	29	Unchanged
Scott Paper Mill, Spain	$E_{op}DEPD$	4–6	–		No attack on cellulose
Munksjo, Sweden	OQE_1PE_2P	2–4	–	–	Unchanged
Ballarpur Paper Industries, India	CE_pH	2–3	–	nd	Unchanged
Crest Brook Forest Industries, Canada	$DE_{op}DED$	–	7%–17% reduction in Kappa factor, increased productivity	nd	Unchanged
Metsa-Sellu, Finland	–	–	12% total active chlorine	nd	Unchanged
Morrum Pulp Mill, Sweden	–	–	22.9% ClO_2	22.9	Unchanged
Donohue St. Felicie Mill, Quebec	–	–	24% reduction in Kappa factor	nd	Good strength

Source: From Springer Science+Business Media: *Advances in Biochemical Engineering/Biotechnology*, vol. 56, 1996, p. 1, Bajpai, P., and Bajpai, P. K; Bajpai, P. 1999. *Biotechnology Progress* 15(2):147–157.

a combination of xylanase with the laccase pretreatment resulted in 35% lesser chlorine dioxide, and 50% lower post color (PC) number without affecting the brightness of the pulp than the control. Further, the biodegradability of organic matter present in the effluents is also increased due to less AOX, which facilitates the microbial growth. Finally, 13.98% higher BOD and 26.39% higher COD values were observed in effluents generated from enzyme-treated pulps. Release of degradation products from lignin by laccase-mediator system and hydrolysis of xylan by xylanase are responsible for the higher COD (Sharma et al. 2014). About 11% higher ISO brightness and 3.6 units lower Kappa number were observed than control when xylanase (*Aureobasidium pullulans*) and laccase (*Trichophyton* sp. LKY-7) were sequentially used followed by P-stage bleaching of hardwood kraft pulp (HWKP) (Yoon and Jung 2014).

Xylanase pretreatment leads to loosening of the lignin carbohydrate complex resulting in more lignin available for the laccase-mediator system to contact. Xylanase pretreatment opens many more sites for oxidized mediator diffusion leading to deeper penetration in the fiber. Therefore, more Kappa number is reduced after xylanase pretreated pulp in comparison to untreated pulp or after laccase or laccase-mediator catalyzed bleaching (Dedhia et al. 2014, Sharma et al. 2014).

Manganese Peroxidase and Lignin Peroxidase

The white-rot fungi possess the capability to bleach kraft pulp including *P. chrysosporium, T. versicolor, P. sordida* YK-624, and *IZU-154* (Machii et al. 2004). During biobleaching of kraft pulp with white-rot fungi, manganese peroxidase (MnP) was found to be a key enzyme due to linear correlation of its activity detected in the culture with brightness increase of the kraft pulp (Paice et al. 1993, Hirai et al. 1994, Katagiri et al. 1995). Mn(II) ion is necessary for production and function of MnP from *P. sordida* YK-624. Unbleached HWKP contains Mn at a concentration of about 50 mg/kg pulp (Paice et al. 1993, Hirai et al. 1994), and *P. sordida* YK-624 utilizes Mn during the biobleaching of HWKP because this fungus can produce MnP and brighten HWKP in a culture containing only HWKP and water (Hirai et al. 1994, 1995).

Kondo et al. (1994) have shown the bleaching efficiency of purified MnP in the presence of Mn^{2+}, Tween 80, malonate, and H_2O_2. Furthermore, MnP treatment is more economic and environmentally safe as compared to laccase-mediator system as it does not require any mediator. However, MnP is not potentially used at large scale as a biobleaching agent. MnP produced by the white-rot fungus *Bjerkandera* sp. strain BOS55 was found to be effective toward lignin oxidation and bleaching of eucalyptus oxygen-delignified kraft pulp in an enzymatic stage, which was followed by chelator, hydrogen peroxide with oxygen pressure-stage sequences (Feijoo et al. 2008). The MnP of white-rot fungus *Bjerkandera* sp. strain BOS55 has been reported to bleach kraft pulp and for delignification. The *Bjerkandera* MnP also significantly reduced the Kappa number of eucalyptus oxygen-delignified kraft pulp by 6% without any Mn addition (Moreira et al. 2001).

Fiber Modifications

Recently, the interest in using enzyme for the fiber modifications has been increased for the improvement in refining/beatability and drainage.

Refining

Pulp and paper industry ranks fourth among most energy-intensive industries (Abdelaziz et al. 2011). The cost of energy consumed accounts for almost half of the production cost of paper (Ozalp and Hyman 2006, Hong et al. 2011). Energy is required in paper mill in the form of heat and electricity in various papermaking operations (Chen et al. 2012). The refining process requires a large part of electrical energy, which constitutes around 20%–30% of the total energy consumed (Lecourt et al. 2010). This corresponds to about 18%–25% of the total manufacturing cost (Bhardwaj et al. 1996, Bajpai et al. 2006).

Enzymes can be added at a pre-refining stage by optimizing different process variables such as temperature, pH, pulp consistency, enzyme dosage, and reaction time for a particular type of fiber/pulp used (Figure 6.14). However, reports are also available for energy reduction by biopulping approach but due to longer reaction time and contamination by growth of other unwanted microorganisms associated with it, a pretreatment stage using enzyme is preferred before refining of fiber (Singh and Bharadwaj 2010, Torres et al. 2012).

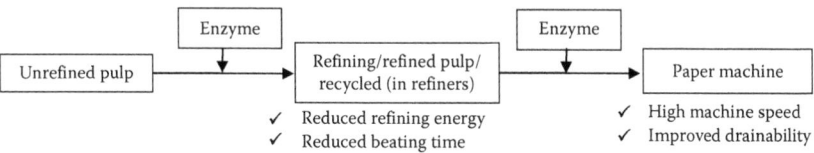

FIGURE 6.14   Basic model of enzyme-assisted refining.

Several enzymes such as cellulase, hemicellulase/xylanase, proteinase, laccase, MnP, amylase, and pectinase are reported in literature showing the effect on refining (Mansfield et al. 1999, Suurnäkki et al. 2000, Wong et al. 2000, Sigoillot et al. 2001, Spiridon et al. 2003, Meza et al. 2006, Torres et al. 2012). The most dynamically investigated enzyme is cellulase and it stands to be a good choice for refining application.

Enzymes are used in paper industry from a long time but gained popularity in the last two decades. In 1942, a patent claimed that hemicellulase enzymes from *Bacillus* and *Aspergillus* species aid refining and hydration of pulp fibers (Diehm 1942). Another patent claimed that the use of cellulase enzyme from *Aspergillus niger* fibrillates pulp (Bolaski 1962). Yerkes (1968) patented for cellulase enzymes from a white-rot fungus to reduce beating or refining time. The interest for cellulase and associated mechanism of enzyme action on cellulose was developed (Lee et al. 1983). After that, Noe et al. (1986) reported external xylanase facilitates fibrillation of bleached chemical pulps and reduced energy demand in the papermaking process. The authors concluded that because of enzyme action, removal of xylan may occur and thus, it resulted in decreased fiber intrinsic strength, while an increased bonding ability between fibers is due to increased fiber flexibility.

As cellulase acts directly on the fiber layer of interest (i.e., S2 layer), most of the cited literature focused cellulase for refining application using fungal, bacterial, and commercial enzymes for different pulp grades (Zhang and Hu 2011, Liu and Hu 2012, Zhang et al. 2013). Revisions were made several times for the role of biotechnology in pulp and paper industry including refining/beating process (Singh and Bhardwaj 2010, Torres et al. 2012, Bajpai 2013). Earlier, it was reported that enzymes can be beneficial in reducing energy requirement at mill scale for various pulp grades (Friermuth et al. 1994, Caram et al. 1996). Pere et al. (1996) reported that the cellulase component CBH I, from *Trichoderma reesei* can be helpful in reducing energy consumption, apparently as a result of selective action on crystalline cellulose. Contrary to that, endoglucanase was reported to be the main component for assistance in refining of pulp (Clark et al. 1997). Several studies were performed using cellulase/xylanase or their mixtures for refining energy reduction in the last decade showing variable results depending on the pulp and enzyme composition used (Bajpai et al. 2006, Tripathi et al. 2008, Gil et al. 2009, Ahmad et al. 2010, Yang et al. 2011).

Nomura (1985) reported that cellulase can facilitate pulp fibrillation without strength loss. Mansfield et al. (1997) used cellulase and xylanase for investigating the nature of changes in fiber and observed substantial reduction in pore volume of the fibers. They reported that enzyme treatment can erode the surfaces of fibers. In another study, endoxylanases treatment on CTMP spruce fibers resulted in improved surface area of the fibers by fibrillation (Lorenzo et al. 2009). Janardhnan and Sain (2006) used fungal enzyme for refining and found significant impact on defibrillation characteristics of the bleached kraft pulp of northern black spruce fibers.

In a toweling tissue producing European mill, enzymatic treatment reduced 30% refiner energy, reduced 20% long fiber use, reduced 50% starch consumption without losing dry strength, increased 2% retention (due to less fines generated), increased 20% stretch, and reduced the steam consumption (due to improved dewatering) (Gill 2008). In another European mill of toweling tissue production, refiner 1 and 2 consumed 60% and 25% less energy, respectively along with 10% lower long fiber consumption after enzymatic treatment. Another toweling tissue production mill shut down one refiner without affecting the strength properties of the product (Bajpai 2012). In Liberty Paper, Becker, MN, USA, the output of paper machine was jumped by 15 tpd along with saving in refining energy using a fungal enzyme product of Dyadic International (Thomas and Murdoch 2006). Fibrezyme®G200 (Dyadic International, Netherlands) was reported to reduce the production cost with improved fiber to fiber bonding and pulp-refining properties (Murdoch, B. 2011. Dyadic International launches high-performance enzyme for pulp and paper industry using new CI production platform. Personal communication).

Enzymes produce a better fibrillation, so paper properties that depend on fibril content are generally affected. Employing the enzymatic approach, significant improvement in sheet density, smoothness, burst, and tensile indices can be obtained. Enzyme-assisted refining reduces pulp-refining energy requirement of the mills. Enzymes are expected to give more benefits to those mills, which are not having captive power generation and/or are limited by refining capacity. Enzymatic treatment also increases the pulp freeness. The enzymatic attack involves peeling mechanisms that removes fibrils and leaves the fibrils less hydrophilic and easier to drain. The increase in drainage may also be attributed to the cleaving of cellulose on the surface of fines. An improved drainability, that is, better dewatering by using enzymes results in decreased steam load for paper drying and reduced steam consumption. Enzymes produce a better fibrillation, so paper properties that depend on fibril content are improved, for example, tensile strength, burst strength, etc. Efficient drainage of pulp furnishes on the wire of paper machines is desired to maximize machine speed. Improved drainage results in shorter time for the drying period; so, paper machine runnability increases and results in enhanced productivity. Enzymes partially eliminate fine fibrils and colloidal materials, mainly those contained in the white water loop due to longer contact time with the residual enzyme activities in water. This produces cleaner recycled water, with optimum fine and fibrils content (Bajpai et al. 2006). Enzymatic treatments are useful in cleaning the process water within the mill. Enzymes partially hydrolyze cellulose debris to low-molecular-weight saccharides that are easily biodegraded in the wastewater treatment system.

Pulp viscosity decreases when cellulases cleave cellulose chains lowering the degree of cellulose polymerization and destroying the fiber integrity. The main challenge in using enzymes to enhance fiber bonding is to increase fibrillation without pulp/paper-quality deterioration. Though promising results have been achieved, yet more intensive studies on enzyme-assisted pulp refining are required to make the process cost-effective for easy adoption by paper mills.

Drainage Improvement

Poor drainability is one of the major problems related with the recycled and nonwoody raw materials. Poor drainability means less drainage of water from the fibrous wet web formed during paper formation process due to the higher relative surface area of recycled fiber's fines than virgin fiber's fines. These fines and microfibrils of the fiber, the main cause of lower drainage rate in recycled fibers, mainly consist of amorphous cellulose (Oksanen et al. 2000, Dienes et al. 2004). Therefore, in comparison to virgin pulp, the use of recycled fiber decreases the productivity of the papermaking because dewatering properties of the pulp strongly affect the energy efficiency of paper machine and runnablity. Thus, the overall cost efficiency of papermaking is also affected. Improved removal of water at paper sheet former and press section allows less energy consumption at the dryer section or increased paper production capacity. Alternately, due to improved drainage, a shorter drying section would decrease investment costs.

The problem of low drainage can be overcome effectively by reducing the effective surface area of the fines and fiber using drainage aids in the wet-end section and/or more intense wet pressing in the press section

(Hubbe 2000, Antunes et al. 2008). Application of drainage aids can deteriorate the quality of paper and high wet press levels decrease the bulk. Therefore, enzymatic treatments may be considered as novel pulp modification and dewatering innovations to make the paper with high-speed paper machines.

Enzymatic treatments have improved dewatering of different paper and board grades (Jackson et al. 1993, Bhardwaj et al. 1995, 1997). Enzymes act on the fiber fines by flocculation or hydrolysis process and remove fibrils from the surface of large fibers. At low enzyme doses, aggregation of fines and small fiber particles with each other improves pulp drainage. At high enzyme doses, hydrolysis of fines dominates over the flocculation process (Jackson et al. 1993, Davies and Henrissat 1995).

Several commercial cellulase enzymes are available for improvement of drainage of secondary fibers applied after refining/beating of the pulp. However, the use of cellulase mixtures can be harmful for some of the pulp properties. The hydrolysis rate of amorphous cellulose is 5–30 times higher than that of crystalline cellulose (Ghana et al. 1993, Ortega et al. 2001, Lynd et al. 2002, Al-Zuhair 2007). Therefore, a well-judged selection of the enzyme component, dose, and the retention time could make it possible to attain desired pulp properties without any detrimental effects on the fibers (Bhat et al. 1991, Eriksson et al. 1998).

According to Verma et al. (2015), monocomponent cellulase (i.e., endoglucanase) may be more efficient for the drainage improvement by selective hydrolysis of additional ultra fines and other dissolved colloidal materials. Endoglucanase treatment improved drainability of recycled pulp by approximately 15%–23% along with better tensile index and smoothness. Higher freeness due to endoglucanase treatment can be utilized for enhancing the operation rate and/or greater dilution can be applied in the head box of the paper machine (Verma et al. 2015). The improvement of pulp drainage of wheat straw pulp treated by cellulase generally occurred at initial 30 min of the enzyme treatment, resulting in a yield reduction of <0.4%. Also, enzymatic treatment improved physical strength properties and paper formation (Gong et al. 2003).

In a mill trial of North America producing 200 g/m² liner, the machine speed was increased with 6%–7% lower steam consumption after enzyme treatment due to improvement in the drainage. While in Asian paper mills, enzymatic treatment saves the steam consumption producing old corrugated containers (OCC) pulp (Shaikh and Luo 2009). In a European mill of towel and tissue production, machine speed was increased from

1650 to 1750 m/min and from 1600 to 1750 m/min for tissue and towel production with a reduction of 12.5% specific refining energy. This effect may be due to the CBH enzyme activity present in the enzyme product (Bajpai et al. 2006).

Deinking

Conventional chemical deinking methods require huge quantities of chemicals resulting in an expensive treatment of wastewater to meet the environmental regulations (Prasad et al. 1993, Woodward et al. 1994). Alternatively, enzymatic approach has been proven as an efficient and eco-friendly option to resolve the problem (Prasad et al. 1993, Putz et al. 1994). A number of enzymes such as cellulase, hemicellulase, pectinase, lipase, esterase, α-amylase, and lignolytic enzymes have already been applied for deinking of various recycled fibers depending on the specific activity of that enzyme, types of waste papers, ink, and coating material. However, cellulases and hemicellulases are the main enzymes applied for deinking (Bajpai and Bajpai 1998).

Researchers have proposed different mechanisms for ink removal from the recycled fibers by enzymes. According to Kim et al. (1991), cellulase enzyme partially hydrolyzes and depolymerizes cellulose polysaccharide chain at fiber surfaces; thus, bonds between fibers become weak enough to free them from one another, which facilitate the dislodgement of ink particles easily from fibers during pulping and separation during washing/flotation. Eom et al. (2007) pointed out that enzymatic action weakens bonds, may be by improved fibrillation or by removal of surface layers of individual fibers. Mere cellulase binding with the cellulose portion may disrupt fiber surfaces in such a way and to a level enough to release ink particles during pulping (Vyas and Lachke 2003). Hemicellulase assists deinking by breaking lignin–carbohydrate complexes to release lignin fragments along with ink particles from fiber surfaces (Treimanis et al. 1999). Cellulase and hemicellulase treatment facilitate the release of lignin to remove ink from newsprint (Bobu and Ciolacu 2007). Cellulases release ink particles for dispersal in suspension by peeling the fibrils from fiber surfaces (Kim et al. 1991) and improve the pulp freeness (Lee et al. 1983). Mechanical action increases susceptibility to enzymatic attack by the distortion of cellulose chains at or near fiber surfaces (Zeyer et al. 1994). Medium consistency is more favorable for enzymatic deinking than low consistency due to more fiber–fiber friction (Jeffries et al. 1994). In contrary to Zeyer et al. (1994), Putz et al. (1994) disagreed with the importance of mechanical action because of unaffected brightness at higher shear forces (at higher consistencies) or for extended times. Enzymes can lose their 3D structure due to distortion at higher shear forces caused by fiber–fiber friction (Bajpai and Bajpai 1998).

Indirectly, enzymes are able to remove microfibrils from the fiber surface and fines resulting into improved freeness (Jeffries et al. 1994). Enzymatic treatment removes fibrous material from ink particles, which improves hydrophobicity of ink particles of nonimpact-printed papers facilitating separation during flotation (Jeffries et al. 1994). Enzymes attack more on easily accessible cellulose chains (Zeyer et al. 1994). Mechanical action is also required for the removal of a significant amount of ink because surface friction on the fiber is able to improve interaction by opening the outermost layers of the fiber, which expose the full cellulose chains. Lipases and esterases degrade the carrier having vegetable oil and thus facilitate the dispersion of pigments in ink (Morkbak et al. 1999, Morkbak and Zimmermann 1998), while amylases are specific for the ink removal from the starch-coated waste papers (Seo et al. 1999).

Application of crude cellulases to recycled pulps could facilitate the deinking process (Kim et al. 1991). In a laboratory-scale deinking experiments, about 73% of deinking efficiency was obtained by the cellulase and hemicellulase of *A. niger* for laser-printed waste papers (Lee et al. 1983). Lignin peroxidase (LiP), manganese-dependent peroxidase (MnP), laccase, and mixture of these enzymes were evaluated for deinking of waste papers toward Kappa number reduction and brightness improvement. These enzymes were produced by three white-rot fungi, that is, *Fomes lividus, Thelephora* sp., and *T. versicolor* (Selvam et al. 2005). Immobilization of cellulase provides better deinking results than soluble cellulase (Zuo and Saville 2005). Deinking ability of lipase was caused by a partial degradation of the binder of the soybean oil-based inks, which facilitates the release of ink particles from the paper (Morkbak and Zimmermann 1998, Morkbak et al. 1999). For the first time, Gübitz et al. (1998) used a combination of magnetic deinking along with the enzymes (enzymatic–magnetic deinking) achieving about 94% deinking efficiency and 2.8% yield loss. This loss was much less than the loss resulted during flotation deinking (15%). The bacterial α-amylase acts on coated colored printed magazine resulting in greatest ink particle reduction (Elegir et al. 2000). Viesturs et al. (1998) suggested that for alkaline papers, coatings and filler materials also facilitate the ink removal from the papers as the majority of inks are localized on these materials. Acidification process of pulp suspension before enzyme treatment dissolves $CaCO_3$ before flotation, which results in efficient detachment and dispersion of toner specks resulting in improved deinking efficiency. Ink films fragmentation was also reported by the enzymatic treatment (Kim et al. 1991, Treimanis et al. 1999). Therefore, it is essential to control the action of enzymes to maintain the ink size in such a range so that the ink particles can be separated easily by flotation or washing processes.

In comparison to crude cellulase, mono-component endoglucanase is more efficient toward the deinking of mixed office waste papers (MOW) (Elegir et al. 2000). Extracellular endoxylanase of *Aspergillus terreus* CCMI 498 and endoglucanase of *Trichoderma viride* CCMI 84 showed better strength and ink removal properties for mixed office waste paper deinking than control (Marques et al. 2003). The enzyme treatment using two extracellular alkali-stable 1,4-α-D-glucan-4-glucanohydrolase fractions (i.e., Endo A and Endo B) of alkali-tolerant *Fusarium* strain resulted in the improvement of brightness with the reduction in ink counts of the recycled papers. Furthermore, higher dose of enzymes favor ink removal along with significant reduction in the strength properties of paper (Magnin et al. 2002). Pure endoglucanases from the fungal strains *Gloeophyllum sepiarium* and *G. trabeum* were used for the deinking of laser-printed wastepaper individually as well as in combinations. It was found that 94% of deinking efficiency was obtained using pure endoglucanases along with improvement in freeness level, slightly decreased strength of intrinsic fiber, and unaffected or slightly improved strength of handsheet (Gübitz et al. 1998). Endoglucanases lacking a cellulose-binding domain could result in superior deinking effects and strength properties (Geng et al. 2003). Crude preparation of cellulase enzyme along with xylanase produced by fungus *Coprinopsis cinerea* and commercial cellulase were found to remove toner ink from the photocopier waste papers having unaffected paper strength properties. Toners are considered as hard to deink due to large toner particles. Enzymatic deinking was found to be affected by point of enzyme addition, dose of enzyme, pulp consistency, and reaction time. Enzymatic treatment was given at a dose of 0.6 IU/g for 60 min at 12% pulp consistency. The enzymatic deinking in comparison to chemical deinking resulted in higher deinking efficiency (+25.4%), brightness (+5.2%), tensile index (+5.6%), burst index (+23.9%), and folding endurance (+15.9%) but reduced tear index (−7.6%) of paper. The pulp freeness was also improved by 19.6% (Table 6.5). The deinking efficiency and freeness of crude enzyme were found to be similar to commercial

TABLE 6.5 Comparison of Enzymatic and Chemical Deinking in Terms of Deinking Performance and Pulp and Handsheet Properties

Properties	Chemical Deinking	Enzymatic Deinking		Control
		(Commercial Enzyme)	(Crude Enzyme of *C. cinerea*)	
Deinking efficiency (%)	75.9 ± 1.0	94.6 ± 1.3	95.2 ± 1.0	43.7 ± 09
ISO brightness (%)	80.4 ± 0.5	78.7 ± 0.7	84.6 ± 0.5	73.1 ± 04
Residual ink (ppm)	77.2 ± 2.3	31.9 ± 1.6	33.2 ± 2.4	185.2 ± 8.3
Yield (%)	77.9 ± 0.8	77.1 ± 0.7	76.1 ± 0.5	79.1 ± 05
CSF (mL)	510	620	610	470
Drainage time (sec.)	6.7 ± 0.7	6.0 ± 0.2	5.6 ± 0.1	7.4 ± 0.2
Tensile index (N.m/g)	28.5 ± 4.9	29.3 ± 3.6	30.1 ± 1.9	26.95 ± 2.45
Folding endurance	0.75 ± 0.12	0.75 ± 0.10	0.87 ± 0.14	0.70 ± 014
Burst index (kPa.m^2/g)	1.34 ± 0.14	1.58 ± 0.36	1.66 ± 0.15	1.41 ± 021
Tear index (mN.m^2/g)	6.75 ± 0.85	5.55 ± 0.88	6.24 ± 0.45	7.11 ± 088
Opacity (%)	88.7 ± 0.62	90.1 ± 1.0	89.3 ± 1.06	90.3 ± 06
Bauer McNett Fiber Classification (wt% Basis)				
+40 mesh	16.2	18.2	19.2	20.5
+100 mesh	53.5	57.5	56.2	57.6
+150 mesh	6.6	6.5	6.5	4.3
+200 mesh	6.6	4	5.5	3.2
Fines	17.1	13.8	12.6	14.4
Effluent Characteristics				
Soluble COD (ppm)	270 ± 8	148 ± 4	189 ± 5	149 ± 5
BOD$_5$ (ppm)	122 ± 5	85 ± 4	105 ± 4	74 ± 3
BOD/COD ratio	0.45	0.57	0.55	0.50

Source: Pathak, P., Bhardwaj, N. K., and Singh, A. K. 2011. *Bioresources* 6:447–463; Pathak, P., Bhardwaj, N. K., and Singh, A. K. 2014. *Appita Journal* 67:291–301.

cellulase with significant improvement in strength properties of handsheets. The lower effluent load during the enzymatic treatment makes this process ecofriendly (Pathak et al. 2014).

In Stora Dalum Deinking Plant, 35% reduction in dirt speck along with 2.2% increase in brightness of final product was achieved after enzymatic application on low-grade office and print-house waste to prepare writing and printing paper. Enzymatic treatment reduced stickies content by ~ >50% (indexed area from 100 to 46 units) with an increase (~8 tons/day, i.e., from 215.5 to 223.7 tons/day) in mill production. The conversion was at least cost neutral (Knudsen et al. 1998). In Moulin Vieux Mill in France, increased brightness by 4–6 points, lower residual ink content, maintained mechanical strength, and no runnability problem on paper machine were obtained with the enzymatic deinking as compared to the mill chemical deinking (Magnin et al. 2002, Saari 2005). According to Tausche (2002), enzyme application reduced the visible/subvisible dirt count by 50% and reduced effective residual ink concentration by 35% in old newsprint/ magazines mill. In an Indian paper mill, enzyme application on multigrade furnish (such as ONP and MOW) resulted in reduced residual ink count, improved brightness, and cost saving by reduction in chemical consumption (saving of 50% sodium hydroxide, 37% sodium silicate and 100% hydrogen peroxide) (Mohammed 2010). Centre Technique du Papier (CTP) has evaluated alkaline cellulase Novozyme 613 toward the deinking of 35% MOW, 35% toner, and 30% wood-free magazines and achieved improved deinking efficiency (from 95% to 99%). It also confirmed that alkaline cellulases can be applied in neutral deinking resulting into lower stickies count (Bajpai 2012).

It is not possible to establish a key activity for ink removal regarding the enzymes for different deinking efficiencies. Although, cellulase and/ or xylanase activities are frequently related to effective deinking trials, but it is difficult to establish a relation about the involvement of each enzyme in the deinking process (Jeffries et al. 1993, Kim et al. 1991, Prasad et al. 1993). The major contribution toward deinking process is of endoglucanases and xylanases. However, the variability in the results is possibly owing to the use of various types of paper samples, printing inks, and enzymes obtained from different microbial sources (Zeyer et al. 1994).

Enzymatic treatment produced almost 50% lower COD loads than conventional deinking (Putz et al. 1994). On the other hand, about 20%–40% higher COD level was also reported in enzymatic process (Magnin et al. 2002, Saari 2005). The process water of enzyme (Novozyme 342) deinking had higher COD, but easily biodegradable than chemical deinking. Sludge production during enzymatic treatment was also found to be low due to lower inorganic content, thus reduced sludge treatment cost (Bobu and Ciolacu 2007).

Enzyme-treated deinked pulp has superior or unaffected strength properties, higher ISO brightness, lower residual ink, no alkaline yellowing, improved pulp freeness, improved drainage, and better machine runnability than chemical deinked pulp. Enzymatic deinking may not require additional dewatering, dispersion, subsequent reflotation, and washing steps, therefore, saves capital costs and electrical energy. Additionally, bleaching chemicals requirement may be lowered for enzymatic deinking.

Generally, enzymatic deinking reduces load on wastewater treatment due to less use of chemicals.

Stickies Control

The presence of contaminants (such as adhesives, coatings, glues, and binders) in the secondary fibers entering the mill is the main problem in most of the recycled fiber-based mills. These materials become a part of the pulp slurry which is called as "stickies." Therefore, it is difficult to remove these contaminants based on their physical characteristics due to its deformable nature which allows extruding these contaminants through screening. Interestingly, specific gravity of stickies is almost equal to water and fiber and therefore, these are accepted by cleaners (specific device designed to allow water and fiber into the system) (Scholz and Tse 1997). The effect of these qualities is that macro-stickies (rejected by a 0.10 mm slotted plate) are accepted into the post screening process (Heise et al. 1998). These macro-stickies are often responsible for filling of felts, plugging of wires, which cause paper defects resulting in increased machine downtime and reduced machine production efficiency. These stickies have been classified into various categories such as polyvinyl acetate, styrene butadiene, pressure-sensitive adhesives, and others (Doshi 1997, Doueck 1997). Due to inefficiency of screens and cleaners toward mechanical removal of macro-stickies from the process, they remain within the system. Conventional chemical fixes the stickies to the fiber to control the stickies (Dykstra 1990, Hall and Nguyen 2000), the dispersing of stickies using solvent and surfactant blends (Hoekstra and May 1998), polymeric stabilization (Magee and Taylor 1994), and combinations of each of these programs. Alternatively, stabilized enzymes such as esterases are also unique method for controlling stickies in recycled paper machine systems, which cleave the ester bonds of the macrostickies making them smaller in size as well as reducing their tackiness. Ren et al. (2013) have showed better cleaning situation at enzyme dose of 2×10^4 U/g of cellulase enzyme, 2×10^3 U/g of amylase and 10^4 U/g of lipase, 7.0 pH, and 50°C temperature. About 30% and 75% stickies reduction in deinked pulp and CTMP have also been reported using lipase- and esterase-type enzymes (Zeng et al. 2009).

Buckman Laboratories, Canada have developed an enzyme Optimyze (esterase) that is found to be very efficient toward stickies control at pH range of 6.5–10 and temperature of 25–60°C (Covarrubias and Eng 2005). In a U.S. mill, producing the coated paperboard from OCC, stickies deposition on paper machine was reduced by 75% using Optimyze enzyme from Buckman, which resulted into approximately \$1.33 million annual return as saving. Using the same enzyme, in another U.S. mill of 1000 tpd production of tissue papers, the shutdown time has been reduced by 60%. While a Brazilian mill having production capacity of 270 tpd of linerboard using 100% OCC achieved a record production (Patrick 2004).

Starch Modification

Because the viscosity of the natural starch is too high for paper sizing, after proper modification, that is, viscosity reduction, starch is currently used for manufacturing of coated papers to improve the gloss and smoothness for better printing quality. This viscosity reduction is traditionally dropped by the use of chemicals. Nowadays, the α-amylase enzyme (as an oxidizing agent) is being used for the viscosity reduction, which is an economical option over chemical route. This low-viscosity and high-molecular-weight starch is used for coating of paper. The starch can be oxidized using suitable α-amylase enzyme at the mill site in batch or continuous process. The purpose of sizing of paper is to improve the quality of paper by the enhancement of the stiffness, strength, and erasability. It also provides smooth coating on the surface and protection to the paper from the damage during processing. Enzymatically hydrolyzed starch is added to the paper surface in the size press. Two rollers transfer the starch slurry on the paper which is picked up by the paper at the temperature range of 45–60°C. Viscosity of the starch should be kept constant to get the reproducible results at this stage. The mills are flexible to vary the starch viscosity for various paper grades. The operation conditions depend on the type and source of enzymes and starches used. Commercially available successful amylase preparations are Amizyme® (PMP Fermentation Products, Peoria, USA), Termamyl®, Fungamyl, BAN® (Novozymes, Denmark) and α-amylase G9995® (Enzyme Biosystems, USA), etc. (Gupta et al. 2003).

Stora Enso Nymolla Mill, Sweden (an integrated mill) producing 325,000 tons pulp and 450,000 tons uncoated fine paper per annum has replaced the hydrogen peroxide-based chemical oxidation with the enzyme modification, which reduces the chemical cost and problem of corrosion of equipment (Svensson 2006).

The α-amylase modification on the tapioca starch showed the modified starch solution of 25% solid content and 13.8 mPa.s viscosity, which provides a better condition for surface sizing. Using this modified starch for surface sizing of newsprint, improvement in the degree of sizing of the paper as well as strength property of paper was observed (Tong et al. 2007).

For the surface-sizing agent for lightweight paper (60 g/m^2), 0.02% (based on dry starch) α-amylase was cooked at 80°C for 20 min, then the temperature was raised to 98°C at a high speed with a holding time for 30 min at the same temperature. The solid content of the product was 9.0% and viscosity was in the range of 5.5–6.5 mPa.s at 60°C. The strength was increased by 23.2% and the cost of surface sizing was also reduced (Feng et al. 2012). The amylase treatment also decreases the equipment corrosion (Wanqi 2010).

Enzymatically modified starches (partially hydrolyzed) are totally free from the AOX products, which are generally produced during the chemical oxidization process using sodium hypochlorite. Under controlled conditions, any type of starch solubilization can be avoided but desired viscosity level will also be maintained due to the highly selective nature of the enzymes. Enzymatic modification is done on the papermaker's site as per the desired viscosity level. The cost of the enzymatically modified starches is considered lower than that of the chemical oxidized starches; therefore, the mills have started using enzymatic modified starch in place of oxidized starch to reduce the cost. Due to the presence of some proteins in the native starches, the brightness of enzymatic modified starch is slightly lowered which is further compensated using brightening agents.

Dissolving Grade Pulp

Dissolving pulp is high-purity cellulose used as a raw material to produce regenerated cellulose and cellulose derivatives. Rayon, a natural man-made fiber used in textile industry, is the main product of regenerated cellulose while cellulose ethers, cellulose acetates and cellulose nitrates, etc. are the cellulose derivatives resulted from different derivatization reactions (Köpcke 2010). Its regeneration and derivatization require its prior solublization in the reagents or solvents, which is a crucial step. The poor accessibility or reactivity of cellulosic pulp presents a major problem in its subsequent processing. In recent years, many researchers have employed cellulase treatment specifically monocomponent endoglucanases to dissolving pulp as well as paper-grade pulp to enhance the pulp reactivity (Henriksson et al. 2005, Engström et al. 2006, Kvarnlöf et al. 2007, Köpcke et al. 2008, Ibarra et al. 2010, Köpcke 2010, Gehmayr and Sixta 2011, Gehmayr et al. 2011, Miao et al. 2014, Wang et al. 2014). There was a marked increase in Fock reactivity of hardwood kraft-based dissolving pulp from 47.67% to 79.90% when the cellulase treatment was subjected at a dose of 2 U/g (Miao et al. 2014). Among the three major types of cellulases, monocomponent endoglucanases have been shown to be very promising in reactivity improvement. Henriksson et al. (2005) and Engström et al. (2006) reported that the reactivity of softwood sulfite pulp was significantly increased with relatively low amounts of endoglucanases used. An increase from 40% to approximately 80% in pulp reactivity was found by Kvarnlöf et al. (2007) when the paper-grade pulp was treated with different types of cellulases. Similar trend of reactivity improvement was exhibited by hardwood-dissolving pulp and softwood-dissolving pulp after being subjected to cellulase treatment. Also, the nonwood pulps were treated with monocomponent endoglucanases resulting in enhanced cellulose reactivity (Ibarra et al. 2010, Köpcke 2010). Cellulase treatment was also found to have a significant role in bioconversion of paper-grade pulp to dissolving pulp. Paper-grade pulp having high hemicellulose content and poor pulp reactivity renders it unsuitable for being used as dissolving pulp. This is attributed to the hemicelluloses which hamper the subsequent dissolution of pulp resulting in poor end product quality.

For this purpose, xylanases have been utilized to lower the hemicellulose content in aid with alkali extraction. Subsequent cellulase treatment of xylan and alkali extraction of pulp then fulfills the required dissolving pulp. This approach has been studied by Ibarra et al. (2010) and Köpcke (2010) for the upgradation of wood and nonwood paper-grade pulps to dissolving pulps. Eucalyptus and sisal pulps when treated in a treatment sequence of xylanase, alkali and endoglucanase provided the reduced hemicellulose content, uniform molecular weight distribution, and improved pulp reactivity meeting the criteria of dissolving pulp as a whole. However, reduced viscosity was also observed that was attributed to cellulase treatment. In addition to cellulases, xylanases have also been used for reactivity improvement of dissolving pulp. Wu et al. (2015) reported that reactivity of bamboo-dissolving pulp obtained from prehydrolysis kraft process was significantly increased at the xylanase dose of 1 IU/g simultaneously lowering the pentosan from 3.42% to below 3.0%. In a study by Ambjörnsson et al. (2014), dissolution of dissolving pulp in NaOH/ZnO solution was increased from 29% to 81% after treatment with 16.7 AXU/g dose of xylanase followed by alkali extraction with 7% NaOH and endoglucanase treatment with 10 ECU/g dose. Gehmayr and Sixta (2011) investigated xylanase treatment and cold caustic extraction (CCE) treatment to attain the target residual level of xylan in *E. globulus* kraft pulp, which was subsequently subjected to TCF bleaching followed by endoglucanase treatment. Endoglucanase treatment was

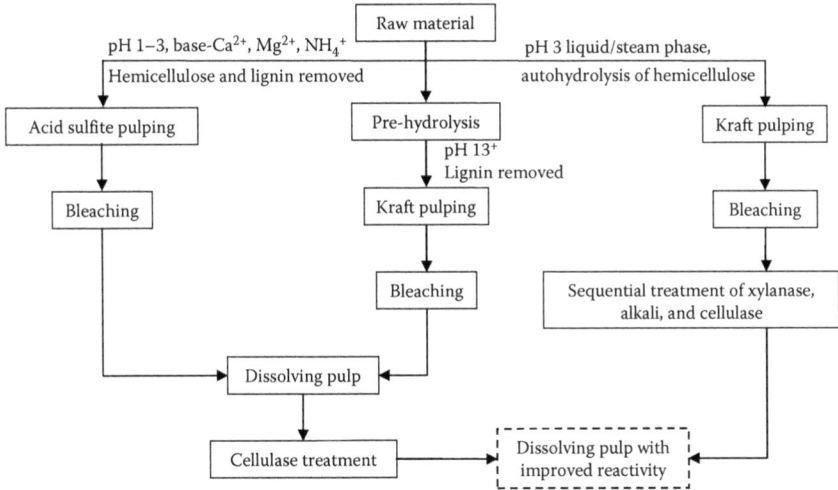

FIGURE 6.15 Process diagram showing the production of dissolving grade pulp by different routes and reactivity improvement using enzymes.

used for adjustment of the final average DP and reactivity increase. This resulted into the production of high-purity dissolving pulp from commercial oxygen-delignified *E. globulus* kraft pulp. Figure 6.15 shows different routes for the production of dissolving grade pulp and reactivity improvement using enzymes.

Therefore, it is evident that xylanase treatment has potential for aiding the hemicellulose removal from either fully bleached kraft pulp or delignified kraft pulp in the process of dissolving pulp production. Also, it can be utilized for the reactivity improvement of dissolving pulp along with endoglucanases whereas the latter has its significant role in the enhancement of pulp reactivity only.

Integrated Biorefinery

Pulp and paper industry is looking for extra sources of revenue generation for improving their viability, profitability along with global competitiveness, where introduction of biorefinery approach can partially fulfill this requirement. The integrated biorefinery is a process, which selectively extracts carbohydrates, lignin, oils, and other materials from biomass for the conversion into fuels and other value-added products. Nowadays, the interest in conversion of biomass to bioethanol has increased because of the limited oil supply and increased concern for green house gas emissions.

The main component of wood is cellulose, hemicellulose, and lignin. Hardwoods have around 48%–54% cellulose, 15%–22% hemicellulose, and 21% lignin. Hemicelluloses have been identified as a source for the production of higher value-added products such as ethanol and acetic acid in an integrated forest products biorefinery (IFBR) based on kraft mill (Van Heiningen et al. 2006). Theoretically, unbleached pulp yield should be 63%–76%, if lignin removal is selective but kraft pulp yields figures around 47%–50%. It is known that during kraft pulping, almost half of the hardwood hemicelluloses get dissolved in black liquor and further the black liquor containing these hemicelluloses sugars is currently being burnt producing steam and electricity. In view of the lower heating value of hemicellulosic sugar than aromatic lignin, this is not an effective utilization of these valuable sugars. Therefore, partial pre-extraction of hemicellulose (as oligomers from wood chips) prior to pulping offers an interesting economic opportunity for the pulp and paper industry to convert these sugars into valuable products (Figure 6.16), otherwise the hemicellulose will be burnt with the black liquor (Van Heiningen et al. 2006). The heating value of hemicellulose is 13.6 MJ/kg, which is just half to that of lignin, that is, 27 MJ/kg. On the other hand, the amount of hemicellulose in black liquor is much lower than that of lignin; therefore the removal of hemicellulose in the pre-extraction stage does not affect heat recovery from black liquor significantly (Gullichsen 2000). Hemicellulose is mainly comprised of pentose sugars (xylose and arabinose) along with hexose sugars (glucose and mannose). This hemicellulose is solubilized

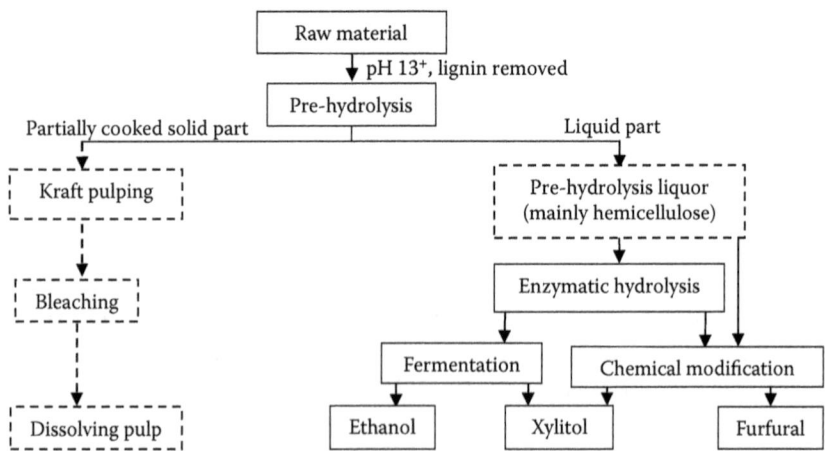

FIGURE 6.16 Process diagram showing an integrated biorefinery concept.

when wood chips are exposed to auto-hydrolysis prior to pulping stage. Thus, so produced pre-hydrolysis liquor mainly contains different types of sugars such as monosaccharides (xylose, arabinose, mannose, galactose, and glucose), oligosaccharides (galactoglucomannan, glucuronoxylan, etc.), and other chemical compounds (acetic acid, furfural, phenolic compounds, etc.) (Marinova et al. 2010).

Hemicellulose prior to pulping can be extracted by different methods, Prehydrolysis: mildly acidic by heating water at 170°C (Lai 1990), Auto hydrolysis: steam (175–220°C) by organic solvents (Lai 1990), Steam explosion: at 200–250°C by explosive discharge (Puls and Saake 2004), Enzymatic hydrolysis: by a group of enzymes (Jeoh 1998), Hot water extraction: high pressure at 140–190°C (Yoon et al. 2006). Pre-extraction of hemicellulose from poplar wood chips was recently reported by Al-Dajani and Tschirner (2008) at highly alkaline conditions and relatively low cooking temperature. Approximately 40–50 kg of hemicellulose from 1 ton of wood could be extracted without affecting net pulp yield, followed by modified kraft pulping. By extraction of hemicellulose of pine chips with pressurized hot water at an H factor of about 500 h, 8% of hemicellulose was removed with minor removal of lignin and cellulose (Yoon et al. 2008). Also, pre-extraction is done in such a way that it does not affect the quality or the yield of pulp. In fermenting hemicelluloses into useful products, it is necessary to hydrolyze the oligosaccharide down to its component monosaccharide constituents. Chemical or enzymatic treatments may be used to accomplish this hydrolysis.

The pre-hydrolysis liquor contains different types of inhibitory compounds (furfural, acetic acid, hydroxyl methyl furfural-HMF, etc.) for fermentation process which are detrimental for both the microbial activity and enzymatic activity. Therefore, it needs to be detoxified prior to bioconversion for the optimum production of valuable compounds through any of the physical, chemical, and biological detoxification methods (Nigam 2002, Agbogbo et al. 2006, Chandel et al. 2011). After the detoxification of prehydrolyzate, it is fermented to convert the sugar into valuable product, that is, ethanol, xylitol. The prehydrolyzate from the softwood contains high level of hexose sugar and it is more toxic than that of hardwood because it usually contains more extractives than hardwood. The prehydrolysate from hardwood contains xylose as the major component which is a five-carbon sugar. Among the different yeasts, *Saccharomyces cerevisiae* is the best choice for the fermentation of prehydrolysate extracted from the softwood as it is quite efficient in converting the hexose sugar into high yield of ethanol and on the same hand it can also tolerate the certain level of toxins. The yeast which can degrade pentose sugars efficiently is *Pichia stipitis,* also known as *Yamadazyma stipitis.* The yeast *Torula utilis* is known to utilize both the pentose and hexose sugars and it can also tolerate the inhibitory compounds present in the prehydrolysate mainly furfural (Bhattacharya et al. 2005). The fermentation of prehydrolysate can be done either by separate hydrolysis and fermentation (SHF) or by simultaneous hydrolysis and fermentation (SSF). In the SHF method the prehydrolysate is first subjected to the enzymatic hydrolysis to convert the complex sugars into simple sugars and then it is followed by the fermentation process for the final conversion of the sugars into ethanol while in the case of SSF both the hydrolysis and the fermentation process occur simultaneously. The ethanol produced after the fermentation needs to be separated and purified by distillation method.

Walton and Van Heiningen (2007) used green liquor for extraction of hemicellulose, extracted liquor fermented with *E. coli* K011 and produced 20 g/L of ethanol in 96 h. Mao et al. (2008) also used the green liquor for extraction of hemicellulose but major disadvantage of this process is that carbohydrate removal is less and extracted liquor also contains inorganic salts originated from green liquor. In another study by Al-Dajani et al. (2009), overall pulp yield was lower for the material pre extracted with hot water, but chemical charges and cooking time could be reduced significantly. Water and kraft white liquor (NaOH, Na_2S, and Na_2CO_3) are used for extractions of hemicellulose from silver birch (*Betula pendula*) wood chips (Helmerius et al. 2010).

Wastewater Treatment

Pulp and paper mills generate various aromatic xenobiotics and pollutants, which are responsible for color of the wastewater. These compounds are toxic for aquatic life. White-rot fungi have been reported for the most detoxification and decolorization of pulp and paper mill wastewater by the production of lignin-degrading enzymes. Decolorization of the effluents by a marine fungal isolate was also reported. A main source of phenolic wastes is alkaline extraction-stage effluent arising during pulp bleaching having over 50% of color load. Conventional treatment methods, such as aerated lagoons and activated sludge plants, are ineffective in removing this color. Moreover, chemical and physical treatments (ultrafiltration, ion exchange, and lime precipitation, coagulant/flocculant methods) for the reduction of pollution are costly. Therefore, laccase and peroxidase enzymes may be a better alternative for biological treatment process. Enzyme-mediated bioremediation processes reduce the pollutants either by the polymerization among toxic pollutants themselves or by copolymerization with other nontoxic substances such as humic materials. These polymerizations facilitate easy removal by adsorption, sedimentation, or filtration (Gianfreda et al. 2006, Arora and Sharma 2010). Laccase-mediator system was also successfully applied in the treatment of paper mill effluent (Minussi et al. 2007). The fungal enzymes such as laccases, lignin peroxidases, and MnPs, are able to degrade the lignin effectively but under extreme environmental conditions (high temperature, alkaline pH, and toxicity), their enzyme efficiency reduces. On the other hand, filamentous structure of fungal hyphae also causes structural obstacle for the biological treatment of effluent emitted from pulp and paper industry (Amr et al. 2009). *P. chrysosporium*, one of the important representatives of white-rot fungi has been a widely studied model for lignin degradation research and LiP production.

Recently, some researchers have reported the ability of bacteria to break down lignin and lignin-derived compounds (Masai et al. 2007, Ahmad et al. 2010, Bugg et al. 2011). *Streptomyces viredosporus* T7A is able to depolymerize lignin, using an extracellular lignin peroxidase enzyme (Ramchandra et al. 1988). *P. putida* mt-2 and *R. jostii* RHA1 both could depolymerize lignin in the absence of hydrogen peroxide, suggesting either the use of oxygen-utilizing laccase enzymes, or extracellular enzymes for hydrogen peroxide generation (Ahmad et al. 2010). Catabolic pathways in *Sphingobium* sp. SYK-6 have also been extensively studied for degradation of several lignin components (Masai et al. 2007).

In majority of studies, bacteria treat pulp and paper mill effluent effectively only in the presence of high-valued secondary carbon (glucose, sucrose) and nitrogen (peptone) sources (Raj et al. 2007, Mishra and Thakur 2010, Chandra and Abhishek 2011). Hence, they do not have practical application. Recently, Kumar and Kumar (2006, 2010) demonstrated batch study where defined consortia of three bacteria could reduce color, BOD, and COD of pulp and paper mill wastewater up to discharge norms within 24 h by utilizing lignin as the sole carbon source. However, these studies are still restricted up to lab scale. Therefore, eco-friendly and economically feasible bacterial treatment technology which can be applied at industrial scale needs to be developed for pulp and paper mill wastewater. Although, activated sludge process is widely used and accepted process for treatment of wastewater from pulp and paper mills throughout the world but this process uses organisms by natural selection. Other newly developed processes like packed bed reactors (Jahren et al. 2002, Malmqvist et al. 2003) or low sludge bioprocesses (Chakrabarti et al. 2012) may have certain advantages for treatment of wastewater using the selective consortia.

The application of enzymes for wastewater treatment in the pulp and paper industry is a new field of interest for researchers with new possibility. Presently, the research activity is being focused on lignin degradation through enzymology. Laccase, lignin peroxidase, and MnP are the most important enzymes, which are used for color reduction of bleaching effluents from pulp and paper mill. In another approach, a mixture of enzymes and microbes can be mixed together to remove recalcitrant and injurious compounds from wastewater (Bajpai and Bajpai 1994).

Shives Control

Shives are small bundles of fibers or woody residue, which have not been separated into individual fibers during the pulping process. They appear as dark splinters in the pulp. Shive count is one of the most important quality criteria for the bleached kraft pulp. Normally, bleaching process is able to remove 95%–99% of the shives. A novel enzyme preparation, Shivex, multicomponent mixture of proteins such as xylanases, is reported to improve the shive removal efficiency by bleaching. According to Tolan et al. (1994), 55% shives can be reduced after Shivex treatment followed by bleaching. Shives removal also improves the bleaching efficiency of pulp and reduction of bleaching chemicals. Shives have been reported to be the most important factor triggering web breaks and reducing paper strength (Gregersen et al. 1999).

Shives control is not facilitated by only the bleach-boosting capability of xylanase but there is some other activity in xylanase or other enzyme/proteins. These enzymes act on the fiber surface of the shives to remove the diffusion barrier improving the bleaching efficiency (Tolan et al. 1994).

Pitch Control

Pitch is termed as the hydrophobic components of wood, primarily triglycerides and waxes present in the extractives. During mechanical pulping processes, these extractives are released from the fibrous part and deposit in the process waters. It has the tendency to deposit on equipment surfaces causing maintenance problems, production loss, and affect papermaking process along with product quality.

Different types of sapstain fungi, basidiomycetes, and molds have also the capability to reduce the wood extractives by growing inside the resin canals, tracheids, and fiber cells. The main sapstain fungi are *Ophiostoma ips, O. piceae, O. piliferum, Leptographium lundbergii, Alternaria alternata,* etc. Basidiomycetes fungal strains *P. chrysosporium, P. subacida, Coriolus versicolor, C. subvermispora, Schizophyllum commune, Pleurotus ostreatus,* etc. have been reported to degrade the extractives from the wood. Cartapip treatment (fungal spores of *O. piliferum,* Clariant, UK) trial in a U.S. mill has shown the reduction in extractives (dichloromethane) of southern yellow pine and increase in burst index after thermomechanical pulping (Chen et al. 1994).

Lipases have been applied to solve pitch problem. Lipase catalyzes the hydrolysis of these triglycerides to fatty acids. These enzymes are not effective at high temperatures during the pulping processes. Therefore, thermo-stable lipases such as Resinase (Novozyme) have recently been developed. Using Resinase, Nanping Paper Mill in China has successfully controlled the pitch outbreaks (Chen et al. 2001). Earlier, application of Resinase A2X enzyme at a plant-scale trial in Yatsushiro mill of Jujo paper Co. was found to be effective toward the reduction of triglycerides content from 18%–25% to 7%–9% (Fujita et al. 1991). Up to 70% of total extractives have been removed using lipase Resinase HT at 80°C (more thermostable than the wild-type enzyme) without affecting the optical properties of the sheet. Mechanical properties appeared to be slightly improved (Blanco et al. 2009). Presently, both the Resinase A2X and Resinase HT enzymes are being applied in the mills of different countries such as United States, Canada, China, Japan, and other far east, which are using the wood species containing high resin content (Bajpai 2012). Nippon Paper Industries, Japan, have developed method for the control of pitch using lipase enzyme of *Candida rugosa* to hydrolyze up to 90% of the wood triglycerides. Shu et al. (2012) have reported that 39% of triglycerides of thermo-mechanical pulp of *Pinus massoniana* were hydrolyzed by the thermostable lipase from *Burkholderia* sp. ZYB002. A combination of lipase with the surfactant may have better possibility to extract a broad range of extractives from the softwood pulp (Dube et al. 2009).

Oxidative enzymes such as laccases are also effective toward the extractives (i.e., fatty acids, resin acids, triglycerides, and free/conjugated sterols) degradation in the presence of redox mediators (such as syringaldehyde, HBT, ABTS, acetosyringone, and p-coumaric acids). Laccases provide dual benefits by degrading lignin as well as the extractives. A laccase enzyme produced by *Pycnoporus cinnabarinus* with HBT has been reported to remove 95%–100% free and conjugated sterols from eucalyptus kraft pulp, 65%–100% reduction of triglycerides, resin acids and sterols from spruce thermomechanical pulp, and 40%–100% degradation of fatty acids, alkanes, and sterols for flax soda pulp (Gutierrez et al. 2006).

The biologically treated (using enzyme and fungi) wood facilitates to solve the pitch-based problems by reducing the extractive content. It also solves the problem of downtime of the paper machine by the reduction of paper defects on paper web, reduction of cleaning frequency of pitch deposits, improvement in pulp drainability, and increased machine runnability. All these improvements indirectly save energy and cost in an eco-friendly manner.

Slime Control

Slime is term used for the deposits of microbial origin forming viscous and mucous layers in process waters. The main problems due to these biofilms are paper defects (spots and holes), web breaks, smell problem as a result of H_2S production, and hygienic quality of paper products (Verma et al. 2014). Slime mainly comprised of bacteria and extracellular polymeric substances (EPS) is produced by the bacteria, wood fibers, and various additives used in papermaking. The main components of EPS are different polysaccharides (such as glucose, galactose, glucosamine, galactosamine, L-rhamnose, and fucose), proteins, nucleic acid, and polymeric lipophilic materials. The main microorganisms responsible for the slime formation are *Desulfovibrio, Clostridium, Aspergillus, Penicillium*, etc. Recently, microbial slimes are controlled using enzymes specific to the numerous saccharide units that make up the exopolysaccharide layer or help in attachment process of biofilm. These enzymes comprise galactosidase, galacturonidase, rhamnosidase, xylosidase, fucosidase, arabinosidase, and α-glucosidase. Such enzyme blends have been found specifically to digest microbial slime and reduce microbial attachment and biofilm. For the better efficacy of enzymes toward slime control, pre-knowledge of exact composition of the slime will be required so that suitable enzyme–substrate combination can be formed depending on the nature of the EPS in a biofilm. These enzymes target the specific compounds present in deposits (starch, slime, pitch, adhesives, latex, and other synthetic binders) (Timothy 2007). In another approach, considerable attention is being given to control formation of biofilms by enzymes. These approaches include the enhanced removal of biofilm, preventing the formation of biofilm and improvement in the efficacy of biocides. Application of enzymes reduces or completely replaces organic microbiocides that are currently being used. Due to the complex problem of biofilming in paper machine (PM) whitewater circuits, a combination of microbiocides, biodispersants, and enzymes appears to be more promising (Verma et al. 2014).

Bacteria produce different forms of levan in liquid systems, mostly attached to surfaces. For the degradation of fructans, two types of enzymes are mainly responsible, that is, hydrolases and transferases. The endo- or exo-hydrolytic enzymes produce a series of oligofrucans or only fructose, respectively. The other enzyme, transferases simultaneously break down the fructose dimers and the *trans*-fructosylation gives rise to difructose anhydride (Timothy 2007).

Juan et al. reported the suitability of enzyme-based biocides at pulp and paper mills (Juan and Victor 2011). The bacterial count at wire pit was reduced to 15 millions and 4–5 millions CFU/g from 60 millions CFU/g using enzymatic biocide at a tissue paper mill just after 5 and 10 days, respectively. While, machine downtime was reduced from 5.6 h/day to 0.4 h/day (93% reduction) on PM-1 and the same was reduced from 2.7 h/day to 0.2 h/day on PM-2 (92% reduction) due to elimination of dirt and slime detachment at paper machine by using the enzymatic biocide at an OCC mill (Juan and Victor 2011).

Although the combined enzyme systems comprising cellulase, α-amylase, and protease have been found efficient toward the treatment of microbially produced extracellular polysaccharides in cooling water and in PM backwater, still, there is requirement of new enzyme having more stability and efficiency through the further developments in enzyme products to further benefit the paper industry in future (Verma et al. 2014).

Reduction of Vessel Picking

Hardwood pulp has fibrous cell and vessel elements. Vessel elements are necessary parts of the hardwood (such as eucalyptus) providing a channel for transporting water and nutrients within a tree. However, they create some problems during papermaking process such as vessel picking problem in offset printing. These vessel elements have poorer bond forming ability than the fiber due to their unique shape and larger size. During offset printing, they are effortlessly picked off from the paper surface due to high splitting force in the printing nip (Rakkolainen et al. 2009). Therefore, they remain attached to the printing blanket, which disturbs further ink transfer process causing printing defect such as appearance of white spots.

TABLE 6.6   Reduction in Vessel Picks after Enzymatic Treatment

Trial	Average Vessel, Picks/cm²	Standard Deviation	Reduction (%)
Control	4.0	0.8	0
Cellulase/xylanase mixture 1	2.0	0.8	50
Cellulase/xylanase mixture 2	1.2	0.9	70

Source: Data from Cooper, III., and Elwood, W. 1998. Process for treating hardwood pulp with an enzyme mixture to reduce vessel element picking. U.S. Patent No. 5,725,732.

Consequently, frequent cleaning of printing blanket is required which causes the increased downtime of the printing press. Vessel picking tendency of paper is mainly affected by the proportion of hardwood in paper and the bond forming ability between fibers and vessels. Vessel picking can be prevented either by affecting the bond forming ability of vessel elements or by removal from pulp. The second option is industrially impractical. Usually, vessel picking is controlled by higher degree of refining of pulp (energy-intensive process), stock, or surface sizing. Although, severe pulp refining breaks down vessel elements resulting in improved bonding ability due to increased vessel fibrillation and flexibility but, detrimental for the drainability of the pulp and bulk of paper. Other alternative methods are selection of suitable wood species, hydrocyclone separation, and enzymatic treatment (Rakkolainen et al. 2009).

As per the claim of U.S. patent OO5725 732A, treatment of hardwood brownstock (unbleached) pulp using a mixture of cellulases and xylanases was found to chemically change the hardwood vessel elements, exposing them prone to break even under normal mill refining, therefore, avoiding extra refining equipment (Cooper and Elwood 1998) (Table 6.6). Another patent from Honshu Paper Co. applied commercial cellulases to enhance the flexibility of hardwood vessels (Uchimoto et al. 1988). Enzyme treatment was also reported to reduce vessel picking by 85% with improved smoothness, tensile strength, and draining time.

Debarking

Bark removal is the first step among the different processes of papermaking. Debarking requires extensive amount of electrical energy for high-quality mechanical and chemical pulp because even small amounts of residual bark content cause darkening of the paper product. During complete debarking, some amount of basic raw material required for papermaking is also lost due to extended mechanical treatment in the drums.

There is a cambial layer, made up of only one layer of living cells, between wood and bark, which produces xylem cells facing inside of the stem and phloem cells facing outside. This cambial layer comprises high content of pectins with low content or devoid of lignin (Katō 1981, Dey and Brinson 1984). The pectin content in cambium cells may vary according to different wood species. The amount of pectic and hemicellulosic substances is also high in the phloem cells. Therefore, pectinase and xylanase enzymes are found to be crucial enzymes in the debarking due to the presence of pectin and hemicellulose, respectively (Viikari et al. 1989). After pretreatment with pectinolytic enzymes, the energy demand in debarking was lowered as much as 80% (Rättö et al. 1993). White-rot fungus, *Phanerochaete gigantea*, may also be used to inoculate a wood log. Fungal growth particularly in the border area between the bark and wood for a sufficient time facilitates the debarking from the balance of the log (Blanchette et al. 1996).

Enzymatic pretreatment prior to debarking causes the significant reduction in energy consumption and saving of raw material. Enzymatic treatment reduces the detachment resistance of the bark, thus facilitates the loosening of bark, which requires less mechanical energy. Poor infiltration of enzymes in the cambium of whole logs is one of the difficulties with enzymatic debarking (Viikari et al. 1989, Rättö et al. 1993). Therefore, it would be advantageous to spray enzyme solution on the poorly debarked logs prior to repeat mechanical debarking.

Nanocellulose Preparation

In recent years, nanocelluloses from abundant and renewable natural resources have attracted wide attention in paper industry due to their extraordinary mechanical properties such as high Young's modulus, tensile strength, and extremely large surface areas due to high aspect ratio, low thermal expansion (Fukuzumi et al. 2009, Nogi et al. 2009, Belbekhouche et al. 2011, Moon et al. 2011). This interesting behavior of nanocelluloses leads to the improvement in physical and mechanical properties of paper. This nanocellulose may be used for papermaking, security papers, food packaging, gas barriers, and coating additives. There are different families of micro/nanocellulose materials: (i) MFC and NFC (micro/nanofibrillated cellulose, microfibrils, and nanofibrils), (ii) NCC (nanocrystalline cellulose), and (iii) BNC (bacterial nanocellulose). Currently, an intense research activity is ongoing within the field of cellulose nanofibrils (CNF) and cellulose microfibrils (CMF). CNF and CMF, also known as nanofibrillated cellulose (NFC) and microfibrillated cellulose (MFC), are two types of cellulose nanomaterials which contain both crystalline and amorphous parts and typically have lengths longer than 1 μm. There is also another type of cellulose nanomaterial called cellulose nanocrystals (CNC). CNC elements are, however, much shorter than CNF or CMF and have therefore not the same ability to form networks as the CNF and CMF elements. Various definitions may be found in the literature, but in the proposal for the new TAPPI Standard, CNF have widths of 5–30 nm and CMF have widths in the range between 10 and 100 nm. In this chapter, however, the name CNF will be used exclusively since there has been a lack of distinction between these two materials in many scientific papers.

MFC and NFC, NCC, and BNC represent about 66%, 34%, and less than 1% of global production in 2012, respectively. Since 2008, a substantial work has been dedicated to the development of nanocellulose. The commercial development is under way and an optimistic estimation of production is around 3500 tons in 2017, a quite moderated production. Among the applications, composites are the most consuming area (40%). The pulp and paper industry is on the second position for application of nanocellulose (Charreau et al. 2013).

Figure 6.17 shows different pretreatment processes for the production of micro/nanofibrillated cellulose. The main drawbacks of nanofibrillated

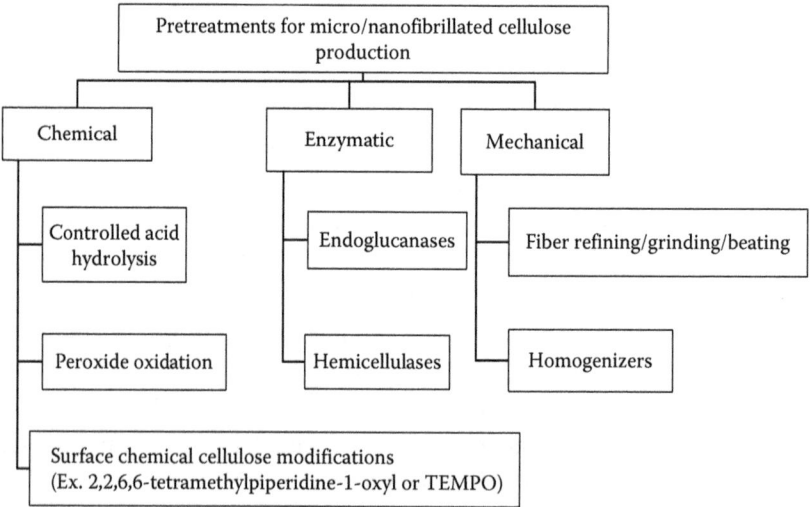

FIGURE 6.17 Different pretreatment processes for the production of micro/nanofibrillated cellulose.

cellulose preparation are energy-consuming manufacture, nonhomogeneous material, and suspensions difficult to pump and to characterize. In order to make cellulose nanofibers commercially competitive, energy consumption during the preparation of NFC can be reduced through different pre-treatments before mechanical treatments, but more energy-efficient processes utilizing low-cost sources are still required.

Nanocrystalline cellulose (NCC) production through acid hydrolysis is hazardous for environmental concern (waste stream treatment) with low yield (30%–40%). Nano-fibrillated cellulose (NFC), mechanical homogenization, or shearing is very energy-consuming process (Energy consumption = 20,000–30,000 kWh/ton or 72–108 GJ/ton) which is approximately 4–5 times higher than the energy stored in wood. NFC can also be produced by chemical pretreatments using 2,2,6,6-tetramethylpiperidine-1-oxyl radical (TEMPO) with significant energy reduction but TEMPO is an extremely expensive chemical and for commercial viability, development of effective recovery processes for TEMPO is needed (Dufresne et al. 1997, Wang et al. 2007). Enzymatic pretreatment before high-shear refining or fibrillation or homogenization is an ecofriendly alternative of chemical pretreatment to facilitate the nanofibrillation via disintegration of cellulose (Henriksson et al. 2007, Zhu et al. 2011). Reduced instrument blockage during high-pressure homogenization is an additional benefit with enzyme pretreatment which facilitates better processing of cellulosic fibers. Un-purified commercial cellulase enzymes shorten the length of the resultant CNF due to synergistic behavior of different components of cellulase (Henriksson et al. 2007). A hyperthermostable endoglucanase enzyme of *Pyrococcus horikoshii* (Ph-GH5) with no xylanase and low cellobiase activities produced CNF from bleached eucalyptus pulp fibers through microfluidization. Treatments with enzymes of Ph-GH5 made easy production of CNF due to the reduced DP of the fibers consuming 30% less reduction of mechanical energy in comparison to control.

FUTURE PROSPECTS FOR ENZYMES AND THEIR APPLICATIONS

Over the last 20 years, enzymes have been produced from the microorganisms present in the nature. The developments in the field of protein engineering, genetic engineering, and bioinformatics along with the availability of sequence data have boosted the possibility to isolate the desired or interesting gene from nature or to introduce small changes to proteins.

Most of the commercially available enzyme products are too expensive to compete with conventional chemicals. Enzyme manufacturing companies are trying to launch efficient products as per the need of papermakers for solving different issues of pulp and paper industry but still there are some constraints. Table 6.7 summarized potential enzymes manufactured by different companies for pulp and paper industry. Enzymes are very susceptible to variations in environmental conditions. Usually enzymes work in a narrow operating range of pH, temperature, and storage time. Therefore, these operating conditions must be accurately controlled to maintain the maximum enzymatic activity. Enzymatic reactions are generally slower in nature and make the process slower. Therefore, it is difficult to retrofit some of the processes into existing operations of pulp and paper mills (Qian and Goodell 2005). Due to the specificity of enzymes, effectiveness of the enzymatic process depends principally on the characteristics of the raw materials/furnish type in comparison to chemical process.

Based on these possibilities, researchers have diverted their mind to tailor new and efficient enzymes attributing toward the specific need of the particular plant. These potential enzymes or microorganisms are evaluated for providing a better solution to the pulp and paper industry. The process of chip pile-based biopulping using fungal strains requires a demonstration for continuing operation at large scale. There is still a requirement of thermostable and alkali-stable enzyme for the paper industry. Nowadays, the focus is on developing the enzyme with improved enzyme properties (alkali and thermostable) for bleaching purpose and improved enzyme performance to make easy application of the enzyme in the existing mill harsh conditions. Cellulase-free xylanase is a prerequisite for the bleaching purpose to avoid any type of cellulose degradation. The identification or development of fast-growing white-rot fungal strains is required for the bleaching purpose in less time. There is immense need to explore the role of other enzymes (besides polygalacturanase and xylanase) for debarking individually or identification of proper composition when used in the mixture.

Therefore, the tailoring of enzyme is an option for the production of the specific type of enzyme. Although tailored enzymes are a rather complex structure consisting of multiple components, they perform singular and synergistic functions. The multiple-component formulations help in efficiency and effectiveness of the desired goal and provide a better fit for the specific attributes of each mill.

TABLE 6.7   Potential Enzymes Manufactured by Different Companies for Pulp and Paper Industry

Company, Headquarters and Website	Main Enzymes Production		Trade Name and Application
AB Enzymes, Germany, www.abenzymes.com	Amylases, xylanases, proteases, lipases, cellulases, pectinases, arabinases, and betaglucanases	Ecopulp	Bleaching
Advanced Enzyme Technologies, Ltd., India, www.advancedenzymes.com	Alphagalactosidase, betaglucosidase, betaxylosidase, betamannosidase, (mannanase), cellulase, hemicellulase, (xylanase), amylases, (alphaamylase, glucoamylase), betaglucanases, xylooligosidase, mannooligosidase, and pectinase	SEBrite 0102S/ 0102L	Blend of a biodegradable long-chain polymeric surfactant along with a lipase
		SEBrite Bleach	Thermostable alkaline xylanase preparation for bleach boosting of virgin pulp
		SEBrite DI Super/ DIL/DI Plus	Cellulase preparation for deinking of recycled paper
		SEBrite PR 810	Blend of a biodegradable long-chain polymeric surfactant and cellulase for deinking of recycled paper
		SEBrite SM 40L/SM 10L	Amylase preparation for starch modification
		SEBrite STR/STR L	Lipase preparation for stickies removal
ANC Enzymes, Singapore, www.ancenzymes.com	Amylases, xylanase, betaglucosidase, mannase, pectinases, proteases, papain, and laccase		
Anil Bioplus, India, www.anilbioplus.com	Amylases, cellulase, catalase, pectinase, proteases, phytase, galactosidases, and glucoamylases	Pulpase DI	Deinking
		Pulpase RF	Refining
		Pulpase BL	Bleaching

(Continued)

TABLE 6.7 (*Continued*) Potential Enzymes Manufactured by Different Companies for Pulp and Paper Industry

Company, Headquarters and Website	Main Enzymes Production	Trade Name and Application	
Anthem Cellutions, India, www.anthemcell.com	Proteases, pectinases, amyloglucosidases, amylases, glucoamylases, cellulases, xylanases, lipases, catalase, and mannase	ArrowPulp	Thermostable xylanases for pulp bleaching
		ArrowDeink	Deinking
		ArrowRefine	Refining
		ArrowFibre	Drainage control
		ArrowSize	Amylase for starch cooking
Aumgene, India, www.aumgene.com	Amylases, amyloglucosidases, cellulases, lipases, glucanases, hemicellulases, proteases, mannanase, xylanase, laccase, and pectinase	Product details not available	
Creative BioMart, USA, www.creativebiomart.net	Mannase, phytase, glucanase, cellulases, pectinases, proteases, amylases, and xylanases	Product details not available	
DuPont (Danisco Genencor), USA, www.dupont.com	Cellulases, proteases, xylanases, betaglucanases, phytases, amylases, hemicellulases, pectinases, and lipases	Product details not available	
Dyadic International Inc., USA, www.dyadic.com	Cellulases, proteases, amylases, xylanases, and endoglucanases	Product details not available	
Enzymatic Deinking Technologies, LLC, USA, www.edt-enzymes.com	Cellulases, endoglucanases, cellobiohydrolases, hemicellulases, mannanases, xylanases, pectinases, lipases, peroxidases, esterase, protease, amylases, betaglucosidases, and laccases	FIBREZYME® G4 FibreZyme G4P SDS FIBREZYME G4 POWDER	Neutral cellulase

(*Continued*)

TABLE 6.7 (*Continued*)   Potential Enzymes Manufactured by Different Companies for Pulp and Paper Industry

Company, Headquarters and Website	Main Enzymes Production	Trade Name and Application
Enzyme Development Corporation, USA, www.enzymedevelopment.com	Cellulases, lipases, esterases, pectinases, proteases, and xylanases	Product details not available
Enzyme Solutions, USA, www.enzymesolutions.com	Amylases, proteases, lipases, xylanases, betaglucanase, glucoamylase, cellulases, mannanase, pectinases, hemicellulases, and betaglycosidase	Product details not available
Enzyme Supplies, UK, www.enzymesupplies.com	Amylases, glucoamylase, cellulases, proteases, xylanases, lipases, and pectinases	Product details not available
EpygenLabs FZ LLC, UAE, www.epygen.com	Cellulases, mannase, alphaamylases, proteases, hemicellulases, phytase, xylanase, and lipases, pectinase	Papyrase® RF Epygen PAPYRASE RF	Enhances swelling and fibrillation of fiber
		Papyrase® DR-	For selective hydrolysis of colloidal organic debris (ultra fines) in virgin and recycled pulp
		Papyrase® IL -	Hemicellulase for fiber modification and deinking
		Papyrase®BBL	An engineered purified xylanase for bleaching
		Papyrase® LIL	Esterase and hemicellulase to control pitch and stickies
		Papyrase® TF-	For fiber development for typical pulp furnish used in tissue making

(*Continued*)

TABLE 6.7 (*Continued*) Potential Enzymes Manufactured by Different Companies for Pulp and Paper Industry

Company, Headquarters and Website	Main Enzymes Production	Trade Name and Application
		Papyrase® WL- Hydrolase for efficient hydrolysis of residual cellulose and hemicellulose in paper mill waste stream
		Papyrase® DN For fiber modification of newsprint furnish
		Papyrase® OCR For fiber modification of OCC furnish in containerboard making
		Papyrase® HTA High temperature stable alpha amylase for liquefaction of gelatinized starch
Leveking Enzymes, China, www.levekingenzymes.com	Lipases, xylanases, alphaamylases, mannase, and protease	LPK-CD05 Deinking
		LPK-CR01 Refining
		LPK-CS01S Starch-based surface sizing
Metgen Oy, Finland, www.metgen.fi	Laccase	MetZyme® LIGNO™ v1 Mechanical pulping
		MetZyme BRILA™ Deinking
		MetZyme POVON™ Bleaching, deinking
		MetZyme Sekalo™ Starch conversion
		MetZyme FORICO™ Removal of residual peroxide

(*Continued*)

TABLE 6.7 (*Continued*) Potential Enzymes Manufactured by Different Companies for Pulp and Paper Industry

Company, Headquarters and Website	Main Enzymes Production	Trade Name and Application	
Novozymes, Denmark, www.novozymes.com	Cellulases, endoglucanases, CBH, hemicellulases, mannanases, xylanases, pectinases, lipases, peroxidases, esterase, protease, amylases, betaglucosidases, and laccases	Product details not available	
Roal Oy, Finland, www.roal.fi	Amylases, xylanases, proteases, lipases, cellulases, pectinases, arabinases, and betaglucanases	Product details not available	
Rossari Biotech, India, www.rossari.com	Xylanases, cellulases, pectinases, amylases, glucanases, and proteases	Rossapase powder	Deinking
		Rossapase Liq New	Deinking
		Rossapase BBP	Bleaching
		Rossapase AFR	Refining
SinoBios, China, www.sinobios.com	Amylases, proteases, pectinase, xylanase, cellulases, alphagalactosidase, mannase, betaglucanases, glucose, oxidase, and lipases	Product details not available	
Specialty Enzymes and Biotechnologies Inc., USA, www.specialtyenzymes.com	Alphagalactosidase, betaglucosidase, betaxylosidase, betamannosidase (mannanase), cellulase, hemicellulase (xylanase), amylases (alphaamylase, glucoamylase), betaglucanases, xylooligosidase, mannooligosidase, and pectinase	SEBrite Bleach	Bleaching
		SEBrite DI	Deinking

(*Continued*)

TABLE 6.7 (*Continued*) Potential Enzymes Manufactured by Different Companies for Pulp and Paper Industry

Company, Headquarters and Website	Main Enzymes Production		Trade Name and Application
Sukahan (Wei-fang) Biotechnology Co. Ltd., China, www.sukahan.com	Xylanase, betaglucanase, proteases, cellulases, pectinases, amylases, lipases, phytase, and glucoamylases	AU-PE89	Pulp bleaching-xylanase
		AU-PE93	Sticky removal
		AU-PEA550S	Starch conversion agent
		AU-PE91	Deinking
		SUKAZYM-TPE90	Tissue specific
		SUKAZYM-OCE90	Recycled pulp and paperboard
Tex Biosciences, India, www.texbiosciences.com	Proteases, lipases, xylanases, pectinase, cellulase, and amylases	Texzyme I	Deinking
		Texzyme J	Bleaching
		Texzyme M	Refining
		Texzyme SM	Starch modification
		Texzyme AS	Stickies control

Source: Company's website date of access September 9, 2015.

Lack of technical knowledge of enzymatic processes, conventional belief among manufacturers, suppliers, and governmental bureaucracy during approval and adaptation of new ideas in many countries tend to delay broader implementation. Therefore, steps should be taken to overcome these hurdles and speed up the harvesting of several benefits offered by the enzymatic approach through education on biotechnological and enzymatic processes. Awareness on biotechnology should be developed through regular arrangement of seminars, workshops, and best-practice schemes with the help of microbiologists, biochemists, and process engineers in association with different paper industries so that openness will increase toward the use of enzymes in industry to create efficient and well-designed networks that are progressive and result oriented.

CONCLUSIONS

The paper industry has attempted different biotechnological routes in various processes. It has been demonstrated that microbial enzymes may augment various pulp and paper processes with less effort and better results including savings in electricity, water, and harsh chemicals, thereby reducing environmental impacts. Some applications have been commercialized. However, many applications are still in the pilot or lab scale, some of these having potential for commercialization in future. A variety of reasons are limiting the commercialization in some cases. Presently, the enzymatic application in pulp and paper industry is a promising and reliable green approach.

ACKNOWLEDGMENTS

The authors are thankful to Mr. Piyush Verma, Ms. Manasi Purwar, and Dr. Rashmi Singh for their support and help.

REFERENCES

Abdelaziz, E. A., Saidur, R., and Mekhilef, S. 2011. A review on energy saving strategies in industrial sector. *Renewable and Sustainable Energy Reviews* 15:150–168.

Agbogbo, F. K., Coward-Kelly, G., Torry-Smith, M., and Wenger, K. S. 2006. Fermentation of glucose/xylose mixtures using *Pichia stipitis. Process Biochemistry* 41:2333–2336.

Ahmad, M., Taylor, C. R., Pink, D. et al. 2010. Development of novel assays for lignin degradation: Comparative analysis of bacterial and fungal lignin degraders. *Molecular Biosystems* 6(5):815–821.

Akhtar, M. 1994. Biochemical pulping of aspen wood chips with three strains of *Ceriporiopsis subvermispora. Holzforschung* 48:199–202.

Al-Dajani, W. W., and Tschirner, U. W. 2008. Pre-extraction of hemicelluloses and subsequent kraft pulping. Part I: Alkaline extraction. *TAPPI Journal* 7:3–8.

Al-Dajani, W. W., Tschirner, U. W., and Jensen, T. 2009. Pre-extraction of hemicelluloses and subsequent kraft pulping. Part II: Acid-and autohydrolysis. *TAPPI Journal* 8:30–37.

Al-Zuhair, S. 2007. The effect of crystallinity of cellulose on the rate of reducing sugars production by heterogeneous enzymatic hydrolysis. *Bioresource Technology* 99:4078–4085.

Ambjörnsson, H. A., Östberg, L., Schenzel, K., Larsson, P. T., and Germgård, U. 2014. Enzyme pretreatment of dissolving pulp as a way to improve the following dissolution in NaOH/ZnO. *Holzforschung* 68:385–391.

Amr, A., Hanafy, E. J., Hassan, E., Abd-Elsalam, H. E., and Elsayed, E. 2009. Molecular characterization of two native Egyptian ligninolytic bacterial strains. *Journal of Applied Sciences Research* 4:1291–1296.

Anfinsen, C. B. 1973. Principles that govern the folding of protein chains. *Science* 141(4096): 223–230.

Antunes, E., Garcia, F. A. P., Ferreira, P., Blanco, A., Negro, C., and Rasteiro, M. G. 2008. Use of new branched cationic polyacrylamides to improve retention and drainage in papermaking. *Industrial and Engineering Chemistry Research* 47:9370–9375.

Arias, M. E., Arenas, M., Rodríguez, J., Soliveri, J., Ball, A. S., and Hernández, M. 2003. Kraft pulp biobleaching and mediated oxidation of a nonphenolic substrate by laccase from *Streptomyces cyaneus* CECT 3335. *Applied and Environmental Microbiology* 69:1953–1958.

Arora, D. S., and Sharma, R. K. 2010. Ligninolytic fungal laccases and their biotechnological applications. *Applied Biochemistry and Biotechnology* 160:1760–1788.

Asgher, M., Bhatti, H. N., Ashraf, M., and Legge, R. L. 2008. Recent developments in biodegradation of industrial pollutants by white rot fungi and their enzyme system. *Biodegradation* 19:771–783.

Bairoch, A. 2000. The enzyme database in 2000. *Nucleic Acids Research* 28:304–305.

Bajpai, P. 1999. Application of enzymes in the pulp and paper industry. *Biotechnology Progress* 15(2):147–157.

Bajpai, P. 2004. Biological bleaching of chemical pulps. *Critical Reviews in Biotechnology* 24(1):1–58.

Bajpai, P. 2012. *Biotechnology for Pulp and Paper Processing*. Springer Science & Business Media, New York.

Bajpai, P. 2013. Pulp and paper bioprocessing. *Encyclopedia of Industrial Biotechnology*. John Wiley and Sons, Inc., 1–17. DOI: 10.1002/9780470054581. eib297.pub2.

Bajpai, P., and Bajpai, P. K. 1994. Biological colour removal of pulp and paper mill wastewaters. *Journal of Biotechnology* 33:211–220.

Bajpai, P., and Bajpai, P. K. 1996. Realities and trends in enzymatic prebleaching. In *Advances in Biochemical Engineering/Biotechnology*, ed. T. Scheper, Springer-Verlag, Berlin, vol. 56, pp. 1–32.

Bajpai, P., and Bajpai, P. K. 1998. Deinking with enzymes: A review. *TAPPI Journal* 81:111–117.

Bajpai, P., Mishra, S. P., Mishra, O. P., Kumar, S., and Bajpai, P. K. 2006. Use of enzymes for reduction in refining energy: Laboratory studies. *TAPPI Journal* 5:25–32.

Battan, B., Kuhar, S., Sharma, J., and Kuhad, R. C. 2007. Biodiversity of hemi-celluloses degrading microorganisms and their enzymes. In *Lignocellulose Biotechnology Future Prospects*, eds. R. C. Kuhad and A. Singh, 121–145. IK International Publishing House Pvt. Ltd., New Delhi.

Beg, Q., Kapoor, M., Mahajan, L., and Hoondal, G. S. 2001. Microbial xylanases and their industrial applications: A review. *Applied Microbiology and Biotechnology* 56(3–4):326–338.

Belbekhouche, S., Bras, J., Siqueira, G. et al. 2011. Water sorption behavior and gas barrier properties of cellulose whiskers and microfibrils films. *Carbohydrate Polymers* 83:1740–1748.

Bhardwaj, N. K., Bajpai, P., and Bajpai, P. K. 1995. Use of enzymes to improve drainability of secondary fibres. *Appita Journal* 48:378–380.

Bhardwaj, N. K., Bajpai, P., and Bajpai, P. K. 1996. Use of enzymes in modifica-tion of fibers for improved beatability. *Journal of Biotechnology* 51:21–26.

Bhardwaj, N. K., Bajpai, P., and Bajpai, P. K. 1997. Enhancement of strength and drainage of secondary fibres. *Appita Journal* 50:230–232.

Bhat, G. R., Heitmann, J. A., and Joyce, T. W. 1991. Novel techniques for enhanc-ing the strength of secondary fibre. *TAPPI Journal* 74:151–157.

Bhat, M. K. 2000. Cellulases and related enzymes in biotechnology. *Biotechnology Advances* 18(5):355–383.

Bhattacharya, P. K., Jayan, R., and Bhattacharjee, C. 2005. A combined biologi-cal and membrane-based treatment of prehydrolysis liquor from pulp mill. *Separation and Purification Technology* 45:119–130.

Blanchette, R. A., Farrell, R. L., and Behrendt, C. J. 1996. U.S. Patent No. 5,518,921. Washington, DC: U.S. Patent and Trademark Office.

Blanco, A., Negro, C., Diaz, L., Saarimaa, V., Sundberg, A., and Holmbom, B. 2009. Influence of thermostable lipase treatment of thermomechanical pulp (TMP) on extractives and paper properties. *Appita Journal* 62:113–117.

Bobu, E., and Ciolacu, F. 2007. Environmental aspects of enzyme deinking. *Professional Papermaking* 4:6–13.

Bolaski, W. 1962. Enzymatic conversion of cellulosic fibers. U.S. Patent No. 3,041,246. Washington, DC: U.S. Patent and Trademark Office.

Bugg, T. D., Ahmad, M., Hardiman, E. M., and Singh, R. 2011. The emerging role for bacteria in lignin degradation and bio-product formation. *Current Opinion in Biotechnology* 22:394–400.

Call, H. P. 1994. Process for modifying, breaking down or bleaching lignin, mate-rials containing or like substances. World Patent Application WO 94/29510.

Caram, F. C., Sarkar, J. M., Didwania, H. P., Espinoza, E., and Benavides, J. C. 1996. Papermill evaluation of a cellulolytic enzyme and polymers for improv-ing the properties of waste paper pulp. *Tappi Proceedings of Papermakers Conference*, Philadelphia, USA, 481.

Chakrabarti, S. K., Gupta, S., Purwar, M., Bhist Shekhar, C., and Vardhan, R. 2012. Combined chemical–biological treatment of pulp and paper mill effluent. *IPPTA Journal* 22:160–163.

Chandel, A. K., Silva, S., and Singh, O. V. 2011. Detoxification of lignocellulosic hydrolysates for improved bioethanol production. In *Biofuel Production—Recent Developments and Prospects*, ed. Marco Aurélio dos Santos Bernardes, INTECH Open Access Publisher, 225–246. DOI: 10.5772/16454.

Chandra, R., and Abhishek, A. 2011. Bacterial decolorization of black liquor in axenic and mixed condition and characterization of metabolites. *Biodegradation* 22:603–611.

Charreau, H., L., Foresti, M., and Vázquez, A. 2013. Nanocellulose patents trends: A comprehensive review on patents on cellulose nanocrystals, microfibrillated and bacterial cellulose. *Recent Patents on Nanotechnology* 7:56–80.

Chen, S., Lin, Y., Zhang, Y., Wang, X. H., and Yang, J. L. 2001. Enzyamatic pitch control at Nanping paper mill. *TAPPI Journal* 84:44–47.

Chen, T., Wang, Z., Gao, Y., Breuil, C., and Hatton, J. V. 1994. Wood extractives and pitch problems: Analysis and partial removal by biological treatment. *Appita Journal* 47:463–466.

Chen, Y., Wan, J., Zhang, X., Ma, Y., and Wang, Y. 2012. Effect of beating on recycled properties of unbleached eucalyptus cellulose fiber. *Carbohydrate Polymers* 87:730–736.

Clark, T. A., Allison, R. W., and Kibblewhite, R. P. 1997. Effects of enzymatic modification on radiata pine kraft fibre wall chemistry and physical properties. *Appita Journal* 50:329–335.

Cooper, III., and Elwood, W. 1998. Process for treating hardwood pulp with an enzyme mixture to reduce vessel element picking. U.S. Patent No. 5,725,732.

Covarrubias, R. M., and Eng, G. H. 2005. Optimyze: Enzymatic stickers control developments. *91st Annual Meeting Pulp and Paper Technical Association of Canada*, Montreal, QC, Canada, Feb. 8–10, 2005, Book A, A107–A116.

Dashtban, M., Schraft, H., Syed, T. A., and Qin, W. 2010. Fungal biodegradation and enzymatic modification of lignin. *International Journal of Biochemistry and Molecular Biology* 1:36.

Davies, G., and Henrissat, B. 1995. Structures and mechanisms of glycosyl hydrolases. *Structure* 3:853–859.

Dedhia, B. S., Vetal, M. D., Rathod, V. K., and Levente, C. 2014. Xylanase and laccase aided bio-bleaching of wheat straw pulp. *The Canadian Journal of Chemical Engineering* 92:131–138.

Demuner, B. J., Pereira Junior, N., and Antunes, A. 2011. Technology prospecting on enzymes for the pulp and paper industry. *Journal of Technology Management and Innovation* 6:148–158.

Dey, P. M., and Brinson, K. 1984. Plant cell-walls. *Advances in Carbohydrate Chemistry and Biochemistry* 42:265–382.

Diehm, R. A. 1942. Process of manufacturing paper. U.S. Patent 2280307. Washington, DC: U.S. Patent and Trademark Office.

Dienes, D., Egyházi, A., and Réczey, K. 2004. Treatment of recycled fibre with *Trichoderma cellulases*. *Industrial Crops and Products* 20:11–21.

Doshi, M. R. 1997. In *Paper Recycling Challenge*, eds. M. R. Doshi and J. M. Dyer, Doshi and Associates, Appleton, pp. 3–6.

Doueck, M. 1997. In *Paper Recycling Challenge*, eds. M. R. Doshi and J. M. Dyer, Doshi and Associates, Appleton, pp. 15–21.

Dube, E., Shareck, F., Hurtubise, Y., Beauregard, M., and Daneault, C. 2009. Enzyme-based approaches for pitch control in thermomechanical pulping of softwood and pitch removal in process water. *EXFOR and Annual Meeting*, 69–74. Montreal, QC, Canada.

Dufresne, A., Cavaillé, J. Y., and Vignon, M. R. 1997. Mechanical behavior of sheets prepared from sugar beet cellulose microfibrills. *Journal of Applied Polymer Science* 64:1185–1194.

Dykstra, G. M. 1990. In *Recycling Paper—From Fiber to Finished Product*, Vol. 2. (M. J. Coleman, Ed), TAPPI Press, Atlanta, pp. 117–124.

Elegir, G., Caldirola, C., and Canetti, M. 2000. Cellulase and amylase assisted enzymatic deinking of mixed office waste 1139–43. *Tappi Pulping/Process and Product Quality Conference*, Boston, MA, United States.

Elegir, G., Sykes, M., and Jeffries, T. W. 1995. Differential and synergistic action of *Streptomyces* endoxylanases in prebleaching of kraft pulps. *Enzyme and Microbial Technology* 17(10):954–959.

Engström, A. C., Ek, M., and Henriksson, G. 2006. Improved accessibility and reactivity of dissolving pulp for the viscose process: Pretreatment with monocomponent endoglucanase. *Biomacromolecules* 7:2027–2031.

Eom, T. J., Kim, K. J., and Yoon, K. D. 2007. Deinking of electrostatic wastepaper with cellulolytic enzymes and surfactant in neutral pH. *Polpu, Chongi Gisul* 39:12–20.

Eriksson, K. E., Ander, P., Henningsson, B., Nilsson, T., and Goodell, B. 1976. U.S. Patent no. 3,962,033. Washington, DC: U.S. Patent and Trademark Office.

Eriksson, K. E., and Cavaco-Paulo, A. 1998. Enzyme applications in fiber processing. In *ACS Symposium Enzyme Applications in Fiber Processing*, San Francisco, California, 1997. American Chemical Society, Distributed by Oxford University Press.

Eriksson, K. E. L., Blanchette, R. A., and Ander, P. 1990. *Microbial and Enzymatic Degradation of Wood and Wood Components*, 89–177. Springer-Verlag, Berlin.

Eriksson, L. A., Heitmann, J. A., and Venditti, R. A. 1998. Freeness improvement of recycled fibres using enzymes with refining. In *Enzyme Applications in Fibre Processing*, 41–54. ACS Symposium Series 687.

Feijoo, G., Moreira, M. T., Alvarez, P., Lú-Chau, T. A., and Lema, J. M. 2008. Evaluation of the enzyme manganese peroxidase in an industrial sequence for the lignin oxidation and bleaching of eucalyptus kraft pulp. *Journal of Applied Polymer Science*, 109:1319–1327.

Feng, G. P., Tian, Z. S., and Li, C. 2012. The application of α-amylase in surface sizing of light weight paper. *China Pulp and Paper Industry* 4:76–77.

Ferraz, A., Guerra, A., Mendonça, R., Masarin, F., Vicentim, M. P., Aguiar, A., and Pavan, P. C. 2008. Technological advances and mechanistic basis for fungal biopulping. *Enzyme and Microbial Technology* 43(2):178–185.

Fonseca, M. I., Fariña, J. I., Castrillo, M. L., Rodríguez, M. D., Nuñez, C. E., Villalba, L. L., and Zapata, P. D. 2014. Biopulping of wood chips with *Phlebia brevispora* BAFC 633 reduces lignin content and improves pulp quality. *International Biodeterioration and Biodegradation* 90:29–35.

Friermuth, B., Garrett, M., and Jokinen, O. 1994. The use of enzymes in the production of release papers. *Paper Technology* 35:21–23.

Fujita, Y., Awaji, H., Matsukura, M., and Hata, K. 1991. Enzymic pitch control in papermaking process. *Kami Pa Gikyoshi* 45:905–921.

Fukuzumi, H., Saito, T., and Isogai, A. 2009. Properties of TEMPO-oxidized cellulose nanofiber film. In *Abstracts of Papers of the American Chemical Society* (Vol. 237). 1155 16th St, NW, Washington, DC 20036 USA: American Chemical Society.

Garmaroody, E. R., Resalati, H., Fardim, P., Hosseini, S. Z., Rahnama, K., Saraeeyan, A. R., and Mirshokraee, S. A. 2011. The effects of fungi pretreatment of poplar chips on the kraft fiber properties. *Bioresource technology* 102(5):4165–4170.

Gehmayr, V., Schild, G., and Sixta, H. 2011. A precise study on the feasibility of enzyme treatments of a kraft pulp viscose application. *Cellulose* 18:479–491.

Gehmayr, V., and Sixta, H. 2011. Dissolving pulps from enzyme treated kraft pulps for viscose application. *Lenzinger Berichte* 89:52–160.

Geng, X., Li, K., Kataeva, I. A., Li, X. L., and Ljungdahl, L. G. 2003. Effects of two cellobiohydrolases, CbhA and CelK, from *Clostridium thermocellum* on deinking of recycled mixed office paper. *Progress in Paper Recycling* 12:6–10.

Ghana, M. F., Teixeira, A. J., and Mota, M. 1993. Cellulose morphology and enzymatic reactivity: A modified solute exclusion technique. *Biotechnology and Bioengineering* 43:581–587.

Gianfreda, L., Iamarino, G., Scelza, R., and Rao, M. A. 2006. Oxidative catalysts for the transformation of phenolic pollutants: A brief review. *Biocatalysis and Biotransformation* 24:177–187.

Gil, N., Gil C., Amaral, M. E., Costa, A. P., and Duarte, A. P. 2009. Use of enzymes to improve the refining of a bleached *Eucalyptus globulus* kraft pulp. *Biochemical Engineering Journal* 46:89–95.

Gill, R. 2008. Advances in use of fibre modification enzymes in paper making. *Conference Aticelca XXXIX Congresso Annuale*, Fabriano, Italy, May 29–30, 2008.

Gong, M. R., Bi, S. L., Xu, S. B., and Wang, Y. J. 2003. Effect of cellulase on the drainage of bleached wheat pulp. *Journal of Cellulose Science and Technology* 11(3):6–11.

Gregersen, O. W., Hansen, Å., and Helle, T. 1999. The influence of shives on newsprint strength. In *Proceedings of 1999 TAPPI International Paper Physics Conference*, San Diego, California, September 26–30, pp. 211–216.

Gübitz, G. M., Lischnig, T., Stebbing, D., and Saddler, J. N. 1997. Enzymatic removal of hemicellulose from dissolving pulps. *Biotechnology Letters* 19(5):491–495.

Gübitz, G. M., Mansfield, S. D., Bohm, D., and Saddler, J. N. 1998. Effect of endoglucanases and hemicellulases in magnetic and flotation deinking of xerographic and laser-printed papers. *Journal of Biotechnology* 65(2):209–215.

Gullichsen, J. 2000. *Chemical Pulping in Papermaking Science and Technology*, Book 6B, Fappet Oy, Helsinki 7.

Gupta, R., Gigras, P., Mohapatra, H., Goswami, V. K., and Chauhan, B. 2003. Microbial α-amylases: A biotechnological perspective. *Process Biochemistry* 38:1599–1616.

Gutiérrez, A., del Rio, J.C., David, I., Jorge, R., Javier, R., Mariela, S., Susana, C., María, J.M., and Ángel, T. 2006. Enzymatic removal of free and conjugated sterols forming pitch deposits in environmentally sound bleaching of eucalypt paper pulp. *Environmental Science and Technology* 40(10):3416–3422.

Hall, J. D., and Nguyen, D. T. 2000. Stickies control using nonionic polymers in systems with low opertating temperatures. *PaperAge* 116: 34–36.

Heise, O. H., Coa, B., Dehm, J. et al. 1998. *TAPPI 1998 Process and Product Quality Conference Proceedings*, TAPPI Press, Atlanta, pp. 183–207.

Helmerius, J., von Walter, J. V., Rova, U., Berglund, K. A., and Hodge, D. B. 2010. Impact of hemicellulose pre-extraction for bioconversion on birch kraft pulp properties. *Bioresource Technology* 101:5996–6005.

Henriksson, G., Christiernin, M., and Agnemo, R. 2005. Monocomponent endoglucanase treatment increases the reactivity of softwood sulphite dissolving pulp. *Journal of Industrial Microbiology and Biotechnology* 32(5):211–214.

Henriksson, M., Henriksson, G., Berglund, L.A., and Lindström, T. 2007. An environmentally friendly method for enzyme-assisted preparation of microfibrillated cellulose (MFC) nanofibers. *European Polymer Journal* 43(8):3434–3441.

Hirai, H., Kondo, R., and Sakai, K. 1994. Screening of lignin-degrading fungi and their ligninolytic enzyme activities during biological bleaching of kraft pulp. *Mokuzai Gakkaishi—Journal of the Japan Wood Research Society* 40:980–986.

Hirai, H., Kondo, R., and Sakai, K. 1995. Effect of metal ions on biological bleaching of kraft pulp with *Phanerochaete sordida* YK-624. *Mokuzai Gakkaishi—Journal of the Japan Wood Research Society* 41:69–75.

Hoekstra, P. M., and May, O. W. 1990. Developments in the control of stickles. In *Recycling Paper—From Fiber to Finished Product*, Vol. 2. (M. J. Coleman, Ed), TAPPI Press, Atlanta, pp. 446–450.

Hofrichter, M., Ziegenhagen, D., Vares, T. et al. 1998. Oxidative decomposition of malonic acid as basis for the action of manganese peroxidase in the absence of hydrogen peroxide. *FEBS Letters* 434:362–366.

Hong, G. B., Ma, C. M., Chen, H. W., Chuang, K. J., Chang, C. T., and Su, T. L. 2011. Energy flow analysis in pulp and paper industry. *Energy* 36:3063–3068.

http://www.bccresearch.com/pressroom/bio/global-market-industrial-enzymes-reach-nearly-$7.1-billion-2018. Accessed on Oct. 12, 2015.

http://www.hexaresearch.com/research-report/enzymes-industry/Accessed on Oct. 12, 2015.

Hubbe, M. A. 2000. Fines management for increased paper machine productivity. *Proceedings of Scientific and Technical Advances in Wet End Chemistry*, Pira International, Leatherhead, UK.

Hunt, C., Kenealy, W., Horn, E., and Houtman, C. 2004. A biopulping mechanism: Creation of acid groups on fiber. *Holzforschung* 58:434–439.

Ibarra, D., Köpcke, V., and Ek, M. 2010. Behaviour of different monocomponent endoglucanases on the accessibility and reactivity of dissolving-grade pulps for viscose process. *Enzyme and Microbial Technology* 47:355–362.

Jackson, S., Heitmann, J. A., and Joyce, T. W. 1993. Enzymatic modifications of secondary fiber. *TAPPI Journal* 76:147–154.

Jaeger, K. E., and Eggert, T. 2004. Enantioselective biocatalysis optimized by directed evolution. *Current Opinion in Biotechnology* 15:305–313.

Jahren, S. J., Rintala, J. A., and Ødegaard, H. 2002. Aerobic moving bed biofilm reactor treating thermomechanical pulping whitewater under thermophilic conditions. *Water Research* 36:1067–1075.

Janardhnan, S., and Sain, M. M. 2006. Isolation of cellulose microfibrils—An enzymatic approach. *Bioresources* 1:176–188.

Jeffries, T. W., Klungness, J. H., Sykes, M. S., and Rutledge-Cropsey, K. 1993. *TAPPI Recycling Symposium* 183–188. TAPPI Press, Atlanta.

Jeffries, T. W., Klungness, J. H., Sykes, M. S., and Rutledge-Cropsey, K. 1994. Comparison of enzyme-enhanced with conventional deinking of xerographic and laser-printed paper. *TAPPI Journal* 77:173–179.

Jeoh, T. 1998. Steam explosion pretreatment of cotton gin waste for fuel ethanol production. Doctoral dissertation. Virginia Polytechnic Institute and State University.

Juan, C., C., and Victor, O. 2011. Green technology: Last developments in enzymes for paper recycling. *TAPPI Papercon Conference*, Northern Kentucky Convention Center in Covington, KY, May 1–4, 1630–1639.

Karigar, C. S., and Rao, S. S. 2011. Role of microbial enzymes in the bioremediation of pollutants: A review. *Enzyme Research*. DOI: 10.4061/2011/805187.

Karmakar, M., and Ray, R. R. 2011. Current trends in research and application of microbial cellulases. *Research Journal of Microbiology* 6(1):41–53.

Kashyap, D. R., Vohra, P. K., Chopra, S., and Tewari, R. 2001. Applications of pectinases in the commercial sector: A review. *Bioresource Technology* 77:215–227.

Katagiri, N., Tsutsumi, Y., and Nishida, T. 1995. Correlation of brightening with cumulative enzyme activity related to lignin biodegradation during biobleaching of kraft pulp by white rot fungi in the solid-state fermentation system. *Applied and Environmental Microbiology* 61(2):617–622.

Katō, K. 1981. Ultrastructure of the plant cell wall: Biochemical viewpoint. In *Plant Carbohydrates I*, eds. w. Tanner and F. A. Loevus, Springer, Berlin, pp. 29–46.

Kim, T. J., Ow, S. S. K., and Eom, T. J. 1991. Enzymatic deinking method of wastepaper. In *Proceedings of TAPPI Pulping Conference*, Atlanta, 1023–1031.

Kirk, T. K., and Jeffries, T. W. 1996. Roles for microbial enzymes in pulp and paper processing. In *ACS Symposium Series*, 655:2–14. Washington, DC: American Chemical Society.

Knudsen, O., Young, J. D., and Yang, J. L. 1998. Mill experience of a new technology at Stora Dalum deinking plant. *PTS-CTP Deinking Symposium 1998*, Munich, Germany, May 5–7, 1998, pp. 17.

Kondo, R., Harazono, K., and Sakai, K. 1994. Bleaching of hardwood kraft pulp with manganese peroxidase secreted from *Phanerochaete sordida* YK-624. *Applied and Environmental Microbiology* 60:4359–4363.

Köpcke, V. 2010. Conversion of wood and non-wood paper-grade pulps to dissolving-grade pulps. Doctoral dissertation, KTH.

Köpcke, V., Ibarra, D., and Ek, M. 2008. Increasing accessibility and reactivity of paper grade pulp by enzymatic treatment for use as dissolving pulp. *Nordic Pulp and Paper Research Journal* 23:363–368.

Kumar, R., and Kumar, A. 2006. US PATENT: US7022511 B2.

Kumar, R., and Kumar, A. 2010. US PATENT: US7736879 B2.

Kvarnlöf, N., Germgård, U., Jönsson, L., and Söderlund, C. A. 2007. Optimization of the enzymatic activation of a dissolving pulp before viscose manufacture. *TAPPI Journal* 6:14–19.

Lai, Y. 1990. Chemistry degradation. In *Wood and Cellulosic Chemistry*, ed. N.-S. David. On and Nobuo Shiraishi, Marcel Dekker, New York, 10:455–523.

Leatham, G. F., Myers, G. C., and Wegner, T. H. 1990a. Biomechanical pulping of aspen chips: Energy savings resulting from different fungal treatments. *Tappi Journal* 73(5):197–200.

Leatham, G. F., Myers, G. C., Wegner, T. H., and Blanchette, R. A. 1990b. Biomechanical pulping of aspen chips: Paper strength and optical properties resulting from different fungal treatments. *Tappi Journal* 73(3):249–255.

Lecourt, M., Meyer, V., Sigoillot, J. C., and Petit-Conil, M. 2010. Energy reduction of refining by cellulases. *Holzforschung* 64:441–446.

Lee, S., Klm, K. H., Ryu, J. D., and Taguchi, H. 1983. Structural properties of cellulose and cellulase reaction mechanism. *Biotechnology and Bioengineering* 25:33–52.

Liu, J., and Hu, H. 2012. The role of cellulose binding domains in the adsorption of cellulases onto fibers and its effect on the enzymatic beating of bleached kraft pulp. *BioResources* 7:878–892.

Lorenzo, M. L., Nierstrasz, V. A., and Warmoeskerken, M. M. C. G. 2009. Endoxylanase action towards the improvement of recycled fibre properties. *Cellulose*16:103–115.

Lynd, L. R., Weimer, P. J., van Zyl, W. H., and Pretorius, I. S. 2002. Microbial cellulose utilization: Fundamentals and biotechnology. *Microbiology and Molecular Biology Reviews* 66:506–577.

Machii, Y., Hirai, H., and Nishida, T. 2004. Lignin peroxidase is involved in the biobleaching of manganese-less oxygen-delignified hardwood kraft pulp by white-rot fungi in the solid-fermentation system. *FEMS Microbiology Letters* 233:283–287.

Madhavi, V., and Lele, S. S. 2009. Laccase properties and applications. *BioResources* 4:1694–1717.

Magee, K. L., and Taylor, J. L. 1994. *TAPPI 1994 Papermakers Conference Proceedings*, 621–628. Book 2. TAPPI Press, Atlanta.

Magnin, L., Lantto, R., and Delpech, P. 2002. Use of enzymes for deinking of wood-free and wood-containing recovered papers. *Progress in Paper Recycling* 11(4):13–20.

Malmqvist, A., Welander, T., Berggren, B., Asselin, C., and Marquis, J. 2003. Removal of chronic toxicity and organic matter from a paper mill effluent in an MBBR process. In *TAPPI International Environmental Conference,* Portland, OR, May 4–7.

Mansfield, S. D., Jong, E. D., Stephens, R. S., and Saddler, J. N. 1997. Physical characterization of enzymatically modified kraft pulp fibers. *Journal of Biotechnology* 57:205–216.

Mansfield, S. D., Wong, K. K. Y., and Richardson, J. D. 1999. Improvements in mechanical pulp processing with proteinase treatments. *Appita Journal* 52:436–440.

Mao, H., Genco, J. M., Yoon, S. H., Van Heiningen, A., and Pendse, H. 2008. Technical economic evaluation of a hardwood biorefinery using the "near-neutral" hemicellulose pre-extraction process. *Journal of Biobased Materials and Bioenergy* 2:177–185.

Marinova, M., Mateos-Espejel, E., and Paris, J. 2010. From kraft mills to forest biorefinery: An energy and water perspective. II. Case study. *Cellulose Chemistry and Technology* 44(1–3):21–26.

Marques, S., Pala, H., Alves, L., Amaral-Collaco, F. M., and Gama, G. F. M. 2003. Characterization and application of glycanases secreted by *Aspergillus terreus CCMI 498* and *Trichoderma viride CCMI 84* for enzymatic deinking of mixed office wastepaper. *Journal of Biotechnology* 100:209–219.

Masai, E., Katayama, Y., and Fukuda, M. 2007. Genetic and biochemical investigations on bacterial catabolic pathways for lignin-derived aromatic compounds. *Bioscience, Biotechnology, and Biochemistry* 71:1–15.

Masarin, F., Pavan, P. C., Vicentim, M. P., Souza-Cruz, P. B., Loguercio-Leite, C., and Ferraz, A. 2009. Laboratory and mill scale evaluation of biopulping of *Eucalyptus grandis* Hill ex Maiden with *Phanerochaete chrysosporium* RP-78 under non-aseptic conditions. *Holzforschung* 63(3):259–263.

Meza, J. C., Sigoillot, J. C., Lomascolo, A., Navarro, D., and Auria, R. 2006. New process for fungal delignification of sugarcane bagasse and simultaneous production of laccase in a vapor phase bioreactor. *Journal of Agricultural and Food Chemistry* 54:3852–3858.

Miao, Q., Chen, L., Huang, L., Tian, C., Zheng, L., and Ni, Y. 2014. A process for enhancing the accessibility and reactivity of hardwood kraft-based dissolving pulp for viscose rayon production by cellulase treatment. *Bioresource Technology* 154:109–113.

Minussi, R. C., Pastore, M. G., and Duran, N. 2007. Laccase induction in fungi and laccase/N-OH mediator systems applied in paper mill effluent. *Bioresource Technology* 98:158–164.

Mishra, M., and Thakur, I. S. 2010. Isolation and characterization of alkalotolerant bacteria and optimization of process parameters for decolorization and detoxification of pulp and paper mill effluent by Taguchi approach. *Biodegradation* 21:967–978.

Mohammed, S. H. 2010. Enzymatic deinking: A bright solution with a bright future. *IPPTA Journal* 22:137–138.

Moon, R. J., Martini, A., Nairn, J., Simonsen, J., and Youngblood, J. 2011. Cellulose nanomaterials review: Structure, properties and nanocomposites. *Chemical Society Reviews* 40:3941–3994.

Moreira, M. T., Sierra-Alvarez, R., Lema, J. M., Feijoo, G., and Field, J. A. 2001. Oxidation of lignin in eucalyptus kraft pulp by manganese peroxidase from *Bjerkandera* sp. strain BOS55. *Bioresource Technology* 78:71–79.

Morkbak, A. L., Degn, P., and Zimmermann, W. 1999. Deinking of soy bean oil based ink-printed paper with lipases and neutral surfactant. *Journal of Biotechnology* 67:29–36.

Morkbak, A. L., and Zimmermann, W. 1998. Deinking of mixed office paper, old newspaper and vegetable oil-based ink printed paper using cellulases, xylanases and lipases. *Progress in Paper Recycling* 7:14–21.

Motta, F. L., Andrade, C. C. P., and Santana, M. H. A. 2013. A review of xylanase production by the fermentation of xylan: Classification, characterization and applications. In *Biochemistry, Genetics and Molecular Biology "Sustainable Degradation of Lignocellulosic Biomass—Techniques, Applications and Commercialization,"* eds. A. K. Chandel and S. S. Silva, 251–283. Intech, Croatia.

Mótyán, J. A., Tóth, F., and Tőzsér, J. 2013. Research applications of proteolytic enzymes in molecular biology. *Biomolecules* 3:923–942.

Nigam, J. N. 2002. Bioconversion of water-hyacinth (*Eichhornia crassipes*) hemicellulose acid hydrolysate to motor fuel ethanol by xylose-fermenting yeast. *Journal of Biotechnology* 97:107–116.

Noe, P., Chevalier, J., Mors, F., and Comtat, J. 1986. Action of xylanases on chemical pulp fibers part ii: Enzymatic beating. *Journal of Wood Chemistry and Technology* 6(2):167–184.

Nogi, M., Iwamoto, S., Nakagaito, A. N., and Yano, H. 2009. Optically transparent nanofiber paper. *Advanced Materials* 21:1595–1598.

Nomura, Y. 1985. Digestion of pulp. Japanese Patent 126,395/85.

Oksanen, T., Pere, J., Paavilainen, L., Buchert, J., and Viikari, L. 2000. Treatment of recycled kraft pulps with *Trichoderma reesei* hemicellulases and cellulases. *Journal of Biotechnology* 78:39–48.

Onysko, K. A. 1993. Biological bleaching of chemical pulps: A review. *Biotechnology Advances* 11:179–198.

Ortega, N., Busto, M. D., and Perez-Mateos, M. 2001. Kinetics of cellulose saccharification by *Trichoderma reesei* cellulases. *International Biodeterioration and Biodegradation* 47:7–14.

Östberg, L. 2012. Some aspects on pulp pre-treatment prior to viscose preparation. Licentiate thesis, Karlstad University Studies, Karlstad.

Ozalp, N., and Hyman, B. 2006. Energy end-use model of paper manufacturing in the US. *Applied Thermal Engineering* 26:540–548.

Paice, M. G., Reid, I. D., Bourbonnais, R., Archibald, F. S., and Jurasek, L. 1993. Manganese peroxidase, produced by *Trametes versicolor* during pulp bleaching, demethylates and delignifies kraft pulp. *Applied and Environmental Microbiology* 59(1):260–265.

Pathak, P., Bhardwaj, N. K., and Singh, A. K. 2010. Enzymatic deinking of office waste paper: An overview. *IPPTA Journal* 22:83–88.

Pathak, P., Bhardwaj, N. K., and Singh, A. K. 2011. Optimization of chemical and enzymatic deinking of photocopier waste paper. *Bioresources* 6:447–463.

Pathak, P., Bhardwaj, N. K., and Singh, A. K. 2014. Enzymatic deinking of photo-copier waste papers using crude cellulase and xylanase of *Coprinopsis cine-rea* PPHRI-4 NFCCI-3027. *Appita Journal* 67:291–301.

Patrick, K. 2004. Enzyme technology improves, efficiency, cost, safety of stickers removal program. *PaperAge* 9:22–26.

Pere, J., Liukkonen, S., Siika-aho, M., Gullichsen, J., and Viikari, L. 1996. *TAPPI Proceedings Pulping Conference*, Nashville, USA, vol. 2, p. 693.

Pere, J., Puolakka, A., Nousiainen, P., and Buchert, J. 2001. Action of puri-fied *Trichoderma reesei* cellulases on cotton fibers and yarn. *Journal of Biotechnology* 89(2–3):247–255.

Piontek, K., Smith, A. T., and Blodig, W. 2001. Lignin peroxidase structure and function. *Biochemical Society Transactions* 29:111–116.

Prasad, D. Y., Heitmann, J. A., and Joyce, T. W. 1993. Enzymatic deinking of colored offset newsprint. *Nordic Pulp and Paper Research Journal* 2:284–286.

Puls, J., and Saake, B. 2004. Industrially isolated hemicelluloses. In *ACS Symposium Series* 864:24–37. American Chemical Society 1999, Washington, DC.

Putz, H. J., Renner, K., Gottsching, L., and Jokinen, O. 1994. Enzymatic deinking in comparison with conventional deinking of offset news. In *Proceedings of Tappi Pulping Conference*, 877–884. Tappi Press, Atlanta.

Qian, Y., and Goodell, B. 2005. Deinking of laser printed copy paper with a medi-ated free radical system. *Bioresource Technology* 96:913–920.

Raj, A., Reddy, M. M., and Chandra, R. 2007. Decolourisation and treatment of pulp and paper mill effluent by lignin-degrading *Bacillus* sp. *Journal of Chemical Technology and Biotechnology* 82:399–406.

Rakkolainen, M., Kontturi, E., Isogai, A., Enomae, T., Blomstedt, M., and Vuorinen, T. 2009. Carboxymethyl cellulose treatment as a method to inhibit vessel picking tendency in printing of eucalyptus pulp sheets. *Industrial and Engineering Chemistry Research* 48:1887–1892.

Ramchandra, M., Don, L., Crawford, D. L., and Hertel, G. 1988. Characterization of an extracellular lignin peroxidase of the lignocellulolytic actinomy-cete *Streptomyces viridosporust*. *Applied and Environmental Microbiology* 54:3057–3063.

Rättö, M., Kantelinen, A., Bailey, M., and Viikari, L. 1993. Potential of enzymes for wood debarking. *TAPPI Journal* 76(2):125–128.

Ren, J., Zhao, C. S., and Yu, D. M. 2013. Research on the technology and mech-anism of inhibiting stickies of blanket by enzyme treatment. *Advanced Materials Research* 750:1373–1376.

Risdianto, H., and Sugesty, S. 2015. Pretreatment of *Marasmius* sp. on biopulping of oil palm empty fruit bunches. *Modern Applied Science* 9(7):1–6.

Riva, S. 2006. Laccases: Blue enzymes for green chemistry. *Trends in Biotechnology* 5:219–225.

Saari, J. 2005. New process engineering for deinking Part 2: The potential of enzymatic deinking for woodfree paper grades. *Paper Technology* (Bury, UK) 46:34–37.

Sánchez, C. 2009. Lignocellulosic residues: Biodegradation and bioconversion by fungi. *Biotechnology Advances* 27:185–194.

Sánchez, O., Alméciga-Díaz, C. J., and Sierra, R. 2011. Delignification Process of Agro-Industrial Wastes an Alternative to Obtain Fermentable Carbohydrates for Producing Fuel. In *Alternative Fuel*, ed. Maximino Manzanera, Intech Open Access Publisher, 111–154. DOI: 10.5772/851.

Scholz, W. S., and Tse, J. 1997. In *Paper Recycling Challenge*, eds. M. R. Doshi and J. M. Dyer, 3–6. Doshi & Associates, Appleton.

Schwanninger, M., Hinterstoisser, B., Gradinger, C., Messner, K., and Fackler, K. 2004. Examination of spruce wood biodegraded by *Ceriporiopsis subvermispora* using near and mid infrared spectroscopy. *Journal of Near Infrared Spectroscopy* 12(6):397–310.

Scott, G. M., Akhtar M., and Lentz, M. J. 1997. Engineering, scale-up, and economic aspects of fungal pretreatment of wood chips. In *Environmentally Friendly Technologies for Pulp and Paper Industry*, eds. R. A. Young and M. Akhtar, 341–383. Wiley, New York.

Selvam, K., Swaminathan, K., Rasappan, K., Rajendran, R., Michael, A., and Pattabi, S. 2005. Deinking of waste papers by white rot fungi *Fomes lividus*, *Thelephora* sp. and *Trametes versicolor*. *Nature, Environment and Pollution Technology* 4:399–404.

Seo, H. I., Ryu, J. Y., Shin, J. H., Song, B. K., and Ow, S. K. 1999. Recycling of Wastepaper (VI)—The effect of starch adhesive on OCC drainage properties and the application of amylase. *Journal—Technical Association of the Pulp and Paper Industry of Korea* 31:25–33.

Shaikh, H., and Luo, J. 2009. Identification, validation and application of a cellulose specifically to improve the runnability of recycled furnishes. *Proceedings 9th International Technical Conference on Pulp, Paper and Allied Industry* (Paperex 2009), New Delhi, India, pp. 277–283.

Sharma, A., Thakur, V. V., Shrivastava, A. et al. 2014. Xylanase and laccase based enzymatic kraft pulp bleaching reduces adsorbable organic halogen (AOX) in bleach effluents: A pilot scale study. *Bioresource Technology* 169:96–102.

Shu, Z., Wu, J., Chen, D., Cheng, L., Zheng, Y., Chen, J., and Huang, J. 2012. Optimization of *Burkholderia* sp. ZYB002 lipase production for pitch control in thermomechanical pulping (TMP) processes. *Holzforschung* 66(3):341–348.

Shul'pin, G. B., Nizova, G. V., Kozlov, Y. N., and Pechenkina, I. G. 2002. Oxidations by the "hydrogen peroxide–manganese (IV) complex–carboxylic acid" system. Part 4. Efficient acid–base switching between catalase and oxygenase activities of a dinuclear manganese (IV) complex in the reaction with H_2O_2 and an alkane. *New Journal of Chemistry* 26:1238–1245.

Sigoillot, J. C., Petit-Conil, M., Herpoel, I., et al. 2001. Energy saving with fungal enzymatic treatment of industrial poplar alkaline peroxide pulps. *Enzyme and Microbial Technology* 29:160–165.

Singh, A., Kuhad, R. C., and Ward, O. P. 2007. Industrial application of microbial cellulases. In *Lignocellulose Biotechnology: Future Prospects*, eds. R. C. Kuhad and A. Singh, 345–358. I. K. International Publishing House, New Delhi.

Singh, R., and Bhardwaj, N. K. 2010. Enzymatic refining of pulps: An overview. *IPPTA Journal* 22:109–115.

Spiridon, I., Duarte, A. P., and Curto, J. 2003. Influence of xylanase treatment on *Pinus pinaster* kraft pulp. *Cellulose Chemistry and Technology* 37:497–504.

Sunna, A., and Antranikian, G. 1997. Xylanolytic enzymes from fungi and bacteria. *Critical Reviews in Biotechnology* 17:39–67.

Suurnäkki, A., Tenkanen, M., Buchert, J., and Viikari, L. 1997. Hemicellulases in the bleaching of chemical pulps. *Advances in Biochemical Engineering/ Biotechnology* 57:261–287.

Suurnäkki, A., Tenkanen, M., Siika-Aho, M., Niku-Paavola, M. L., Viikari, L., and Buchert, J. 2000. *Trichoderma reesei* cellulases and their core domains in the hydrolysis and modification of chemical pulp. *Cellulose* 7(2):189–209.

Svensson, G. 2006. Alternative enzymatic conversion of surface sizing starch at Nymolla Mill. http://www.chemeng.lth.se/exjobb/E256.pdf. Accessed on Oct. 31, 2015.

Tausche, J. 2002. Mill-scale benefits in enzymatic deinking. *7th Pira International Recycling Technology Conference*, Belguim, Brussels, paper, 7, 1–3.

Thakur, V. V., Jain, R. K., and Mathur, R. M. 2012. Studies on xylanase and laccase enzymatic prebleaching to reduce chlorine-based chemicals during CEH and ECF bleaching. *BioResources* 7(2):2220–2235.

Thomas N., and Murdoch, B. 2006. Speed, production get bio-boost at liberty paper: A bio-refining enzyme helps the company run faster and cleaner. *Paper* 360:17.

Thurston, C. F. 1994. The structure and function of fungal laccases. *Microbiology* 140:19–26.

Timothy, K. 2007. Electrolytic bromine: A green biocide for cooling towers. *Proceedings of the Water Environment Federation, Industrial Water Quality* 7:546–552.

Tolan, J. S., Guenette, M., Thibault, L., and Winstanley, C. 1994. The use of novel pretreatment enzymes to improves the efficiency of shives removal by bleaching. *Pulp and Paper Canada* 95(12):493.

Tong, G. L., Zhang, X. L., and Jing, Y. 2007. A study on improving the surface sizing performance of starch treated with amylase enzymolysis. *China Pulp and Paper Industry* 28(12):57–359.

Torres, C. E., Negro, C., Fuente E., and Blanco, Á. 2012. Enzymatic approaches in paper industry for pulp refining and biofilm control. *Applied Microbiology and Biotechnology* 96:327–344.

Treimanis, A., Leite, M., Eisimonte, M., and Viesturs, U. 1999. Enzymatic deinking of laser-printed white office wastepaper. *Chemical and Biochemical Engineering Quarterly* 13:53–57.

Tripathi, S., Sharma, N., Mishra, O. P., Bajpai, P., and Bajpai, P. K. 2008. Enzymatic refining of chemical pulp. *IPPTA Journal* 20:129–132.

Uchimoto, I., Endo, K., and Yamagishi, Y. 1988. (Honshu Paper Co.). Kouyouyu parupu no kaishitsu houhou (Improvement method of hardwood pulp). Japanese Patent 63–135597.

Van Heiningen, A. R. P., Yoon, S. H., Zou, H., Jiang, J., and Goyal, G. 2006. U.S. Patent application; 11(640): 820.

Verma, P., Bhardwaj, N. K., and Varadhan, R. 2014. Microbial life in paper machine: Prevention and control. *IPPTA Journal* 26(3):44–48.

Verma, P. K., Bhardwaj, N. K., and Singh, S. P. 2015. Selective hydrolysis of amorphous cellulosic fines for improvement in drainage of recycled pulp based on ratios of cellulase components. *Journal of Industrial and Engineering Chemistry* 22:229–239.

Vermelho, A. B., Noronha, E. F., Filho, E. X., Ferrara, M. A., and Bon, E. P. S. 2013. Diversity and biotechnological applications of prokaryotic enzymes. In *The Prokaryotes*, eds. E. Rosenber, E. F. DeeLong, S. Lory, E. Stackebrandt, and F. Thompson, 213–240. Springer-Verlag, Berlin.

Viesturs, U., Leite, M., Eisimonte, M., Eremeeva, T., and Treimanis, A. 1998. Biological deinking technology for the recycling of office wastepapers. *Bioresource Technology* 67:255–265.

Viikari, L., Ranua, M., Kantelinen, A., Sundquist, J., and Linko, M. 1986. Bleaching with enzymes. *Proceedings of 3rd International Conference on Biotechnology in the Pulp and Paper Industry*, 67–9. STFI, Stockholm.

Viikari, L., Rättö, M., and Kantelinen, A. 1989. *Finish Patent Applied*. 896:291.

Vyas, S., and Lachke, A. 2003. Bio-deinking of mixed office waste paper by alkaline active cellulases from alkalotolerant *Fusarium* sp. *Enzyme and Microbial Technology* 32:236–245.

Wall, M. B., Stafford, G., Noel, Y., Fritz, A., Iverson, S., and Farrell, R. L. 1996. Treatment with *Ophiostoma piliferum* improves chemical pulping efficiency. In *Biotechnology in the Pulp and Paper Industry. Recent Advances in Applied and Fundamental Research*, eds. E. Srebotnik and K. Messner, Facultus Universitatsverlag, Vienna, 205–210.

Walton, S. L., and Van Heiningen, A. R. 2007. Biological conversion of hemicelluloses extracted from hardwood: Enabling co-production of ethanol and pulp in an integrated forest bio-refinery. In *The 29th Symposium on Biotechnology for Fuels and Chemicals,* Denver, CO, April 29–May2.

Wang, B., Sain, M., and Oksman, K. 2007. Study of structural morphology of hemp fiber from the micro to the nanoscale. *Applied Composite Material* 14:89–103.

Wang, H., Pang, B., Wu, K., Kong, F., Li, B., and Mu, X. 2014. Two stages of treatments for upgrading bleached softwood paper grade pulp to dissolving pulp for viscose production. *Biochemical Engineering Journal* 82:183–187.

Wanqi, W. M. L. 2010. The application of amylase for surface sizing in culture paper. *East China Pulp and Paper Industry* 5:030.

Wesenberg, D., Kyriakides, I., and Agathos, S. N. 2003. White-rot fungi and their enzymes for the treatment of industrial dye effluents. *Biotechnology Advances* 22:161–187.

Wong, D. W. 2009. Structure and action mechanism of ligninolytic enzymes. *Applied Biochemistry and Biotechnology* 157:174–209.

Wong, K. K., Richardson, J. D., and Mansfield, S. D. 2000. Enzymatic treatment of mechanical pulp fibers for improving papermaking properties. *Biotechnology Progress* 16:1025–1029.

Woodward, J., Stephan, L. M., Koran, L. J., Wong, K. K. Y., and Saddler, J. N. 1994. Enzymatic separation of high-quality uninked pulp fibers from recycled newspaper. *Bio/Technology* 12:905–908.

Wu, C., Zhou, S., Li, R., Wang, D., and Zhao, C. 2015. Reactivity improvement of bamboo dissolving pulp by xylanse modifications. *Bioresources* 10(3): 4777–4970.

Yang, G., Lucia, L. A., Chen, J., Cao, X., and Liu, Y. 2011. Effects of enzyme pretreatment on the beatability of fast-growing poplar APMP pulp. *BioResources* 6:2568–2580.

Yerkes, W. D. 1968. Process for the digestion of cellulosic material by enzymatic action of *Trametes suaveolens*. U.S. Patent No. 3,406,089. Washington, DC: U.S. Patent and Trademark Office.

Yoon, C., and Jung, H. 2014. Bleaching of kraft pulp with xylanase and laccase-mediator system. *Journal of Korea TAPPI* 46:1–10.

Yoon, S. H., Macewan, K., and van-Heiningen, A. 2006. Pre-extraction of southern pine chips with hot water followed by kraft cooking. In *TAPPI Engineering, Pulping, and Environmental Conference* 5:157.

Yoon, S. H., Macewan, K., and van-Heiningen, A. 2008. Hot water pre extraction from loblolly pine in an integrated forest products biorefinery. *TAPPI Journal* 7(7): 22–27.

Zeng, X., Fu, S., Yu, J., Li, K., Zhan, H., and Li, X. 2009. Study on the degradation of stickies in the pulps by complex enzymes. *China Pulp Paper* 28:1–4.

Zeyer, C., Joyce, T. W., Heitmann, J. A., and Rucker, J. W. 1994. Factors influencing enzyme deinking of recycled fiber. *TAPPI Journal* 77:169–177.

Zhang, Z. J., and Hu, H. R. 2011. Action mechanism analysis of enzymatic refining for bleached simao pine kraft pulp. *Advanced Materials Research* 236:1425–1430.

Zhang, Z. J., Chen, Y. Z., Hu, H. R., and Sang, Y. Z. 2013. The beatability-aiding effect of *Aspergillus niger* crude cellulase on bleached simao pine kraft pulp and its mechanism of action. *BioResources* 8:5861–5870.

Zhu, J. Y., Sabo, R., and Luo, X. 2011. Integrated production of nano-fibrillated cellulose and cellulosic biofuel (ethanol) by enzymatic fractionation of wood fibers. *Green Chemistry* 13:1339–1344.

Zuo, Y., and Saville, B. A. 2005. Efficacy of immobilized cellulase for deinking of mixed office waste. *Journal of Pulp and Paper Science* 31:3–6.

CHAPTER 7

Rhizobacteria

Tools for the Management of Plant Abiotic Stresses

Anjali Singh, Ajay Shankar, Vijai Kumar Gupta, and Vishal Prasad

CONTENTS

ABSTRACT

Plants are continuously exposed to a plethora of threats originating from different biotic and abiotic stressors, which have escalated over time due to change in global climate pattern as well as human interferences and the subject of stress is imperative in influencing plant growth and crop production all around the world. Diverse methods like use of plants with natural tolerance or plants with modified tolerance by use of stress-related genes are available for alleviation of such stresses. Nevertheless, the use of rhizospheric microorganisms having plant growth-promoting traits has also been proved to be effective under stressful conditions. Such microbes can symbiotically or non-symbiotically alleviate the effects of stress and enhance plant growth. Therefore, diverse microbial species and strains are isolated, screened, tested, and used as microbial inoculums worldwide. The vast influence of rhizospheric microbes toward the growth and production of crop plants can be environmentally and economically significant. This chapter highlights the management and mitigation of various abiotic stressors and their effects on plants by exploring the opportunities available with the rhizobacterial microbes.

INTRODUCTION

A healthy plant faces a number of abiotic environmental stresses, in its edaphic environment such as flooding, salinity, heavy metal toxicity, drought, chilling, and heat. These stresses have a negative impact on the growth and development of plants, decrease crop yield, and also reduce soil fertility (Nadeem et al., 2014; Gupta et al., 2015; Sulmon et al., 2015). On the other hand, an ever-increasing world population and climatic variability are anticipated to severely enhance the worldwide appeal for farmable land, a resource that is already in high demand (Barrow et al., 2008; Derr and Tringe, 2014). It is expected that global population will count 9 billion by 2050, and the global food supply will have to be increased by 70% to meet the briskly growing requirement of the increasing population (Editorial, 2010). The necessity of providing food for the world's burgeoning population while repelling abiotic stresses is a bigger challenge today, and it has given an imperative significance in plant and soil productivity research (Dimpka et al., 2009; Ahemad and Kibret, 2014; Glick, 2014; Nadeem et al., 2014). Under such situation, it needs appropriate technology to improve crop productivity and soil health through interactions of soil microorganisms and plant roots under stressful conditions (Egamberdieva, 2012; Sarma et al., 2012). Even though various approaches have been tested for the mitigation of abiotic stresses on plant growth, adoption of biological methods containing soil rhizobacteria has been confirmed to be efficient for the mitigation of divergent stresses (Dimpka et al., 2009; Miransari, 2012; Nadeem et al., 2014; Gupta et al., 2015). To fulfill the rising appeal of food supply, improved adoption of crop plants to abiotic stresses as well as geographical enlargement of agricultural lands by making impractical lands usable for the production of crop is an essentiality. Thus, both improvements in productivity of crops as well as the farming of additional farmland are required for this enlargement in agricultural production. In this regard beneficial soil microbes may become very accessible and administer an expeditious resolution to the problem (Egamberdieva, 2009; Bashan et al., 2012; Sarma et al., 2012; Kang et al., 2014; Gusain et al., 2015; Sulmon et al., 2015). Diverse kinds of soil bacteria are attracted by the exudates of plant roots. They occupy the rhizosphere of many plant species and provide benefit to the plants by enhancing plant growth and lowering disease development. Rhizobacteria are primarily used for enhancing crop yield and preserving soil productivity (Saharan and Nehra, 2011; Bashan et al., 2012; Glick, 2014). They are not only useful in agriculture but they also have potential to solve environmental problems including abiotic stresses. Recent studies have shown that some strains of plant growth-promoting rhizobacteria (PGPR) evoke tolerance to abiotic stresses, such as drought, salinity, nutrient excess or deficiency (Yang et al., 2009; Derr and Tringe, 2014; Nadeem et al., 2014; Sulmon et al., 2015).

In this chapter, an attempt has been made to discuss the effect of different abiotic stresses on plants and the role of rhizobacteria in mitigating the adverse effects of these abiotic stresses on plants. The molecular mechanisms used by these rhizobacteria in regulating these responses and how they can be used for sustainable agricultural productions have also been incorporated.

IMPACTS OF ABIOTIC STRESSES ON PLANTS

Abiotic stresses are perceived as the major intimidation to natural resources globally. The leading factors that reinforce these abiotic stresses are climate change and human interferences. These facts symbolize severe threats to sustainable food production by affecting the production of crops. Increased occurrences of abiotic stresses have evolved into major cause for sluggishness of primary crop yields (Christensen et al., 2007; Saharan and Nehra, 2011). Plants are sessile and are continually exposed to changes in environmental circumstances. When these changes are hasty and intense, plants commonly perceive them as stresses. During a stress situation the plant faces certain physiological and metabolic disorders such as nutritional and hormonal imbalance, accelerated production of ethylene, scarcity of water, ion toxicity, oxidative stress, changes in metabolic processes, membrane foul-up, decline of cell division and enlargement, and genotoxicity (Munns et al., 2002; Mayak et al., 2004; Glick, 2007; Dimpka et al., 2009; Ahemad and Kibret, 2014; Nadeem et al., 2014; Gupta et al., 2015; Kang et al., 2015). Some of the considerable plant mechanisms that are troubled by abiotic stresses are root inhibition, impact on nodulation, biochemical disorder such as epinasty, senescence, chlorophyll demolition, and leaf abscission. Physiological changes include closure of stomata, regulation of the synthesis of compounds, such as chlorophyll and phenolic compounds and enhanced respiratory enzymes (Barassi et al., 2006; Govindasamy et al., 2008; Dimpka et al., 2009; Chakraborty et al., 2011; Farooq et al., 2014; Egamberdieva and Lugtenberg, 2014; Sulmon et al., 2015).

There are reports showing declines in productivity of wheat and paddy crops in different parts of South Asia due to accelerating water stress and air temperature and decrease in the number of rainy days. The normal temperature in the Indian sub-continent has increased by 0.57°C in the last 100 years and it is estimated that it is inclined to increase further to a maximum of 2.5°C by 2050 and 5.8°C by 2100 (Grover et al., 2011). The necessity of irrigation water in arid and semi-arid regions is projected to rise by 10% with every 1°C rise in temperature. In addition high temperature, drought, elevated CO_2, more floods after extreme rainfall events, cyclones, heat and cold waves, salinity are the other influential natural calamities that construct severe economic catastrophe and are expected to be endorsed as a consequence of global warming. These determinants are likely to cause severe destructive effects on growth and yield of crops as well as promulgate uncompromising burden on our farmland and other natural resources. An accelerated level of ethylene during stress condition is a major reason of reduced plant growth and development because it is considered as a stress hormone as it is released in response to both edaphic and adaphic environmental stresses (Tank and Saraf, 2010). Ethylene, a plant hormone is gaseous in nature, plays an important role in abiotic as well as biotic stresses and is therefore immediately anticipated and oozed throughout the plant (Bleecker and Kende, 2000). However, a limited amount of ethylene is necessary for plant growth and development, but at a higher concentration it had several adverse impacts on plants such as root growth inhibition, which affects the overall growth and development of plants (Nadeem et al., 2012). Different types of abiotic stresses are responsible for an increase in ethylene concentration during that particular stress condition. Flood-like situation is also accountable for accelerating ethylene concentration, due to increase in the activeness of 1-aminocyclopropane-1-carboxylic acid (ACC)-synthase and ACC-oxidase in root and shoot appropriately. There are several reports showing that soil flooding increased ACC-oxidase activity in plants and resulted in accelerated ethylene synthesis (Govindasamy et al., 2008; Grover et al., 2011; Barnawal et al., 2012; Nadeem et al., 2014). Under anaerobic conditions, synthesis of

ACC occurs in plant roots and then transported to shoot, whereby oxidizing it produces ethylene and causes anomalous growth such as leaf epinasty (Else and Jackson, 1998; Nadeem et al., 2012). Most of the plants are also sensitive to fluctuations in temperature, which leads to changes in plant hormonal balances and ultimately affects plant growth and development. Chilling and heat both enhance ethylene concentration and there prevails a decisive interrelationship between temperature and ethylene concentration. In case of drought, another major abiotic stress that curbs yield of crops and affects both cellular and molecular level. This plant water shortfall has been mainly associated with an accelerated level of ethylene, which leads to aberrant growth such as restriction of root and shoot growth, abscission, and premature senescence of different parts of plants. The raised production of ethylene during drought stress also causes diversification in rectitude of plasma membrane, decreased leaf dry weight, and content of pigment. Salinity is another major abiotic stress, which is responsible for affecting plant growth and development by disturbing different plant physiology and biochemical processes such as misbalancing protein metabolism and hormone balance, enzyme activities, ion toxicity of salts, hindering nutrient uptake and making ion inadequacy, and bothers symbiotic interaction between legumes and *Rhizobium* (Upadhyay et al., 2011; Egamberdieva and Lugtenberg, 2014; Nadeem et al., 2014). Salinity also disrupts photosynthesis by enhancing photorespiration, altering the normal homeostasis of cells and creates an accelerated production of reactive oxygen species (ROS) and ethylene (Miller et al., 2010; Carmen and Roberto, 2011). Seed germination and seedling growth are mostly affected by salinity (Egamberdieva et al., 2011). The heavy metals pose another abiotic threat to plants and their concentration in soil is gradually rising due to their inclusion through different agricultural and industrial sources such as chemical fertilizers, use of pesticides, impeachment of sewage and municipal waste, mining and smelting of metals, electroplating, energy and fuel production, gas exhaust, improper treatment of industrial wastes, leakage of landfill leachate, mining wastes, incomplete collection of used batteries, accidental spills, and military activities (Kim et al., 2001; Miransari, 2012; Turan et al., 2012). The high level of these contaminants disturbs plant growth and creates hindrance of root growth in contaminated soil. High ethylene concentration during heavy metal stress not only restricts root growth but also depletes nutrients such as iron and reduced fixation of carbon dioxide (Burd et al., 2000; Selvakumar et al., 2012; Nadeem et al., 2014). The stresses induced by contaminants present in air also cause enhanced ethylene synthesis in plants (Dimpka et al., 2009; Nadeem et al., 2012).

It is noticeable from the discussions above that abiotic environmental stresses are inimical for growth of plants. The divergent abiotic stresses influence the growth and development of plants by inducing detrimental effects on the morphology as well as on the physiological and biochemical mechanisms. Under stress conditions, many such mechanisms are affected directly, while others are indirectly (Nadeem et al., 2014). So any method to control the accelerated production of ethylene will be of considerable importance in promoting the growth of plants under stress conditions. It has been shown that this can be conceivably accomplished by rhizobacteria having ACC-deaminase to preserve plants from the injurious effects of environmental stresses (Mayak et al., 2004; Turan et al., 2012; Glick, 2014; Nadeem et al., 2014; Kang et al., 2015).

Rhizobacteria

Plant-associated microbial communities have appreciable capabilities to negotiate many of the abiotic stress effects on plants (Mayak et al., 2004; Tank and Saraf, 2010; Marasco et al., 2012; Derr and Tringe, 2014). Entire plants and almost all tissues within the plant are populated by a diversity of microorganisms, many of which endeavor benefits to the host, enhancing uptake of nutrients, protecting from pathogen attack, and enhancing growth of plants under unfavorable environmental conditions. In return, these microorganisms secure shelter from the ambience environment and connection to a carbon-rich food supply (Yang et al., 2009; Turner et al., 2013). The root soil amalgamate comprises a compelling microcosm known as the rhizosphere, where microorganisms, plant roots, and soil components interact (Azcon et al., 2013). The rhizosphere, a physical, chemical, and biological environment where diversified microbial activities are known to take place, considerably leverage plant health and soil fertility, and microbial improvements in such environment can help the host plant to reciprocate to abiotic stress conditions (Barea et al., 2005; Azcon et al., 2013). Rhizobacteria are beneficial and free-living microorganisms located near the plant root and strive to make a significant impact on plants ranging from direct influence to indirect effects (Figure 7.1) against abiotic environmental stresses (Kloepper et al., 1980; Kohler et al., 2006; Saharan and Nehra, 2011; Jackson et al., 2012; Nadeem et al., 2012; 2014; Ahemad and Kibret, 2014; Glick, 2014). So the use of these bacteria on plants or seeds promotes plant growth and protects the plants from

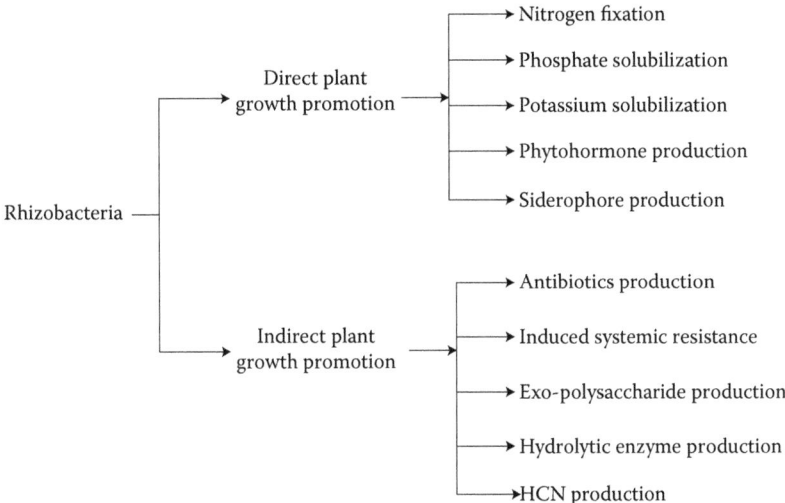

FIGURE 7.1 Schematic representation of different plant growth-promoting mechanisms used by rhizobacteria.

harmful effects of drought, salinity, temperature, and heavy metals stress (Kloepper et al., 1980; Dimpka et al., 2009; Lugtenberg and Kamilova, 2009; Azcon et al., 2013; Egamberdieva and Lugtenberg, 2014; Nadeem et al., 2014).

Some of the rhizobacteria belonging to different genera are *Pseudomonas, Bacillus, Serratia, Azospirillum, Klebsiella, Azotobacter, Arthrobacter, Enterobacter, Burkholderia, Paenibacillus*, etc. (Zahir et al., 2008; Belimov et al., 2009; Dimpka et al., 2009; Ahemad and Kibret, 2014; Gupta et al., 2015). The alleviation of abiotic environmental stresses is primarily due to phytohormones such as auxins and gibberellins, which are propagated by rhizobacteria (Patten and Glick, 1996; Dimpka, et al., 2009). These phytohormones embellish plant growth either directly or in concert with other bacterial secondary metabolites (Dimpka et al., 2009). Other compounds that are produced by rhizobacteria include enzymes, osmolytes, nitric oxide, siderophore, antibiotics, and organic acids, which are also responsible for plant growth by different mechanisms (Chakraborty et al., 2006; Dimpka et al., 2009). Figure 7.1 presents a schematic representation of various direct and indirect mechanisms used by rhizobacteria for plant growth promotion. The considerable tool used by rhizobacteria having ACC-deaminase for enhancing plant growth under stress conditions is the decreasing of ethylene concentration in the plant by hydrolyzing ACC, an instantaneous precursor of ethylene biosynthetic pathway (Nadeem et al., 2014). These rhizobacteria use ACC as a point of supply of their nitrogen that results in low ACC levels in the plant root, which eventually reduces the ethylene level in plants. The mitigation of harmful impacts of higher ethylene results in an enhancement of plant growth under stress conditions (Mayak et al., 2004; Nadeem et al., 2014). With careful investigations it has been identified that several plant growth-promoting bacteria encompass the enzyme ACC-deaminase. When PGPR which consists of the enzyme ACC-deaminase are associated with the seed coat of a growing seedling, they may perform as a sink for ACC and thus as a tool for securing that the concentration of ethylene does not inflate to the point where root growth is injured (Grichko and Glick 2001). Adjustment of microorganisms to stress is an intricate multilevel regulative operation in which several genes are involved (Srivastava et al., 2008). In certain contemporary species living under intense environments, better metabolic mechanisms such as enzymatic activities and membrane strength take place at high temperature or high salt concentration appropriately, while other microorganisms prosper various adaptive mechanisms to encounter the stress (Egamberdieva and Lugtenberg, 2014). Some bacteria such as *Pseudomonas* endure under stress conditions due to the excretion of exopolysaccharides (EPS), which insulate microorganisms from water stress and fluctuations in water potential by increasing water confinement and determining the carbon sources dispersion in the microbial environment (Sandhya et al., 2009). EPS acquire exclusive water holding and sealing properties, and thus play an essential role in the construction and stabilization of aggregates of soil and regulation of nutrients and flow of water across plant roots through the formation of biofilms (Vacheron et al., 2013; Zelicourt et al., 2013).

Plant–Microbe Interactions

Soil-grown plants are engaged in a sea of microbes and divergent profitable microorganisms such as plant growth-promoting bacteria can vitalize growth of plants and deliberate intensified resistance to abiotic stresses (Lugtenberg and Kamilova, 2009). The establishment of valuable plant–microbial interactions needs the reciprocal appreciation and an extensive composition of the reactions at both the plant and the microbial side. A substantial part of carbohydrates produced by plants at the time of photosynthesis is passed to microbes associated with plant roots. Since significant amounts of phosphate, nitrate, and some other minerals required for plant growth are generally not accessible in free form or present in finite amounts in the soil, root-associated useful microbes are crucial companions, which make phosphate and nitrate accessible to plants (Zelicourt et al., 2013).

Certain microorganisms produce phytohormones, such as indole acetic acid (IAA) and gibberellic acid, which persuade increased root growth and by that advantage to embellish uptake of nutrients (Egamberdieva and Kucharova, 2009; Saharan and Nehra, 2011; Barea et al., 2013). Plants have the capability to achieve a condition of induced systemic resistance (ISR) to pathogens after inoculation with rhizobacteria. Together with plant roots, rhizobacteria can bloom the plant-inherited immune system and deliberate resistance to a large number of pathogens with the slightest effect on growth and yield (Zelicourt et al., 2013). Various rhizobacteria, including *Pseudomonas fluorescens*, *Pseudomonas putida*, *Serratia marcescens*, *Bacillus pumilus*, and *Azospirillum brasilense* colonize roots and secure on a wide array of plant species, including crops, vegetables, and even trees, against abiotic stresses and foliar diseases in greenhouse and field experiments (Van Loon, 2007). Nitrogen fixation, production of siderophores, phosphorus solubilization, production of enzyme ACC-deaminase, plant growth regulators, chitinase and glucanase, and organic acids are the mechanisms used by microbes to stimulate plant growth (Glick et al., 2007; Berg 2009; Hayat et al., 2010; Nadeem et al., 2014). In addition, many of these rhizobacterial strains can also improve plant tolerance against drought, salinity, heavy metal toxicity, and flooding, and thus facilitate plants to endure unfavorable environmental conditions (Mayak et al., 2004; Glick et al., 2007; Zahir et al., 2008; Sandhya et al., 2009; Grover et al., 2011). Reports establish that plants treated with the rhizobacterium *Paenibacillus polymyxa* have shown an enhanced resistance to drought and further bacterial attack, an effect associated with the expression of the erd15 gene (Timmusk and Wagner, 1999).

RHIZOBACTERIA AS A TOOL FOR THE MANAGEMENT OF ABIOTIC STRESSES

Crop productivity is subjected to number of stresses and potential yields are seldom achieved with a diverse array of abiotic stresses active in the environment. The present challenges such as global climate change, water and soil pollution, less water availability, and urbanization add up to the situation. All crops grown under natural environment are subjected to one or the other stress. Extensive research is going on worldwide to develop strategies for coping with abiotic stresses through development of tolerant varieties, shifting the crop calendars, production of transgenics, etc. One of the recently gaining practices of counteracting the adverse effects of abiotic stresses on plant growth includes the implementation of stress-tolerant rhizobacteria with natural growth-promoting ability in such conditions.

In one study the application of novel PGPR strain *Serratia nematodiphila* PEJ1011 was found to be a good substitute to synthetic fertilizers (Kang et al., 2015). The potential of growth promotion of this strain was also sustained by the improved plant growth characteristics under normal and cold-stressed environments. It was observed that this increase in growth of inoculated plants was accomplished through analogous increment in growth hormone gibberellic acid under normal and cold-stressed circumstances, and vice versa in case of stress hormones such as jasmonic acid and salicylic acid. The *S. nematodiphila* PEJ1011 inoculation continued stress alleviation by the enhanced level of stress-reactive abscisic acid (ABA) in cold conditions and declined level in normal situation in comparison to non-inoculated control plants (Kang et al., 2015). Salinity was found to considerably reduce seed germination, growth, and root length of eggplant (Fu et al., 2010). However, the seeds inoculated with *Pseudomonas* sp. DW1 exhibited an increase in the percentage of germination over its non-inoculated seeds under saline conditions (Fu et al., 2010). Salinity significantly reduced K^+ concentration, accelerating Na^+ concentration, thus depreciating the ratio of K^+/Na^+. It was observed that under salinity conditions, shoot Ca^{2+} concentration of eggplant increased when inoculated with *Pseudomonas* sp. DW1 compared with non-inoculated plants. In this study, it was also found that inoculating regimens with *Pseudomonas* sp. DW1 had no impact on shoot Na^+ concentration in 0.57 and 1 g NaCl kg soil^{-1}, but there were noticeable decline in inoculated treatments in comparison to non-inoculated ones at 2 and 3 g NaCl kg soil^{-1}. Under saline conditions, SOD activity was reduced and POD activity was accelerated significantly, while there was no significant difference in CAT activity. However eggplant inoculated with *Pseudomonas* sp. DW1 had a little positive effect on SOD activity in the leaves of eggplants, and there was no significant change in CAT and POD activities in comparison to non-inoculated plants under salinity (Fu et al., 2010). In another study, metal-antagonistic and ACC-deaminase-producing endophytic bacteria with higher ACC-deaminase activity were isolated from Cu-tolerant plants and were used as the bioinoculant for the persuasive phytoremediation of Cu-polluted environment (Zhang et al., 2011). The isolated endophytic bacteria showed an intuitive ability of articulating resistance against numerous heavy metals and also exhibited some plant growth-promoting activities. Among the three metal-resistant and ACC-deaminase-producing strains, that is, *Ralstonia* sp. J1-22-2, *Pantoea agglomerans* Jp3-3, and *Pseudomonas thivervalensis* Y1-3-9, the utmost Cu uptake in rape crop was recognized after inoculation with strain *Pantoea agglomerans* Jp3-3 (Zhang et al., 2011). An additional compassionate liaison between the ACC-deaminase-producing endophytic bacteria and the plant is essential for the betterment of effective phytoremediation of Cu-polluted soils. Thus these isolates may offer to administer a new endophytic bacterial system for expedited phytoremediation of Cu-polluted environment (Zhang et al., 2011). In one study showing the usability of microbes for abiotic stress tolerance the inoculation with an isolate of *Pseudomonas* sp. strain AKM-P6 bearing 97% homology to *Pseudomonas aeruginosa* on the basis of 16S rDNA sequencing produced enhanced tolerance in sorghum seedlings against elevated temperatures (Ali et al., 2009). This tolerance was observed to be produced by inducing several physiological and biochemical changes in the treated plants. The AKM-P6-inoculated sorghum seedlings showed enhanced rate of survival and growth at elevated temperatures in comparison to uninoculated seedlings. The inoculation was observed to improve the levels of cellular metabolites such as proline, chlorophyll, sugars, amino acids, and proteins, as well as lower the level of membrane injury at elevated temperatures. Some high molecular weight protein molecules were also observed to be synthesized in the leaves of inoculated plants under elevated temperatures (Ali et al., 2009).

In another instance showing salinity tolerance in groundnut by inoculation of *Bacillus licheniformis* strain A2 it was observed that this strain improved the overall growth of groundnut with an increase in fresh biomass, root length, and total length under salinity treatment (Goswami et al., 2014). Besides this the strain A2 was also observed to have two important plant growth promotion traits, one of IAA production and second of phosphate solubilization (Goswami et al., 2014). A list of some of the potential rhizospheric microorganisms used to combat several abiotic stresses in crop plants depicting the results of their inoculation is shown in Table 7.1.

Different strains of *Rhizobium* and *Pseudomonas* may be used on various crops to determine their efficiency in imparting salt tolerance to host plants. Both *Rhizobium* and *Pseudomonas* alleviated salt stress in

TABLE 7.1   Various Rhizobacteria Conferring Abiotic Stress Tolerance in Crop Plants

Stress	Microorganisms	Crop	Results of Inoculation of Bacteria to Plants	References
Drought	Pseudomonas fluorescence, Bacillus cereus	Rice (Oryza sativa L.)	Improved plant growth and activity of antioxidant defense systems	Gusain et al. 2015
	Bacillus sp. Klebsiella sp.	Pepper (Capsicum annuum L.)	Enhanced photosynthetic activity and biomass synthesis	Marasco et al. 2012
	Variovorax paradoxus	Pea (Pisum sativum)	Synthesis of ACC-deaminase	Belimov et al. 2009
	Bacillus sp.	Lettuce (Lactuca sativa)	Enhanced photosynthetic activity and biomass synthesis	Arkhipova 2007
	Pseudomonas putida, Achromobacter piechaudii	Tomato (Lycopersicum esculentum), Pepper (Capsicum annuum)	Plant maintained their growth under water deficit condition, decreased ethylene concentration	Mayak et al. 2004
Salinity	Pseudomonas chlororaphis	Cotton (Gossypium hirsutum)	Modulation of IAA, increment in root and shoot growth along with biomass	Egamberdieva et al. 2015
	Burkholderia cepacia, Promicromonospora sp., Acinetobacter calcoaceticus	Cucumber (Cucumis sativus)	Decreased electrolytic leakage and Na^+ concentration, increased water potential	Kang et al. 2014
	Canola (Brassica napus)	Pseudomonas putida UW4	Conferred salt resistance and enhanced plant growth	Cheng et al. 2012
	Bacillus cereus	Mung bean (Vigna radiata), Chickpeas (Cicer arietinum), Rice (Oryza sativa)	Increased phosphate solubilization and enzyme activity, promoted seedling growth	Chakraborty et al. 2011
	Lettuce (Lactuca sativa)	Azospirillum	Increased tolerance against salinity	Barassi et al. 2006

(Continued)

TABLE 7.1 (*Continued*) Various Rhizobacteria Conferring Abiotic Stress Tolerance in Crop Plants

Stress	Crop	Microorganisms	Results of Inoculation of Bacteria to Plants	References
Flooding	Tulsi (*Ocimum sanctum*)	*Achromobacter* sp., *Serratia ureilytica, Herbaspirillum* sp., *Ochrobactrum rhizosphaerae*	Protected from flooding induced detrimental changes	Barnawal et al. 2012
	Tomato (*Lycopersicum esculentum*)	*Enterobacter cloacae, Pseudomonas putida*	Plant showed healthy growth against flooding stress	Grichko and Glick 2001
Temperature	Pepper (*Capsicum annuum* L.)	*Serratia nematodiphila* PEJ 1011	Increased endogenous gibberellins content, up regulated endogenous ABA level	Kang et al. 2015
	Tomato (*Solanum lycopersicum*)	*Flavobacterium* sp., *P. frederiksbergensis*	Reduction in stress responsive ethylene	Subramanian et al. 2015
	Grapevine (*Vitis vinifera*)	*Burkholderia phytofirmans*	Protected plants from heat and frost through increase in the levels of starch, proline and phenols	Barka et al. 2006
Heavy metal	Maize (*Zea mays*)	*Proteus mirabilis*	Reduced the negative consequences of oxidative stress caused by heavy metal toxicity	Islam et al. 2014
	Wheat (*Triticum aestivum*)	*Pseudomonas aeruginosa*	Improved necessary nutrient availability, elicited antioxidant defense system, lowering Zn metal uptake	Hussain et al. 2013
	Sunflower (*Helianthus annuus*)	*Micrococcus, Klebsiella* sp.	Enhanced Cd mobilization, Promoted root elongation, increased dry weight	Prapagdee et al. 2013

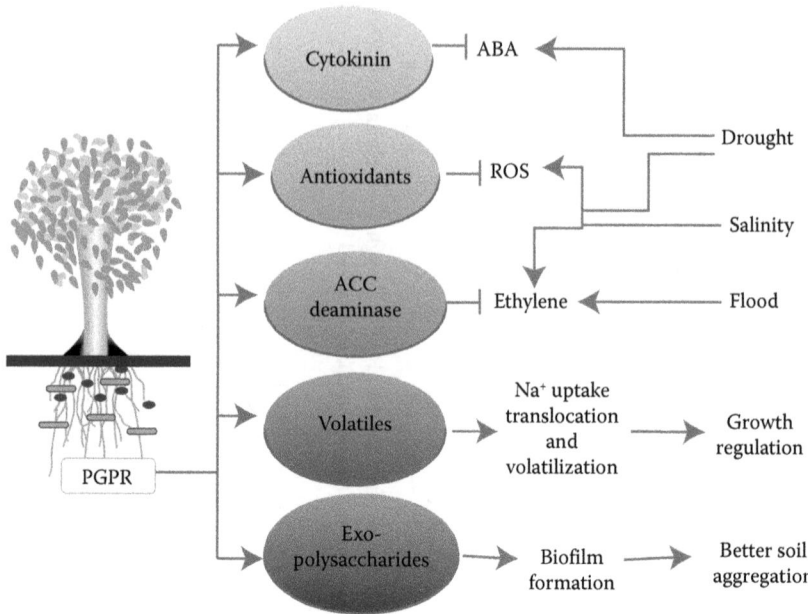

FIGURE 7.2 Illustration depicting uses of plant growth-promoting bacteria in the mitigation of abiotic stresses.

maize by increasing the total uptake of K, P, and Ca without affecting Na accumulation as compared with salt treatment, and the tolerance effect depended on the plant cultivar (Bano and Fatima, 2009). Generally, coinoculation was found to be better than single inoculation. Their mode of action appears to affect the production of proline as osmoregulants, protein production, reduction in electrolyte leakage, and increase in diameter over plant height.

In an experiment demonstrating the role of rhizobacteria to combat chilling stress in tomato it was observed that introduction of an exogenous ACC-deaminase gene in ACC-deaminase negative bacteria helped to overcome the chilling stress in treated plants. A significant reduction in ethylene levels, ACC accumulation, and 1-aminocyclopropane-1-carboxylate oxidase (ACO) activity in tomato plants inoculated with acdS expressing mutants was observed in comparison to plants inoculated with wild-type strains, clearly demonstrating the effect of bacterial ACC-deaminase with respect to chilling stress (Subramanian et al., 2015). The real-time polymerase chain reaction (RT-PCR) analysis has shown that induction of exogenous ACC-deaminase activity not only decreased the expression of ethylene-responsive transcription factor (ETF-13) and ACC-oxidase, but also increased the expression of cold-induced genes *Lycopersicum esculentum* CRT-repeat binding factor-1 (LeCBF1) and LeCBF3. These mutants of psychrotolerant bacteria provide a potential tool to be used for plant growth promotion and also to improve plant resistance against chilling stress conditions. A study on *Bacillus* strains showing tolerance to various abiotic stresses such as temperature, pH, and salinity produced very useful information regarding genetic variability in plant growth promotion (Kumar et al., 2014). Although, these strains were isolated from the same environmental source, they showed heterogeneity in their phenotypic and genetic characters. The studied isolates were categorized into four groups: *Bacillus amyloliquefaciens*, *Bacillus megaterium*, *Bacillus pumilus*, and *Bacillus subtilis* on the basis of molecular characterization. The results obtained were very informative regarding the development of bioinoculants for sustainable agriculture especially for increasing maize crop production under different stress conditions. Some of the possible adverse abiotic stresses that can be alleviated in plants by the applications of PGPR are depicted in Figure 7.2.

CONCLUSIONS

Abiotic stresses pose an unremitting threat to the growth and productivity of crop plants. PGPR provide a very economical and an environment friendly candidate for overcoming the adverse effects of several abiotic stresses by conferring tolerance in plants as well as by promoting their growth and productivity. There are several opportunities available in terms of a diverse array of rhizospheric microorganisms available presently and a large number being discovered everyday with novel properties of conferring tolerance to plants. Continued efforts on research and development in this field have a lot to give and these simple creatures having potential for stress mitigation when combined with modern tools of biology can produce radical changes in crop productivity patterns.

REFERENCES

Ahemad, M. and Kibret, M. 2014. Mechanisms and applications of plant growth promoting rhizobacteria: Current perspective. *Journal of King Saud University—Science* 26:1–20.

Ali, S.Z., Sandhya, V., Grover, M., Reddy, G., and Venkateswarlu, B. 2009. Alleviation of drought stress effects in sunflower seedlings by the exopolysaccharides producing *Pseudomonas putida* strain GAP-P45. *Biology and Fertility of Soils* 46:17–26.

Arkhipova, T.N. 2007. Cytokinin producing bacteria enhance plant growth in drying soil. *Plant and Soil* 292:305–315.

Azcon, R., Medina, A., Aroca, R., and Ruiz-Lozano, J.M. 2013. Abiotic stress remediation by the arbuscular mycorrhizal symbiosis and rhizosphere bacteria/*yeast interactions*. In *Molecular Microbial Ecology of the Rhizosphere*. ed. Feans J.de. Bruijn, 2:991–1002. John Wiley & sons, Hoboken, NJ.

Bano, A. and Fatima, M. 2009. Salt tolerance in Zea mays (L) following inoculation with *Rhizobium and Pseudomonas*. *Biology and Fertility of Soils* 45:405–413.

Barassi, C.A., Ayrault, G., Creus, C.M, Sueldo R.J., and Sobrero. M.T. 2006. Seed inoculation with *Azospirillum* mitigates NaCl effects on lettuce. *Scientia Horticulturae* 109:8–14.

Barea, J.M., Pozo, M.J., Azcón, R., and Azcón-Aguilar, C. 2005. Microbial cooperation in the rhizosphere. *Journal of Experimental Botany* 56:1761–1778.

Barea, J.M., Pozo, M.J. Azcón, R., and Azcón-Aguilar, C. 2013. Microbial interactions in the rhizosphere. *Molecular Microbial Ecology of the Rhizosphere* 1–2:29–44.

Barka, E.A., Nowak, J., and Clement, C. 2006. Enhancement of chilling resistance of inoculated grapevine plantlets with a plant growth-promoting rhizobacterium; *Burkholderia phytofirmans* strain PsJN. *Applied Environmental Microbiology* 72:7246–7252.

Barnawal, D., Bharti, N., Maji, D., Chanotiya, C.H., and Kalra, A. 2012. 1-Aminocyclopropane-1-carboxylic acid (ACC) deaminase-containing rhizobacteria protect *Ocimum sanctum* plants during waterlogging stress via reduced ethylene generation. *Plant Physiology and Biochemistry* 58:227–235.

Barrow, J.R., Lucero, M.E., Reyes-Vera, I., and Havstad, K.M. 2008. Do symbiotic microbes have a role in plant evolution, performance and response to stress? *Communicative & Integrative Biology* 1:69–73.

Bashan, Y., Salazar, B.G., Moreno, M., Lopez B.R., and Linderman, R.G. 2012. Restoration of eroded soil in the Sonoran Desert with native leguminous trees using plant growth promoting microorganisms and limited amounts of compost and water. *Journal of Environmental Management* 102:26–36.

Belimov, A.A., Dodd, I.C., Hontzeas, N., Theobald, J.C., Safronova, V.I., and Davies, W.J. 2009. Rhizosphere bacteria containing 1-aminocyclopropane-1-carboxylate deaminase increase yield of plants grown in drying soil via both local and systemic hormone signaling. *New Phytologist* 181:413–423.

Berg, G. 2009. Plant–microbe interactions promoting plant growth and health: Perspectives for controlled use of microorganisms in agriculture. *Applied Microbiology and Biotechnology* 84:11–18.

Bleecker, A.B. and Kende, H. 2000. Ethylene: A gaseous signal molecule in plants. *Annual Review of Cell and Developmental Biology* 16:1–18.

Burd, G.I., Dixon, D.G., and Glick, B.R. 2000. Plant growth-promoting bacteria that decrease heavy metal toxicity in plants. *Canadian Journal of Microbiology* 46:237–245.

Carmen, B. and Roberto D. 2011. Soil bacteria support and protect plants against abiotic stresses. In *Abiotic Stress in Plants Mechanisms and Adaptations*, ed. A.K. Shanker and B. Venkateswarlu, Pub. In Tech, Rijeka, Croatia, 143–158.

Chakraborty, U., Chakraborty B., and Basnet, M. 2006. Plant growth promotion and induction of resistance in *Camellia sinensis* by *Bacillus megaterium*. *Journal of Basic Microbiology* 46:186–195.

Chakraborty, U., Roy, S., Chakraborty, A.P., Dey, P., and Chakraborty, B. 2011. Plant growth promotion and amelioration of Salinity stress in crop Plants by a salt-tolerant bacterium. *Recent Research in Science and Technology* 3:61–70.

Cheng, Z., Woody, O.W., Mc Conkey, B.J., and Glick. B.R. 2012. Combined effects of the plant growth-promoting bacterium *Pseudomonas putida* UW4 and salinity stress on the *Brassica napus* proteome. Applied Soil Ecology 61:255–263.

Christensen, J.H., Hewitson B., Busuioc, A., Chen, A., Gao, X., Held, I., Jones, R. et al. 2007. Regional climate projections. In *Climate Change 2007: The Physical Science Basis. Contribution of Working Group I to the Fourth Assessment Report of the Intergovernmental Panel on Climate Change*, ed. S. Solomon, D. Qin, M. Manning, Z. Chen, M. Marquis, K.B. Averyt, M. Tignor, and H.L. Miller, Cambridge University Press, Cambridge, United Kingdom and New York, NY, USA.

Derr, D.C. and Tringe, S.G. 2014. Building the crops of tomorrow: Advantages of symbiont-based approaches to improving abiotic stress tolerance. *Frontiers in Microbiology* 5:1–6.

Dimpka, C., Weinand, T., and Asch, F. 2009. Plant–rhizobacteria interactions alleviate abiotic stress conditions. *Plant Cell & Environment* 32:1682–1694.

Editorial. 2010. How to feed a hungry world. *Nature* 466:531–532.

Egamberdieva, D. 2009. Alleviation of salt stress by plant growth regulators and IAA producing bacteria in wheat. *Acta Physiologiae Plantarum* 31:861–864.

Egamberdieva, D. 2012. Pseudomonas chlororaphis: A salt-tolerant bacterial inoculant for plant growth stimulation under saline soil conditions. *Acta Physiologiae Plantarum* 34:751–756.

Egamberdieva, D., Jabborova, D., and Hashem, A. 2015. *Pseudomonas* induces salinity tolerance in cotton (*Gossypium hirsutum*) and resistance to Fusarium root rot through the modulation of indole-3-acetic acid. *Saudi Journal of Biological Sciences* 22:773–779.

Egamberdieva, D. and Kucharova, Z. 2009. Selection for root colonizing bacteria stimulating wheat growth in saline soils. *Biology and Fertility of Soils* 45:561–573.

Egamberdieva, D., Kucharova, Z., Davranov, K., Berg, G., Makarova, N., Azarova, T., Chebotar, V., et al. 2011. Bacteria able to control foot and root rot and to promote growth of cucumber in salinated soils. *Biology and Fertility of Soils* 47:197–205.

Egamberdieva D. and Lugtenberg B. 2014. Use of plant growth promoting rhizobacteria to alleviate salinity stress. In *Use of Microbes for the Alleviation of Soil Stresses*, ed. M. Miransari, 1:73–96. Springer, New York, NY.

Else, M.A. and Jackson, M.B. 1998. Transport of 1-aminocyclopropane 1-carboxylic acid (ACC) in the transpiration stream of tomato (*Lycopersicon esculentum*) in relation to foliar ethylene production and petiole epinasty. *Australian Journal of Plant Physiology* 25:453–458.

Farooq, M., Hussain, M., and Siddique, K.H.M. 2014. Drought stress in wheat during flowering and grain filling periods. *Critical Reviews in Plant Sciences* 33:331–349.

Fu, Q., Liu, C., Ding, N., Lin, Y., and Bin, G. 2010. Ameliorative effects of inoculation with the plant growth-promoting rhizobacterium Pseudomonas sp. DW1 on growth of eggplant (*Solanum melongena* L.) seedlings under salt stress. *Agricultural Water Management* 97:1994–2000.

Glick, B.R. 2014. Bacteria with ACC deaminase can promote plant growth and help to feed the world. *Microbiological Research* 169:30–39.

Glick, B.R., Cheng, Z., Czarny, J., Cheng, Z., and Duan J. 2007. Promotion of plant growth by ACC deaminase-producing soil bacteria. In *New Perspective and Approaches in Plant Growth Promoting Soil Bacteria*, P.A.H.M. Bakker, J.M. Raaijmakers, G. Bloemberg, M. Hofte, P. Lemanceau, and B.M Cooke. Springer, Netherlands, 329–339.

Goswami, D., Dhandhukia, P., Patel, P., and Thakker, J.N. 2014. Screening of PGPR from saline desert of Kutch: Growth promotion in *Arachis hypogea* by *Bacillus licheniformis* A2. *Microbiological Research* 169:66–75.

Govindasamy, V., Senthilkumar, M., Kumar, U., and Annapurna, K. 2008. *PGPR-Biotechnology for Management of Abiotic and Biotic Stresses in Crop Plants.* I.K. International Publishing House Pvt. Ltd, New Delhi, India 26–48.

Grichko, P.V. and Glick, B.R. 2001. Ethylene and flooding stress in plants. *Plant Physiology and Biochemistry* 39:1–9.

Grover, M., Ali, S. Z., Sandhya, V., Rasul, A., and Venkateswarlu, B. 2011. Role of microorganisms in adaptation of agriculture crops to abiotic stresses. *World Journal of Microbiology and Biotechnology* 27:1231–1240.

Gupta, G., Parihar, S.S., Ahirwar, N.K., Snehi, S.K., and Gupta, V.S. 2015. Plant growth promoting rhizobacteria (PGPR): Current and future prospects for development of sustainable agriculture. *Journal of Microbial and Biochemical Technology* 72:96–102.

Gusain, Y.S., Singh, U.S., and Sharma, A.K. 2015. Bacterial mediated amelioration of drought stress in drought tolerant and susceptible cultivars of rice (*Oryza sativa* L.). *African Journal of Biotechnology* 14:764–773.

Hayat, R., Ali S., Amara, U., Khalid, R., and Ahmed, I. 2010. Soil beneficial bacteria and their role in plant growth promotion: A review. *Annals of Microbiology* 60:579–598.

Islam, F., Yasmeen, T., Riaz, M., Arif, M.S., Ali, S., and Raza, S.H. 2014. Proteus mirabilis alleviates zinc toxicity by preventing oxidative stress in maize (*Zea mays*) plants. *Ecotoxicology and Environmental Safety* 110:143–152.

Jackson, L.E., Bowles, T.M., Hodson, A.K., and C. Lazcano. 2012. Soil microbial-root and microbial-rhizosphere processes to increase nitrogen availability and retention in agroecosystems. *Current Opinion in Environmental Sustainability* 4:517–522.

Kang, S.M., Khan, A.L., Waqas, M., You, Y.H., Hamayun, M., Goo, G.J., Shahzad, R., Choi, K.S., and Lee, I.J. 2015. Gibberellin-producing *Serratia nematodiphila* PEJ1011 ameliorates low temperature stress in *Capsicum annuum* L. *European Journal of Soil Biology* 68:85–93.

Kang, S.M., Khan, A.L., Waqas, M., You Y.H., Kin, J.H., Kim, J.G., Hamayun, M., and Lee I.J. 2014. Plant growth promoting rhizobacteria reduce adverse effects of salinity and osmotic stress by regulating phytohormones and antioxidants in *Cucumis sativus*. *Journal of Plant Interaction* 9:673–682.

Kim, S.O., Moon, S.H., and Kim, K.W. 2001. Removal of heavy metals from soil using enhanced electrokinetic soil processing. *Water Air and Soil Pollution* 125:259–272.

Kloepper, J.W., Leong, J., Teintze, M., and Schroth, M.N. 1980. Enhanced plant growth by siderophores produced by plant growth promoting rhizobacteria. *Nature* 286:885–886.

Kohler, J., Caravaca, F., Carrasco, L., and Roldan, A. 2006. Contribution of *Pseudomonas mendocina* and *Glomus intraradices* to aggregates stabilisation and promotion of biological properties in rhizosphere soil of lettuce plants under field conditions. *Soil Use Management* 22:298–304.

Kumar, A., Kumar, A., and Pratush, A. 2014. Molecular diversity and functional variability of environmental isolates of *Bacillus species. SpringerPlus* 3:1–11.

Lugtenberg, B. and Kamilova F. 2009. Plant growth promoting rhizobacteria. *Annual Review of Microbiology* 56:541–556.

Marasco, R., Rolli, E., Ettoumi, B., Vigani, G., Mapelli, F., and Borin, S., 2012. A drought resistance-promoting micro biome is selected by root system under desert farming. *PLoS ONE* 7:1–14.

Mayak, S., Tirosh, T., and Glick, B.R. 2004. Plant growth-promoting bacteria confer resistance in tomato plants to salt stress. *Plant Physiology and Biochemistry* 42:565–572.

Miller, G., Susuki, N., Ciftci-Yilmaz, S., and Mittler, R. 2010. Reactive oxygen species homeostasis and signaling during drought and salinity stresses. *Plant, Cell and Environment* 33:453–467.

Miransari, M. 2012. Microbial products and soil stresses. In *Bacteria in Agrobiology: Stress Management*, ed. A.K. Maheshwari, Springer-Verlag, Berlin, 65–75.

Munns, R., Husain, S., Rivelli, A.R., James, R.A., Condon, A.G., Lindsay, M.P., Lagudah, E.S., Schachtman, D.P., and Hare, R.A. 2002. Avenues for increasing salt tolerance of crops, and the role of physiologically based selection traits. *Plant and Soil* 247:93–105.

Nadeem, S.M., Ahmad, M., Zahir, A., and Ashraf, M. 2012. Microbial ACC-deaminase biotechnology: Perspectives and applications in stress agriculture. In *Bacteria in Agrobiology: Stress Management*, ed. A.K. Maheshwari, Springer-Verlag, Berlin, 141–185.

Nadeem, S.M., Ahmad, M., Zahir, Z.A., Javaid, A., and Ashraf, M. 2014. The role of mycorrhizae and plant growth promoting Rhizobacteria (PGPR) in improving crop productivity under stressful environments. *Biotechnology Advances* 32:429–448.

Patten, C.L. and Glick, B.R. 1996. Bacterial biosynthesis of indole-3-acetic acid. *Canadian Journal of Microbiology* 42:207–220.

Prapagdee, B., Chanprasert, M., and Mongkolsuk, S. 2013. Bioaugmentation with cadmium-resistant plant growth-promoting rhizobacteria to assist cadmium phytoextraction by *Helianthus annuus*. *Chemosphere* 92:659–666.

Saharan, B.S. and Nehra, V. 2011. Plant growth promoting rhizobacteria: A critical review. *Life Science & Medicine Research* 2011: LSMR-21.

Sandhya, V., Ali, S.Z., Grover, M., Reddy, G., and Venkateswarlu, B. 2009. Alleviation of drought stress effects in sunflower seedlings by the exopolysaccharides producing *Pseudomonas putida* strain GAP-P45. *Biology and Fertility of Soils* 46:17–26.

Sarma, B.K., Yadav, S.K., Singh, D.H., and Singh, H.B. 2012. Rhizobacteria mediated induced systemic tolerance in plants: Prospects for abiotic stress management. In *Bacteria in Agrobiology: Stress Management*, ed. A.K. Maheshwari, Springer-Verlag, Berlin, 225–238.

Selvakumar, G., Panneerselvam, P., and Ganeshamurthy, A.N. 2012. Bacterial mediated alleviation of abiotic stress in crops. In *Bacteria in Agrobiology: Stress Management*, ed. A.K. Maheshwari, Springer-Verlag, Berlin, 205–224.

Srivastava, S., Yadav, A., Seem, K., Mishra, S., Choudhary, V., and Nautiyal, C.S. 2008. Effect of high temperature on *Pseudomonas putida* NBRI0987 biofilm formation and expression of stress sigma factor RpoS. *Current Microbiology* 56:453–457.

Subramanian, P., Krishnamoorthy, R., Chanratana, M., Kim, K., and Sa, T. 2015. Expression of an exogenous 1-aminocyclopropane-1-carboxylate deaminase gene in psychrotolerant bacteria modulates ethylene metabolism and cold induced genes in tomato under chilling stress. *Plant Physiology and Biochemistry* 89:18–23.

Sulmon, C., Baaren, J.V., Hurtado, F.C., Gouesbet, G., Hennion, F., Mony, C., Renault, D. et al. 2015. Abiotic stressors and stress responses: What commonalities appear between species across biological organization levels? *Environmental Pollution* 202:66–77.

Tank, N. and Saraf, M. 2010. Salinity-resistant plant growth promoting rhizobacteria ameliorates sodium chloride stress on tomato plants. *Journal of Plant Interaction* 5:51–58.

Timmusk, S. and Wagner, E.G.H. 1999. The plant-growth-promoting rhizobacterium *Paenibacillus polymyxa* induces changes in Arabidopsis thaliana gene expression: A possible connection between biotic and abiotic stress responses. *Molecular Plant–Microbe Interactions* 12:951–959.

Turan, M., Esitken, A., and Sahin, F. 2012. Plant growth promoting rhizobacteria as alleviators for soil degradation. In *Bacteria in Agrobiology: Stress Management*, ed. A.K. Maheshwari, Springer-Verlag, Berlin, 41–63.

Turner, T.R., James, E.K., and Poole, P.S. 2013. The plant microbiome. *Genome Biology* 14:1–10.

Upadhyay, S.K., Singh, J.S., and Singh, D.P. 2011. Exopolysaccharide-producing plant growth-promoting rhizobacteria under salinity condition. *Pedosphere* 21:214–222.

Vacheron, J., Desbrosses, G., Bouffaud, M.L., Touraine, B., Loccoz, Y.M., Muller, D., Legendre, L., Dye, F.W., and Combaret, C. P. 2013. Plant growth promoting rhizobacteria and root system functioning. *Frontiers in Plant Science* 4:1–19.

Van Loon, L.C. 2007. Plant responses to plant growth-promoting rhizobacteria. *European Journal of Plant Pathology* 119:243–254.

Yang, J., Kloepper, J., and Ryu, C.M. 2009. Rhizosphere bacteria help plants tolerate abiotic stress. *Trends in Plant Sciences* 14:1–4.

Zahir, Z.A., Munir, A., Asghar, H.N., Shahroona, B., and Arshad, M. 2008. Effectiveness of rhizobacteria containing ACC-deaminase for growth promotion of peas (*Pisum sativum*) under drought conditions. *Journal of Microbiology and Biotechnology* 18:958–963.

Zelicourt, A.D., Yousif, M.A., and Hirta, H. 2013. Rhizosphere microbes as essential partners for plant stress tolerance. *Molecular Plant* 6:242–245.

Zhang, Y.F., He, L.Y., Chen, Z.J., Wang, Q.Y., Qian, M., and Sheng, X.F. 2011. Characterization of ACC deaminase-producing endophytic bacteria isolated from copper-tolerant plants and their potential in promoting the growth and copper accumulation of *Brassica napus*. *Chemosphere* 83:57–62.

CHAPTER 8

Betulin Biotransformation toward Its Antitumor Activities

A Brief Overview

Dhirendra Kumar and Kashyap Kumar Dubey

CONTENTS

ABSTRACT

In this chapter, we have covered the general biology of betulin and betulinic acid, its distribution in the plant kingdom, biological and chemical
synthesis of betulin derivatives, and their pharmacologically important
activities to available bioprocess initiatives for their synthesis. Mainly we
have focused on specific activities based of betulin and its derivatives due
to change in their specific "R" group(s) chain at either C3, C20, or C28 or
another site in its lupeol skeleton. Betulinic acid (3b-hydroxy-lup-20(29)-
en-28-oic acid), a derivative of betulin is also a lupane-type pentacyclic
triterpene widely distributed with betulin in the plant kingdom, is a precursor for the synthesis of many derivatives such NVX207, B10, DSB, LH15,
LH55, and many more. A variety of biological activities have been recognized to betulin and its derivatives such as anti-human immunodeficiency
virus (HIV), anticancer, antibacterial, antihelminthic, anti-inflammatory,
and antimalarial (Figure 8.1). Betulin extensively isolated from birch bark
plants, selectively inhibits the growth of several different human cancer
cell lines. The poor effectiveness of betulin restricted its clinical development, because of being deficient in toxicity in animal studies even at high

LH15

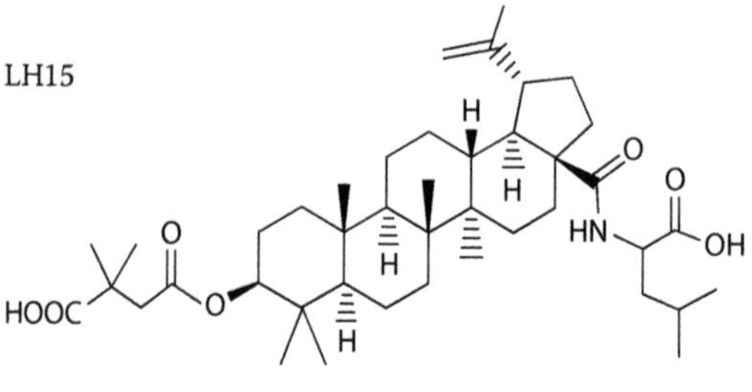

(LH15)-N-[3β-O-(3'3' dimethyl succinyl)-lup20(29)-
en-28-oyl]-L-Leucine

LH55

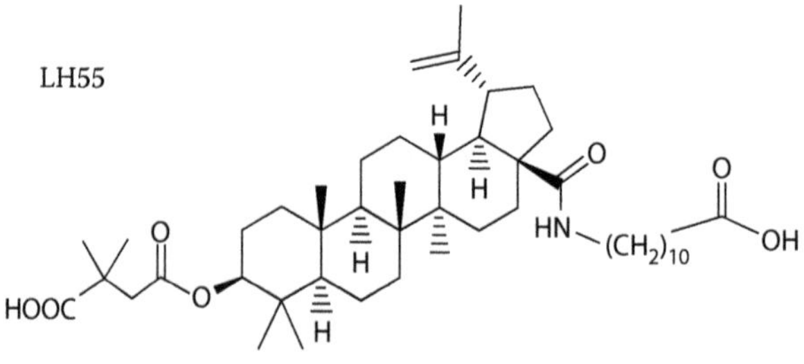

(LH55)-N-[3β-O-(3'3' dimethyl succinyl)-
lup20(29)-en-28-oyl]-11-aminodecaoic acid

FIGURE 8.1   Betulinic acid derivative with dual action mechanism against HIV.

concentrations. We also focused on betulin and betulinic acid as precursors for pharmacologically active molecules as their derivatives often are more effective than both of these. The biochemistry and chemotherapeutic potential of betulin and its analogs for human welfare, including, their formation by rearrangement from betulin derivatives are also the focus of the study. Newest method of action has been recognized for some of the most promising anti-HIV derivatives of betulin which made them potentially useful additive tools to the current anti-HIV therapies which hold great promise for their diagnosis and treatment.

GENERAL BIOLOGY OF BETULIN AND BETULINIC ACID

Plants have been the most common and reliable source for traditional medicines to precursor molecule of modern drugs throughout the world. Since ancient time to modern era, plants have been the best source of medication for even a small injury to serious wounds during a war. Herbs as a whole and/or different parts of the plants represent some of the most important part of traditional medicine all over the world; as the civilization progressed their uses and availability of their synthetic drugs advanced, but plants remained even an integral part of health care throughout the globe, especially in developing nations. In countries such as India, China, Thailand, Japan, Egypt, Russia, and Greece there have been a traditional great system of herbal medicine recognized by their civilization. Approximately 422,000 plant species have been reported (4160 are threatened) in which more than 52,000 plants are used for their known pharmacological properties (Govaerts 2001; Bramwell 2002). *Betula alba*, *B. utilis*, *B. pendula* are distributed throughout Siberian northern mediterranean areas and to temperate regions of Asia France, Iraq, Turkey, Native to Europe, and Northern Asia (*PDR for herbal medicines*, Medical Economic Co. Montvale, New Jersey, 2000, 178–179). *Betula pendula* is commonly recognized as white birch which possesses alkaloids, terpenoids, tannins, and many different secondary metabolites among which betulin and betulinic acid are most significant from a pharmacological perspective. Betulin (3b,28-dixydroxy-20(29)-lupene) and betulinic acid (3b-hydroxy-lup-20(29)-en-(28)-oic acid) are pentacyclic lupane-type triterpenes, both are found widely in the bark extracts of the *Betula* spp. plant in varying amounts. "Isoprenoid" are the building blocks of terpenes. The terpenes are the largest groups of phytochemicals consisting of more than 40,000 individual compounds, with many new compounds adding up over time. Terpenes are mostly from plant sources; however, they are also synthesized by other organisms, such as bacteria and yeast as part of primary or secondary metabolism. Terpenes are synthesized by the isoprenoid units, that is, based on the number of building blocks; terpenes are classified into different classes, such as monoterpenes (e.g., geraniol carvone, d-limonene, and perillyl alcohol), diterpenes (e.g., retinol and *trans*-retinoic acid), triterpenes (e.g., betulin, betulinic acid, oleanic acid, and ursolic acid), and tetraterpenes (e.g., α-carotene, β-carotene, lutein, and lycopene) (Peñuelas and Munné-Bosch 2005; Liby et al. 2007; Withers and Keasling 2007; Rabiand Bishayee 2009). The biosynthesis of triterpenes includes an amalgamation of triterpene hydrocarbon, cyclization of squalene, and precursor of all steroids (Phillips et al. 2006). They can be further classified into subgroups such as lupanes, sqalenes, oleananes, ursanes, cucurbitanes, cycloartanes, dammaranes, euphanes, friedelanes, holostanes, tirucallanes, isomalabaricanes, limonoids, lanostanes, hopanes, protostanes, and many more new classes (Setzer and Setzer 2003; Petronelli et al. 2009; Mullauer et al. 2010). Betulin and betulinic acid are widely distributed throughout the plant kingdom that includes *Betula alba* and other *Betula* spp., rosemary, thyme, apples, oregano, lavender, figs, olives, mistletoe, cranberries, and many more (Ovensná et al. 2004; Neto 2007; Gerhauser 2008; Laszczyk 2009; Rabi and Bishayee 2009).Though *Betula* spp. have been the largest source of betulin that leads to the chemical synthesis of betulinic acid but literature has already shown that these can be isolated from different plants as well (Table 8.1). Previously, terpenes were not considered more effective from a pharmacological point of view but after the recognition of antitumor activity of betulinic acid by researchers it ignited the interest of a large numbers of researchers considering its low toxicity with a larger range of pharmacological activity. Owing to their several medicinal properties such as sedative, analgesic, anti-inflammatory, antipyretic, cardiotonic, hepatoprotective, wound healing, terpenes have been widely used as traditional medicine in many countries such as Russia, China, India, and other South Asian countries. In the traditional medicine system various plant parts are used as decoctions for curing a variety of common diseases. Day by day reports are advocating the significance of triterpenes toward their lower or even zero toxicity on normal cells and remarkable cytotoxicity against a variety of cancer cells (Laszczyk 2009; Petronelli et al. 2009). The latest finding have not only confirmed the conventional properties of triterpenes, but in addition also highlighted a range of other biological activities including antitumor or anti-angiogenic, antiviral, antimicrobial,

TABLE 8.1   Distribution of Betulin and Betulinic Acid in Plants with Their Extraction Source and Solvent Used

Sl No.	Plant Name	Plant Part	Solvent Used	References
1	*Betula verrucosa Ehrh.*	Stem Bark	Ethanol	Banskota et al. (2000)
2	*Combretum quadrangulare*	Leaves	Methanol	De Melo et al. (2009)
3	*Clusia nemorosa L.*	Roots	n-hexane + ethanol	Kwon et al. (2003)
4	*Quisqualis fructus*	Leaves	Methanol	Chandramu et al. (2003)
5	*Vitex negundo*	Leaves	Methanol	Tezuka et al. (2000)
6	*Orthosiphon stamineus*	Leaves and twigs	Methanol	Siddiqui et al. (2000)
7	*Eucalyptus camaldulensis*	Leaves	Methanol	Yi et al. (2000)
8	*Tetracentron sinense*	Stem Bark	Methanol	Kim et al. (2000)
9	*Physocarpus intermedium*	Stem Bark	Methanol	Kim et al. (2002)
10	*Ilex macropoda*	Twigs	Methanol	Prakash et al. (2003)
11	*Caussarea paniculata*	Twigs	Methanol + dichloromethane	Enwerem et al. (2001)
12	*Berlinia grandiflora*	Stem bark	Methanol + hexane + ethyl ether	
13	*Tetracera boiviniana*	Leaves	Methyl ethyl ketone	Schühly et al. (1999)
14	*Ziziphus joazeiro*	Stem bark	Dichloromethane	Steele et al. (1999)
15	*Upaca nidida*	Root bark	Ethanol	Recio et al. (1995)
16	*Diospyros leucomelas*	Root bark	Ethanol	Krogh et al. (1999)
17	*Ipomea pescaprae*	Root bark	Ethanol	Fujioka et al. (1994)
18	*Syzgium claviforum*	Leaves	Ethanol	Mukherjee et al. (1997)
19	*Nelumbo nucifera*	Rhizomes	Methanol	
20	*Anemone raddeana*	Leaves	Ethanol	de Oliveira et al. (2002)
21	*Doliocarpus schottianus*	Leaves and wood	Ethanol	Setzer et al. (2000)
22	*Syncarpia glomulifera*	Stem bark	Chloroform	Nyasse et al. (2009)
23	*Uapaca acumilata*	Stem bark	Ethyl acetate	Ren and Omori (2012)
24	*Platanus occidentalis* (sycamore bark)	Stem bark	Hot water + organic solvent	

antioxidant, anti-allergic, antipruritic, and spasmolytic activity (Sultana and Ata 2008; Shah et al. 2009). Nowadays because of marvelous reports published on the pharmacological significance of triterpenes particularly on betulin derivatives, researchers' increasing interest made it a novel promising molecule of the time. Structural modification of natural compounds has been the basic and most reliable criteria for the synthesis of newer derivatives with a broad spectrum of activities. Research has shown that semi-synthetic derivatives are considered to be the most potent anti-inflammatory and anticarcinogenic agents (Liby and Yore 2007). Many triterpenes are reported under evaluation of phase I trials for antitumor activity with demonstrated antitumor effectiveness in preclinical animal models of cancer (Laszczyk 2009; Petronelli et al. 2009). Betulinic acid, as a derivative of betulin, is regarded as one of the important pentacyclic triterpene used for a drug screened by the National Cancer Institute shown diverse pharmacological properties (Pisha et al. 1995). Betulin can be extracted upto 30% of dry weight while betulinic acid is available only 0.5% of that. Betulin, specially with betulinic acid is widely distributed in the plant kingdom, and its various derivatives are used for many significant biological properties such as antibacterial, antimalarial, antcarminative, anti-leishmanial, antioxidant, and above all antitumor and as anti-human immunodeficiency virus (HIV; Yogeeswari and Sriram 2005). Betulin is a pentacyclic triterpene with a lupeol ring as its backbone is a precursor molecule for the synthesis of betulinic acid through an intermediate compound betulone by chemical synthesis (Figure 8.2) (Yogeeswari 2005). Generally, betulinic acid is found in very low amounts of upto 0.5%–1.5% w/w from birch bark extraction; so betulin is widely used for its synthesis. The cyclization of squalene leads to the formation of pentacyclic triterpenes such as betulin and betulinic acid, which varies from plant to plant and in different species. The highest extracted amount of betulin was reported to be more than 30% from birch outer bark.

DISTRIBUTION OF BETULIN AND BETULINIC ACID IN THE PLANT KINGDOM

The silver birch grows naturally from Xinjiang province in China, Mongolia, Siberia, to Western Europe, to Kazakhstan, and southward to the mountains of the Caucasus and Northern Iran, Iraq, and Turkey. It is also native to Northern Morocco and has become naturalized in some other parts of the world such as the Himalayan Mountains. Its light seeds are easily blown by the wind, a pioneer species, one of the first trees to

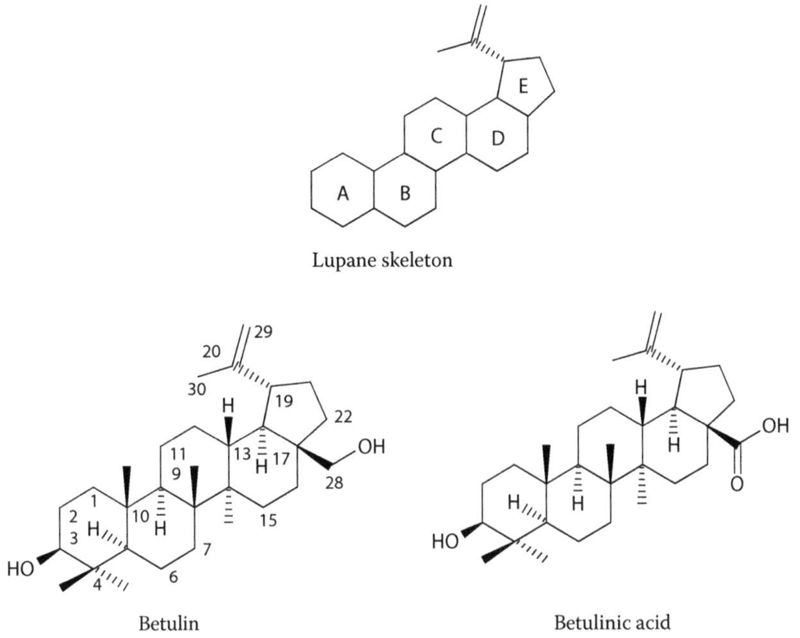

Lupane skeleton

Betulin

Betulinic acid

FIGURE 8.2   Structure of betulin and betulinic acid with their lupane skeleton.

sprout on bare land or after a forest fire. It has great adaptability in polluted regions and needs plenty of light and does best on dry, acidic soils of mountain sides. Its tolerance to pollution makes it suitable for planting in industrial areas and exposed sites (Franiel and Babczynska 2011). It has been introduced into North America where it is known as the European white birch, and is considered invasive to Kentucky, Maryland, Washington, and Wisconsin.

It is naturalized and locally invasive in parts of Canada. The closely related *Betula platyphylla* in Northern Asia and *Betula szechuanica* of Central Asia are also treated as varieties of silver birch by some botanists, as *B. pendula* var. *platyphylla* and *B. pendula* var. *szechuanica*, respectively (Hunt 1993). *Betula pendula* is distinguished from *B. pubescens*, in having hairless, warty shoots (if hair without warts are present then it is downy birch), more triangular leaves with double serration on the margins (single serration and more ovoid leaves in downy birch), and whiter bark often with scattered black fissures (while less in numbers and gray-colored fissured, in downy birch). We can distinguish them also according to their genetic makeup such as silver birches are diploid whereas downy birch is tetraploid. Their hybrids are reported but very rare, and being triploid, are sterile. The two have differences in habitat requirements, with silver birch is found mainly on dry, sandy soils, and downy birch more common on wet, poorly drained sites such as clay soils and peat bogs. Silver birch also demands slightly more summer warmth as compared with downy birch, which is significantly available in the cooler parts of Europe. Owing to increasing variation in plant systematic studies and confusion in the nomenclature of similar individual plants plant taxonomist changed the name *Betula alba to Betula pendula, although also known as Betula pubescens.* Synonyms include *Betula pendula* var. *carelica* (Merckl.) Hämet-Ahti, *Betula pendula* var. *lapponica* (Lindq.) *Betula verrucosa* Ehrh., *Betula fontqueri* Rothm, *Betula pendula* var. *laciniata* (Wahlenb.), *Betula aetnensis* Raf., *Betula montana* V.N. Vassil, *Betula talassica* Poljakov. The rejected name *Betula alba* L. also applied in part to *B. pendula*, though also to *Betula pubescens* (Govaerts 1996; Govaerts and Frodin 1998).

BIOCHEMICAL SYNTHESIS OF BETULIN DERIVATIVES AND THEIR PHARMACOLOGICALLY IMPORTANT ACTIVITIES

Betulin in its native form is not that much active, than its derivatives such as betulinic acid and many others. To obtain different derivatives of betulin and betulinic acid various methods are followed that include esterification, oxidation, hydroxylation, acetylation, glucosidation biotransformation, and many more as per the modification needed. Generally C3, C20, and C28 positions are more prone to be transformed but other than these other modifications were also observed. Researchers attracted more toward the amino derivatives of betulinic acids because of their higher anti-HIV activity followed by the publication of the patent. In another published report, researchers focused on the synthesis of the inverted peptides with the same scheme providing the interaction between the chloroanhydride of 3-O-betulic acid with the excess of 1,7 di-aminoheptane as the first stage (Soler et al. 1996). Esterified betulin derivatives allobetulin and halogen derivatives of betulin were also described (Kashiwada et al. 1996; Flekhter et al. 2002a). Derivatives with acid conjugates such as dimethyl succinic betulinic acid (DSB), dimethyl glutaric acid, and oxapenta-dicarboxylic acids were patented (Kashiwada et al. 1996). Modified derivatives have shown very promising activities against life-threatening diseases such as HIV and cancer. Succinic acid, glutaric acid, and campholic acid used as catalyzing agents for the synthesis of mono-, di-, and tri-acylates of betulin, respectively, were found to be very promising anti-HIV agents (Holz-Smith et al. 2001).

Microbial Catalysis as a Tool for Betulin Biotransformation

It is observed by the chemistry of various steroids and terpenes that biological oxidation can give products at a higher yield and under safer mode which even cannot get through chemical synthesis. As reported earlier, betulinic acid and it derivatives have shown significant effect on various cancer cell lines, which encourage researchers to find more derivatives of the same either through chemical or biological means. In a study *Bacillus megaterium ATCC 13368* due to the presence of P450 mono-oxygenase enzyme gave four metabolites, recognized as 3-oxo-lup-20(29)-en-28-oic acid, 3-oxo-11a-hydroxy-lup-20(29)-en-28-oic acid, 1b-hydroxy-3-oxo-lup-20(29)-en-28-oic acid, and 3b,7b,15a-trihydroxy-lup-20(29)-en-28-oic acid which have shown good anticancer activity on Mel I (lymph node) and Mel II (pleural fluid) human melanoma cell lines (Chatterjee et al. 2000). The biotransformation carried through *Aspergillus ochraceus* and *Mucor rauxii* gave the modified product with C3, C28, and C19 positions (Flekhter et al. 2002a). Studies on transformation through *Chaetomium longirostre* also gave new derivatives of betulin (Akihisa et al. 2002).

Betulin Biotransformation through Glycosylation

In the biological system, the addition of sugar molecules to nonsugar molecules is known as glycosylation and the product is called as glycosides. It was observed that the betulin by itself does not show good biological activity but its glycoside derivative gives better results on application. Glycosides of betulin (yield up to 17%) was obtained by the reaction of betulin with a acetobromo-glucose, using cadmium carbonate ($CdCO_3$) as a catalyst. The major products were anomeric glycoside obtained by the glycosylation of betulin 3-O-acetate. The synthesis of 2-desoxy- a-D, 2-desoxy- a-L, and 2-6 deoxy-a-L arabino-exopyranoside of betulin were catalyzed by glycosylation of 3 and 28 monoacetates of betulin under acidic conditions by glycol acetate (Flekhter et al. 2002a); a series of O,S, and N betulinic acid derivatives were patented for its specific pharmacological significance.

RED–OX Mechanism for Betulin as a Tool for Betulin Biotransformation

Betulin when treated with mono-perphalic acid gave epoxide derivatives, while osmium tetra-oxide gave 20–29 dihydroxy betulin and 3b,-28-dixylroxylup20 (29)-ene-30-al was obtained as an oxidized product when treated with selenium di-oxide. Betulin and its acetylated derivative give

3-O-acetyl betulinic aldehyde and betulonic aldehyde in the presence of strong oxidizers such as chromium tri-oxide. While obtaining betulonic aldehyde under DMSO and cobalt chloride in combination with Pd (OAc), a high yield (93%) was reported (Krasutsky et al. 2001; Flekhter et al. 2002a). Acetyl betulinic acid was easily obtained from betulin aldehyde under the reaction of N-hydroxy-phthimide,2,2′ azobisisobutyronitrile,2,2,6,6 tetramethyl-1-piperidiumoxyl. While studying *in situ*, regeneration of the betulin aldehyde was reported during the selective oxidation through SHO while betulin was the precursor. Chromium containing reagents such as pyridinium-dichromate (PDC) and pyridinium-chlorochromate (PCC) efficiently formed betulonic aldehyde on oxidation of betulin. Betulonic acid was obtained as a main product during the oxidation of betulin in acetic acid with chromic acid (Le Bang Shon 1999). The available litera-ture indicates that platinum, palladium, and nickel were in practice for the hydrogenation of double bond in betulin and betulinic acid. The reac-tion of the 3-O-acetyl -18-19-dehydromethyl betulinate gave the mixture of four compounds, because of the reduction in the presence of lithium in ethylene diamine. Although 3b, 28, 30-trihydroxy lupane was synthesized by the hydroboration of betulin, while on another hand methyl betulinate was transformed to the 3-keto-29-carboxylic acid (Dinda et al. 1995).

Betulin Biotransformation through Modification in Ring "A"

A lactone was synthesized from the "A" ring modification in the methyl ester of dihydrobetulinic acid under the oxidation of a mixture of hydro-gen peroxide with SeO_2. The lactams are generally synthesized from methyl ester of the betulonic acid during heating with hydroxyl amine-O-sulphonic acids in a solution of HCO_2H. The oximes of the reaction of H_2SO_4 form the oleanane series lactams and acetonitrile. Bayer–Willinger reaction was used to opening ring "A" in betulin, whereas lactone as an oxidized product of allobetulone formed with the action of per-acetic acid and methyl ester of seco-acid is formed through its methanolysis. Photo catalysis was also used in the modification of "A" ring in 28-monoace-tate of betulin and dihydro betulin with I_2pHi (AOc)$_2$, where 4 demethyl derivatives of betulin were formed through seco-aldehyde of the previ-ous photolysis reaction (Deng and Snyder 2002). The formation of keto-aldehyde and tri-ketones were described by ozonolysis through opening of "A" ring (Flekhter et al. 2002a). Condensation reaction at C-2 position of allobetulin and betulonic acid was also reported (Flekhter et al. 2000b). Allobetulone and 3 keto betulin gave diosphenol as a product on oxidation reaction ring "A." Another "A" ring derivative 1,2,3, thiodiazole was synthesized from 3-semicarbazone of betulinic acid on Hard–Mewory reaction (Flekhter et al. 2002a). Sulfur and nitrogen containing heterocyclic derivatives were synthesized through classical methods. The formation of dimethyl carbonyl compounds was reported while allobetulone was condensed with ethyl formate, diethyl oxalate, and ethyl trifluroacetate, when treated with hydrazine, hydroxylamine, and methyl hydrazine giving respective heterocyclic compounds.

ANTITUMOR ACTIVITIES THROUGH BETULIN AND ITS DERIVATIVES

Plants and their products have been the best source for treatment of diseases. Betulin and its most widely used derivative betulinic acid act as newer pioneer anticancer drugs. Investigations on pioneer anticancer drugs such as vinblastin and vincristin from *Vinca roseu,* have shown the vanishing of microtubules and emergence of crystal structures upon vinca alkaloid treatment (Risinger et al. 2009). The mechanism of these drugs was identified as the destabilization of microtubules, which blocks mitotic spindle formation that leads to G2/M arrest and apoptosis (Martin and Pasquier 2008). First, taxol from *Taxus buccata* was shown to promote microtubule assembly which is now approved by the FDA to treat various types of ovarian cancers (Chabner and Roberts 2005). Another significant molecule is colchicine and its one of the significant derivatives 3-demethylated thiocolchicin (Dubey et al. 2008, 2013; Dubey and Behera 2011). Naturally colchicine is obtained from *Colchicum autumnale L* and *Gloriosa superba L.* plants but in very low amounts (Ellington and Bastida 2003), colchicoside, 3-demethylated colchicine (3-DMC), thio-colchicocide are more significant and broad-spectrum derivatives of colchicine have shown improved curative features as antitumor and anti-inflammatory drugs. Reports are also available for their use in the clinical behavior of definitive forms of leukemia and solid tumors (Rosner et al. 1981; Solet et al. 1993; Ellington and Bastida 2003). The antitumor property of betulin and its derivatives have been of major interest since it was first documented for anticancer studies (Yasukawa et al. 1991). In a systematic study for screening of more than 2500 plant extracts reports suggested about its antimelanoma activity. It showed remarkable *in-vitro* cytotoxic activity against MEL-1 and MEL-2 cell lines (derived from lymph node and pleural fluid, respectively) with IC_{50} from 0.5 to 4.5 µg/m while other tumor cell lines were found to be resistant comparatively. The antimelanoma activity of betulinic acid was confirmed by membrane blubbing and shrinking of cell together with the detection using flow cytometry analysis of MEL-2 cells. Further *in-vivo* studies also showed highly effective tumor inhibition on nude mice at intraperitoneal upto 500 mg/kg body weight and were selected in rapid access to intervention development (RAID) Programme of National Cancer Institute (USA). A pioneer work that confirmed the effect of betulin derivative (i.e., betulinic acid) was in an *in-vivo* ovarian carcinoma xenograft mouse model (Zuco et al. 2002). Further it was also reported to be effective on, head and neck squamous cellular carcinoma. Studies also indicated toward betulinic acid prospective for the treatment of hematological malignancies (Thurnher 2003). It was also found that murine leukemia cell line L1210 was sensitive to betulinic acid on exposure to specific duration and pH (Noda et al. 1997). It was also reported that risk stratification, leukemia type, patient age, and sex were the significant factors for apoptosis induction by betulinic acid derived from betulin (Ehrhardt et al. 2004). Betulinic acid also induced apoptosis in the anti-leukemic treatment resistant human chronic myelogenous leukemia (CML) cell line K-562 (derived from the blast crisis stage) without affecting the levels of Bcr-Abl (Raghuvar et al. 2005). On the other hand, betulinic acid as a most common and widely used derivative of betulin is found effective but need more specific work against solid, widespread tumor types such as colon, breast, prostate, and lung cancer. Work done so far, to know the effect of betulin and betulinic acid or/with its derivatives on tumors are summarized in Figure 8.3. Although the exact mechanism of the action of betulin and its derivatives are not confirmed, the processes of apoptosis induction in cancer cells of tumors are most widely acceptable and are the suggested mechanisms. Although many researchers worked on betulinic acid-induced apoptosis, their views differ altogether especially on the role of Bcl-2; instead of that other pathways and targets have also been suggested to be involved in betulinic acid-induced cytotoxic effects. Different research groups have suggested different targets to explain the pathways and their role in betulinic acid derived cell death. These include the proteasome (Chintharlapalli et al. 2007; Huang et al. 2007), cell cycle regulation (Fulda 1998a,b; Sawada et al. 2004; Rieber et al. 2006), transcription factor NF-kB, and enzymes such as acetyl-CoA acetyltransferase, kinases, topoisomerase I/II, and aminopeptidase N (CD 13). One of the recent studies suggested that betulinic acid and cisplatin are promising antitumor agents, which induce apoptotic cell death of cancer cells. They synthesized a new series of betulinic acid–cisplatin conjugates and assayed cytotoxicity and selectivity against five different tumor cell lines (Emmerich

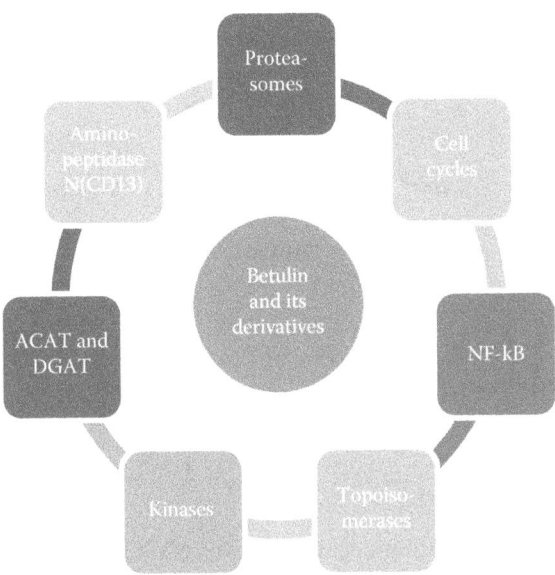

FIGURE 8.3   Molecular targets for betulin and its derivatives (betulinic acid).

et al. 2014). The plan was to join two structural units, both connected with apoptosis induction. A dose-dependent anti-proliferative action and the effect of these structural variations on anticancer activity were studied and discussed at micro-molar concentrations. Many of the derivatives have shown significant antitumor activity; 3-O-acetylbetulinic (2-(2-amino ethyl) amino ethyl) amide has shown the best antitumor activity (IC_{50} = 1.30–2.24 μM). Fascinatingly, betulinic acid–cisplatin conjugates were less cytotoxic than the precursors. Recently glioblastoma cell lines (U251MG, U343MG, and LN229) were used to study the radiobiological and cellular action of newly synthesized molecules NVX207 and B10 betulinic acid derivative (Bache et al. 2014). As shown in Figure 8.4, modified betulinic acid derivatives, NVX207 and B10 were an effective tool against tumors. It was based on comparative IC_{50} value of betulinic acid with NVX 207 and B10 that has shown 2.9- and 1.3-fold higher cytotoxicity, respectively. Decreased cell migration and protein expression levels of survivin, cleavage of poly(ADP-ribose) polymerase (PARP) were observed using both derivatives. Weak radiation sensitivity enrichment was seen in U251MG cells after treatment with both BA derivatives. The enrichment factors at an irradiation dose of 6 Gy after treatment with 5 μM NVX-207 and 5 μM B10 were 1.32 (p = 0.029) and 1.55 (p = 0.002), respectively. In comparison to

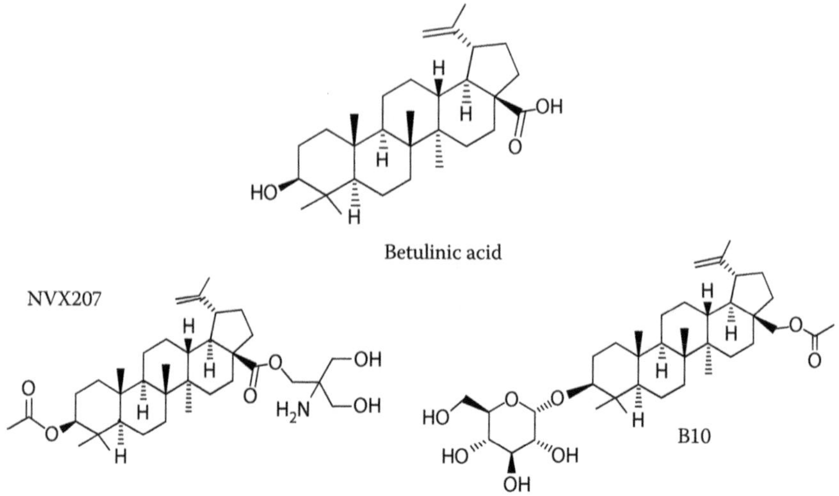

FIGURE 8.4 Modified betulinic acid derivatives, NVX207 and B10 as an effective tool against cancer. (From Bache, M. et al. *Int. J. Mol. Sci.* 2014, 15, 19777–19790.)

betulinic acid, neither B10 nor NVX-207 had additional effects under hypoxic conditions. Results suggested that under normal oxygen conditions NVX-207 and B10, the derivatives of betulinic acid improve the effects of radiotherapy on human malignant glioma cells. The therapeutic potential of betulinic acid on bone metastases and skeletal complications in breast cancer patients were investigated (Park et al. 2014). They suggested the protective and therapeutic potential of betulinic acid on cancer-associated bone diseases. It is a pioneering finding indicating the effect of betulinic acid on breast cancer cells, osteoblastic cells, and osteoclasts in the vicious cycle of osteolytic bone metastasis. Betulinic acid reduced cell viability and the production of parathyroid hormone-related protein (PTHrP), a major osteolytic factor, in MDA-MB-231 human metastatic breast cancer cells stimulated with or without tumor growth factor-β. A raise in the receptor activator of nuclear factor-kappa B ligand (RANKL)/osteoprotegerin ratio was blocked by betulinic acid by lowering RANKL protein expression in PTHrP-treated human osteoblastic cells. Betulinic acid has shown inhibition of RANKL-induced osteoclastogenesis in murine bone marrow macrophages and decreased the production of resorbed area in plates with a bone biomimetic synthetic surface by suppressing the secretion of matrix metalloproteinase (MMP)-2, MMP-9, and cathepsin K in RANKL-induced osteoclasts. Furthermore, estrogen deprivation was observed during oral administration of betulinic acid inhibited bone loss in mice intra-tibially inoculated with breast cancer cells and in ovary-ecto-mized mice as supported by the restored bone morphometric parameters with serum bone turnover markers. Taken collectively, these results pro-posed that betulinic acid may have the potential to prevent bone loss in patients with bone metastases and cancer treatment-induced estrogen insufficiency. A pioneer study on athymic nude mice, bearing MCF-7 breast adenocarcinoma xenografts was taken as the *in-vitro* cytotoxic and *in-vivo* antitumor model (Damle et al. 2013). The antitumor activity of betulinic acid was studied at 50 and 100 mg/kg body weights, whereas cytotoxic activity of MCF-7 cells with IC50 value of 13.5 μg/mL was studied by MTT assay. Betulinic acid treatment has shown significant reduction in tumor size by 77% and 52% tumor size (100 and 50 mg/kg body weight, respec-tively); in addition decreased angiogenesis, proliferation, and invasion in betulinic acid treated mice were also highlighted through histo-pathologi-cal studies. Through topomer CoMFA, some 35 derivatives of betulinic acid were prepared and analyzed against HT29 human colon cancer cells (Ding et al. 2013). The contour maps have shown that bulky and electron-donat-ing groups could be encouraging for activity at the C-28 position, and a moderately bulky and electron-withdrawing group near the C-3 position would improve this activity. Few of the betulin derivatives were designed and synthesized as per the modeling result, while groups such as maleyl, phthalyl, and hexahydrophthalyl (bulky electronegative groups) were directly introduced at the C-28 position. They also found consistency with predicted and actual IC_{50} values of the given analogs against HT29 cells, proving that the present topomer CoMFA model is successful and that it could potentially forward the synthesis of new betulinic acid derivatives with high anticancer activity. Five tumor cell lines were tested against three newly synthesized derivatives where 28-O-hexahydrophthalyl betulin has shown the greatest anticancer activities and its IC_{50} values were lower than other tumor cell line except DU145.In an investigation *Prunella vulgaris* was used for betulinic acid and ursolic acid extraction, which were respon-sible for the anti-estrogenic effects, suggesting their potential application against estrogen-dependent tumors (Kim et al. 2014). In this study, *Prunella vulgaris* constituents were isolated and their individual anti-estrogenic effects were tested. Betulinic acids, ursolic acid, rosmarinic acid, caffeic acid, oleanolic acid, hyperoside, and rutin were isolated from the flower stalks of *P. vulgaris* var. lilacina Nakai (Labiatae) which showed anti-estro-genic effects as a decreased level in the mRNA of GREB1, revealed significant anti-estrogenic effects of betulinic acid and ursolic acid, on estrogen receptors (ERs). They also demonstrated the suppression of estrogen response element (ERE)-dependent luciferase activity and expression of estrogen-responsive genes in response to exposure to estradiol, which also supported the suppressive role of these compounds in estrogen-induced signaling. On the other hand, none of them were capable of suppressing estrogen signaling in cells ectopically over-expressing estrogen receptor α (ERα). Both mRNA and protein levels of ERα were reduced by treatment with betulinic acid and ursolic acid recommended they inhibit estrogen signaling by suppressing the expression of ERα. In a study researchers targeted to determine the regulation mechanism and role of lamin B1 expression in betulinic acid-based therapy for human pancreatic cancer pathogenesis, (Li et al. 2013). The increased level of Lamin B1 leads to the formation of pancreatic cancer to show their connection with the clinico-pathologic characteristics of pancreatic cancer, during identifying betulinic acid target genes through cDNA microarray and utilized tissue microarray to establish the expression levels of lamin B1 in pancreatic cancer tissues. As a result betulinic acid action significantly downregulates lamin B1 in pancreatic cancer in both *in-vitro* culture and xenograft models. *In-vitro* and *in-vivo* models were used in finding the biological impacts of altered lamin B1 expression and the mechanisms underlying lamin B1 overexpression in human pancreatic cancer. Overexpression of lamin B1 was pronounced in human pancreatic cancer, and increased lamin B1 expression was directly linked with low-grade segregation, improved occurrence of distal metastasis, and reduced forecast of patients with pancreatic cancer. Knockdown of lamin B1 also significantly attenuated rise and invasion of pancreatic cancer cells. First time a landmark report was available on Bax/Bak-independent cytochrome-C release induced by betulinic acid in human nasopharyngeal carcinoma cells (Liu and Luo 2012). Usually most of the chemotherapeutic agents kill melanoma, leukemia, lung, colon, breast, prostate, and ovarian cancer cells via induction of caspase activation dependent apoptosis. Here they reported conventional apoptosis by betulinic acid in CNE2 cells, a cell line originated from nasopharyngeal carcinoma cells. Overexpression of Bcl-2 and Bcl-xL could partially prevent apoptosis caused by betulinic acid. In addition, Bax was not activated during the induction of apoptosis. Bax/Bak knockdown and wild-type CNE2 cells have shown the same kinetics of cytochrome-C release. It has also been shown that betulinic acid provides cytochrome-C release from damaged mitochondrial permeability transition pores (mPTPs) and suggested that betulinic acid could be an effective anticancer representative in human nasopharyngeal carcinoma. The cytotoxicity of betulinic acid in two breast cancer cell lines MCF-7 and T47D and their modification in p53 status were evaluated. This indicates p53-independent apoptotic pathway, because response of both p53 mutant and wild-type cell lines was found unchanged after action with pifithrin-α, an inhibitor of p53. Cells were considerably protected when treated with tocopherol, suggesting the involvement of membrane-centered lipid peroxidation-induced system in betulinic-acid-mediated apoptosis (Tiwari et al. 2014).

CONCLUSION AND FUTURE DIRECTIONS

Betulin, as a precursor for betulinic acid and other modified new molecule from this, has, basically a triterpene, emerged as a promising plant-based molecule that selectively destroys cancer cells with a specific approach of action while safe to normal cells (Zuco et al. 2002). As the literature shows, betulinic acid shows broad antitumor effects and it is supposed that there may be more targets that could be investigated in the future for more site specific and better treatment. Due to the advancement in medical science and awareness about tumors, human mortality has been decreased but still tumors are one of the most common causes of death. From various studies via cell culture assays and animal models of tumors as mentioned in this chapter it is clear that betulin holds great prospect not only in the cure of a broad range of tumors but also in preventing these diseases. Betulin and its tailored derivative have shown a wide choice of biological actions with inhibition of inflammation, drop of oxidative stress, regulation of cell cycle, inhibition of tumor proliferation, encouragement of apoptosis, and above all relations with tumor micro-environment via modulation of multiple signal transduction pathways. A landmark study pointed toward the genetic complexity of specific neoplastic diseases showing mutations in about 200 genes detected in breast and colon cancers. Further studies are required to investigate the full potential of these multifunctional molecules in combinatorial studies. Different findings suggest that even the lack of effectiveness for an individual agent at low concentrations, combinations of two or more compounds could be much more effective. Combining numerous reaction groups and/or molecules with betulin skeleton also proposes the chance of reducing their doses and consequently dropping unwanted adverse effects. In view of these advantages, betulin and its derivatives may be used in combination with other chemotherapeutic drugs and radiation therapy to enhance their remedial value while preventing the chemotherapy- and radiotherapy-linked unnecessary side effects. However, more studies are required to validate these finding as shown in Figure 8.5, such as the betulin macrocyclized derivative as HIV 1, maturation inhibitor). Available text highlights that though there are a good number of *in-vitro* studies demonstrating the cytotoxicity of betulin derivatives against various tumor cell lines, only a few are evaluated in preclinical animal models. Perhaps one of the causes for the restricted number of preclinical antitumor studies on betulin and its derivatives includes lack of *in-vivo* studies on derivatives which have already shown efficacy in cell culture systems. The reason behind is that most of these are insoluble in aqueous media restraining their bioavailability in the body which is very vital for *in-vivo* efficacy. To conquer this problem one approach is to enhance water solubility via structural modification of naturally occurring compounds to give more polar composites. Several other possibilities of improving the hydrophilicity of betulin and other terpenes include design and generation of formulations containing nanoparticles, liposomes, cyclodextrin complexes, colloids, and micelles (Guo et al. 2003; Chen et al. 2005; Kang et al. 2007). As presented here we have highlighted many unique biological properties of betulin with its derivatives with an objective of evaluating its clinical potential in the prevention and cure of tumors. Nevertheless, a considerable amount of work needed to be done which includes identification of novel target proteins and pathways in which they function for further drug interactions, and development of selective endpoint and obtaining biomarkers for analyzing positive clinical results. Long-term well-designed clinical trials are also required as well as low-cost biotransformation and modification strategies should also be developed. In short, we strongly suggest that

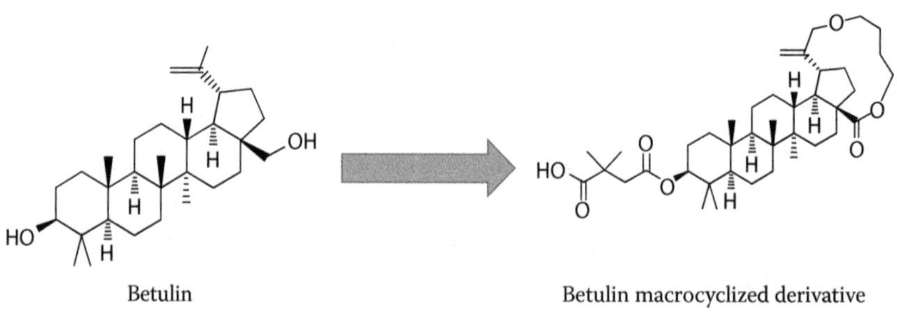

Betulin   Betulin macrocyclized derivative

FIGURE 8.5   Betulin macrocyclized derivative as HIV 1, maturation inhibitor. (From Jun tang et al. *Med. Chem. J.* **2014**, 8, 23–27.)

betulin with its derivatives are potential candidates for effective chemo preventive and chemotherapeutic strategies to reduce the load of tumors in mankind.

ACKNOWLEDGMENTS

The authors are grateful to the University Grants Commission, New Delhi, India, for providing the financial support. They also acknowledge Maharshi Dayanand University, Rohtak (India) for providing the laboratory and supporting facilities. The authors are also grateful to Dr. P. Shukla, the Editor of the Book (Cat No. K27546) and to CRC Press/Taylor & Francis Group publishers for invitation for submitting a chapter in this book.

REFERENCES

Akihisa, T.; Takamine, Y.; Yoshizumi, K. Microbial transformations of two lupane-type triterpenes and anti-tumor-promoting effects of the transformation products. *J Nat Prod.* 2002, 65, 278–82.

Bache, M.; Bernhardt, S.; Vordermark, D. Betulinic acid derivatives NVX-207 and B10 for treatment of glioblastoma—An *in Vitro* study of cytotoxicity and radio sensitization. *Int J Mol Sci* 2014, 15, 19777–19790.

Banskota, AH.; Tezuka, Y.; Adnyana, IK.; Xiong, Q.; Hase, K.; Tran, KQ.; Tanaka, K.; Saiki, I.; Kadota, S. Hepatoprotective effect of *Combretum quadrangulare* and its constituents. *Biol Pharm Bull.* 2000, 23, 456–460.

Bramwell, D. How many plant species are there? *Plant Talk* 2002, 28, 32–34.

Chabner, BA.; Roberts, TG.; Jr. Timeline: Chemotherapy and the war on cancer. *Nat Rev Cancer.* 2005, 5, 65–72.

Chandramu, C.; Manohar, RD.; Krupadanam, DG.; Dashavantha, RV. Isolation, characterization and biological activity of betulinic acid and ursolic acid from *Vitex negundo L. Phytother Res.* 2003, 17, 129–134.

Chatterjee, P.; Kouzi, SA.; Pezzuto, JM.; Hanamm, MT. Biotransformation of the antimelanoma agent betulinic acid by *Bacillus megaterium* ATCC 13368. *Appl Environ Microbiol* 2000, 66, 3850–3855.

Chen, Y.; Liu, J.; Yang, X.; Zhao, X.; Xu, H. Oleanolic acid nanosuspensions: Preparation, in-vitro characterization and enhanced hepatoprotective effect. *J Pharm Pharmacol* 2005, 57, 259–264.

Chintharlapalli, S.; Papineni, S.; Ramaiah, SK.; Safe, S. Betulinic acid inhibits prostate cancer growth through inhibition of specificity protein transcription factors. *Cancer Res* 2007, 67, 2816–2823.

Damle, AA.; Pawar, YP.; Narkar, AA. Anticancer activity of betulinic acid on MCF-7 tumors in nude mice. *Indian J Exp Biol.* 2013, 51, 485–89.

De Melo C.; Queiroz MG.; Arruda filho AC.; Rodrigues AM.; De sousa D.; Almeida, JG. et al. Betulinic acid, a natural Pentacyclic Triterpenoid, prevents abdominal fat accumulation in mice fed a high-fat diet. *J Agric Food Chem* 2009, 57, 8776–8781.

Deng, Y.; Snyder, JK. Preparation of a 24-nor-1,4-dien-3-one triterpene derivative from betulin: A new route to 24-nortriterpene analogues. *J Org Chem.* 2002, 67, 2864–2873.

de Oliveira, BH.; Santos, CA.; Espíndola, AP. Determination of the triterpenoid, betulinic acid, in *Doliocarpus schottianus by HPLC. Phytochem Anal.* 2002, 13, 95–98.

Dinda, B.; Hajra, AK.; Das, SK. Reactions on naturally-occurring triterpene. *Ind J Chem.* 1995, 34, 624–628.

Ding, W.; Sun, M.; Luo, S. 3D QSAR study of betulinic acid derivatives as antitumor agents using topomer CoMFA: Model building studies and experimental verification. *Molecules.* 2013, 18, 10228–10241.

Dubey, KK.; Behera, BK. Statistical optimization of process variables for the production of an anticancer drug (colchicine derivatives) through fermentation: At scale-up level. *New Biotechnol.* 2011, 28, 79–85.

Dubey, KK.; Jawed, A.; Haque, S. Enhanced bioconversion of colchicine to regiospecific 3-demethylated colchicine (3-DMC) by whole cell immobilization of recombinant *E. coli* harboring P450 BM-3 gene. *Process Biochem* 2013, 48, 1151–1158.

Dubey, KK.; Ray, AR.; Behera, BK. Production of demethylated colchicine through microbial transformation and scale-up process development. *Process Biochem.* 2008, 43, 251–257.

Ehrhardt, H.; Fulda, S.; Fuhrer, M.; Debatin, K.; Jeremias, I. Betulinic acid-induced apoptosis in leukemia cells. *Leukemia.* 2004, 18, 1406–1412.

Ellington, E.; Bastida, J.; Viladomat, F.; Codina, C. Supercritical carbon dioxide extraction of colchicine and related alkaloids from seeds of *Colchicum autumnale L. Phytochem Anal.* 2003, 14, 164–169.

Emmerich, D.; Vanchanagiri, K.; Baratto, LC. Synthesis and studies of anticancer properties of lupane-type triterpenoid derivatives containing a cisplatin fragment. *Eur J Med Chem.* 2014, 75, 460–466.

Enwerem, NM.; Okogun, JI.; Wambebe, CO.; Okorie, DA.; Akah, PA. Antiheliminthic activity of the stem bark extracts of *Berlina grandiflora* and one of its active principles, Betulinic acid. *Phytomedicine.* 2001, 8(2), 112–114.

Flekhter, OB.; Nigmatullina, LR.; Baltina, LA.; Karachurina, LT.; Galin, FZ.; Zarudii, FS. et al. Synthesis of betulinic acid from betulin extract and study of the antiviral and antiulcer activity of some related terpenoids. *Pharm Chem J* 2002b, 36, 484–487.

Flekhter, OB.; Tolstikov, GA.; Pliasunova, OA. Synthesis and pharmacological activity of betulin dinicotinate, *Bioorg Khim.* 2002a, 28, 543–50.

Franiel, I.; Babczynska, A. The growth and reproductive effect of *Betula Pendula Roth* in heavy metal polluted area. *Polish J Env Stud.* 2011, 20, 1097–1101.

Fujioka, T.; Kashiwada, Y.; Kilkuskie, RE.; Cosentino, LM.; Ballas, LM.; Jiang, JB.; Janzen, WP.; Chen, IS.; Lee, KH. Anti-AIDS agents, 11. Betulinic acid and platanic acid as anti-HIV principles from *Syzigium claviflorum*, and the anti-HIV activity of structurally related triterpenoids. *J Nat Prod.* 1994, 57, 243–247.

Fulda, S.; Scaffidi, C.; Susin, SA.; Krammer, PH.; Kroemer, G.; Peter, ME.; Debatin, K. Activation of mitochondria and release of mitochondrial apoptogenic factors by betulinic acid. *J Biol Chem*. 1998b, 273, 33942–33948.

Fulda, S.; Susin, SA.; Kroemer, G.; Debatin, K. Molecular ordering of apoptosis induced by anticancer drugs in neuroblastoma cells. *Cancer Res*. 1998a, 58, 4453–4460.

Gerhauser, C. Cancer chemopreventive potential of apples, apple juice, and apple components. *Planta Med*. 2008, 74, 1608–1624.

Govaerts, R. Proposal to reject the name *Betula alba* (Betulaceae). *Taxon* 1996, 45, 697–698.

Govaerts, R. How many species of seed plants are there? *Taxon*. 2001, 50, 1085–1090.

Govaerts, R.; Frodin, DG. 1998. *World Checklist and Bibliography of Fagales*. Royal Botanic Gardens, Kew, London.

Guo, M.; Zhang, S.; Song, F.; Wang, D.; Liu, Z.; Liu, S. Studies on the non-covalent complexes between oleanolic acid and cyclodextrins using electrospray ionization tandem mass spectroscopy. *J Mass Spectrom*. 2003, 38, 723–731.

Holz-Smith, SL.; Sun, I.; Chen, CH. Role of human immunodeficiency virus (HIV) type 1 envelope in the anti-HIV activity of the betulinic acid derivative IC9564. *Antimicrob. Agents Chemother*. 2001, 45, 60–66.

Huang, L.; Ho, P.; Chen, CH. Activation and inhibition of the proteasome by betulinic acid and its derivatives. *FEBS Lett*. 2007, 581, 4955–4959.

Hunt, D. *Betula*. Proceedings of the IDS Betula Symposium, International Dendrology Society, Sussex, UK, 1993, p. 51.

Kang, HS.; Park, JE.; Lee, YJ.; Chang, IS.; Chung, YI.; Tae, G. Preparation of liposomes containing oleanolic acid via micelle-to-vesicle transition. *J Nanosci Nanotechnol*. 2007, 7, 3944–3948.

Kashiwada, Y.; Hashimoto, F.; Cosentino, LM. Betulinic acid and dihydrobetulinic acid derivatives as potent anti-HIV agents. *J. Med Chem*. 1996, 39, 1016–17.

Kim, DK.; Nam, IY.; Kim, JW.; Shin, TY.; Lim, JP. Pentacyclic triterpenoids from *Ilex macropoda*. *Arch Pharm Res*. 2002, 25, 617–620.

Kim, HI.; Quan, FS.; Kim, JE.; Inhibition of estrogen signaling through depletion of estrogen receptor alpha by ursolic acid and betulinic acid from *Prunella vulgaris var. lilacina*. *Biochem Biophys Res Commun*. 2014, 451, 282–287.

Kim, YK.; Yoon, SK.; Ryu, SY. Cytotoxic triterpenes from stem bark of *Physocarpus intermedius*. *Planta Med*. 2000, 66, 485–496.

Krasutsky, P.; Carlson, RM.; Vitaliy, V.; Nesterenko, VV. Method for Manufacturing of Betulinic Acid. *U.S. Patent*. 2001, 6, 271, 405.

Krogh, R.; Kroth R; Berti C; Madeira AO; Souza MM; Cechinel-Filho V; Delle-Monache F; Yunes RA. Isolation and identification of compounds with antinociceptive action from *Ipomoea pescaprae* (L.) R. Br. *Pharmazie*. 1999, 54, 64–66.

Kwon, HC.; Min, YD.; Kim, KR.; Bang, EJ.; Lee, CS.; Lee, KR. A new acylglycosyl sterol from *Quisqualis fructus*. *Arch Pharm Res*. 2003, 26, 275–278.

Laszczyk, MN. Pentacyclic triterpenes of the lupane, oleanane and ursane group as tools in cancer therapy. *Planta Med*. 2009, 75, 1549–1560.

Le Bang Shon. Sintez betulinovoy kisloty razrabotka yeyo liposomalnoy formy (The synthesis of betulinic acid development of its liposomal form): Abstract of Chemical Sciences Dissertation, Moscow, 1999, 26.

Li, L.; Du, Y.; Kong, X. Lamin B1 is a novel therapeutic target of betulinic acid in pancreatic cancer. *Clin Cancer Res.* 2013, 19, 4651–4661.

Liby, KT.; Yore, MM.; Sporn, MB. Triterpenoids and rexinoids as multifunctional agents for the prevention and treatment of cancer. *Nat Rev Cancer.* 2007, 7, 357–369.

Liu, Y.; Luo, W. Betulinic acid induces Bax/Bak-independent cytochrome-C release in human nasopharyngeal carcinoma cells. *Mol Cells.* 2012, 5, 517–524.

Martin, L.; Pasquier, E. Drug-induced livedo and the EGFR pathway. *Eur J Dermatol.* 2008, 18, 601–603.

Mukherjee, P.K.; Saha, K.; Das, J.; Pal, M.; Saha, BP. Studies on the anti-inflammatory activity of rhizomes of *Nelumbo nucifera*. *Planta Med.* 1997, 63, 367–369.

Mullauer, FB.; Kessler, JH.; Madema, JP. Betulinic acid, a natural compound with potent anticancer effects. *Anti-Cancer Drugs.* 2010, 21, 215–227.

Neto, CC. Cranberry and its phytochemicals: A review in *in vitro* anticancer studies. *J Nutr.* 2007, 137, 1865–1935.

Noda, Y.; Kaiya, T.; Kohda, K.; Kawazoe, Y. Enhanced cytotoxicity of some triterpenes toward leukemia L1210 cells cultured in low pH media: Possibility of a new mode of cell killing. *Chem. Pharm. Bull.* 1997, 45, 1665–1670.

Nyasse, B.; Nono, JJ.; Nganso, Y.; Ngantchou, I.; Schneider B. *Uapaca* genus (Euphorbiaceae), a good source of betulinic acid. *Fitoterapia* 2009, 80, 32–34.

Ovensná, Z.; Vachalková, A.; Horváthová, K.; Táthová, D. Pentacyclic triterpenoic acids: Newchemoprotective compounds. *Neoplasma.* 2004, 51, 327–333.

Park, SY.; Kim, HJ.; Kim, KR. Betulinic acid, a bioactive pentacyclic triterpenoid, inhibits skeletal-related heavy met events induced by breast cancer bone metastases and treatment. *Toxicol Appl Pharmacol.* 2014, 275, 152–62.

Peñuelas, J.; Munné-Bosch, S. Isoprenoids: An evolutionary pool for photoprotection. *Trends Plant Sci.* 2005, 10, 166–169.

Petronelli, A.; Pannitteri, G.; Testa, U. Triterpenoids as new promising anticancer drugs. *Anti-Cancer Drugs* 2009, 20, 880–892.

Phillips, DR.; Rasbery, JM.; Bartel, B.; Masuda, SP. Biosynthetic diversity in plant triterpene cyclization. *Curr Opin Plant Biol.* 2006, 9(3), 305–314.

Pisha, E.; Chai, H.; Lee, IS.; Chagwedera, TE.; Farnsworth, NR.; Cordell, GA. et al. Discovery of betulinic acid as a selective inhibitor of human melanoma that functions by induction of apoptosis. *Nat Med.* 1995, 1, 1046–1051.

Prakash, CVS.; Schilling, JK.; Johnson, RK.; Kingston, DG. New cytotoxic lupane triterpenoids from the twigs of *Coussarea paniculata*. *J Nat Prod.* 2003, 66, 419–422.

Rabi, T.; Bishayee, A. Terpenoids and breast cancer chemoprevention. *Breast Cancer Res Treat.* 2009, 115(2), 223–239.

Raghuvar Gopal, DV.; Narkar, AA.; Badrinath, Y.; Mishra, KP.; Joshi, DS. Betulinic acid induces apoptosis in human chronic myelogenous leukemia (CML) cell line K-562 without altering the levels of Bcr-Abl. *Toxicol Lett.* 2005, 155, 343–351.

Recio, MC.; Giner, RM.; Máñez, S.; Gueho, J.; Julien, HR.; Hostettmann, K.; Ríos, JL. Investigations on the steroidal anti-inflammatory activity of triterpenoids from *Diospyros leucomelas*. *Planta Med.* 1995, 61, 9–12.

Ren, H.; Omori, S. A simple preparation of betulinic acid from sycamore bark. *J Wood Sci.* 2012, 58, 169–173.

Rieber, M.; Rieber, MS.; Signalling responses linked to betulinic acid-induced apoptosis are antagonized by MEK inhibitor U0126 in adherent or 3D spheroid melanoma irrespective of p53 status. *Int J Cancer.* 2006, 118, 1135–1143.

Risinger, AL.; Giles, FJ.; Mooberry, SL. Microtubule dynamics as a target in oncology. *Cancer Treat Rev.* 2009, 35, 255–261.

Rosner, M.; Capravo, H.; Jacobson, AE.; Atwell, L. Biological effects of modified colchicine. Improved preparation of 2- DMC, 3MC and (1)-colchicine and rearrangement of the position of the double bond in dehydro-7-deacetamidecolchicine. *J. Med. Chem.* 1981, 24(2), 57–261.

Sawada, N.; Kataoka, K.; Kondo, K.; Arimochi, H.; Fujino, H.; Takahashi, Y. Betulinic acid augments the inhibitory effects of vincristine on growth and lung metastasis of B16F10 melanoma cells in mice. *Br J Cancer.* 2004, 90, 1672–1678.

Schühly, W.; Heilmann, J.; Calis, I; Sticher, O. New triterpenoids with antibacterial activity from *Zizyphus joazeiro*. *Planta Med.* 1999, 65, 740–743.

Setzer, WN.; Setzer, MC. Plant-derived triterpenoids as potential antineoplastic agents. *Mini Rev. Med Chem.* 2003, 3, 540–556.

Setzer, WN.; Setzer, MC.; Bates, RB.; Jackes, BR. Biologically active triterpenoids of *Syncarpia glomulifera* bark extract from Paluma, north Queensland, Australia. *Planta Med.* 2000, 66, 176–177.

Shah, BA.; Qazi, GN.; Taneja, SC. Boswellic acids: A group of medicinally important compounds. *Nat. Prod Rep.* 2009, 26, 72–89.

Siddiqui, BS.; Sultana, I.; Begum, S. Triterpenoidal constituents from *Eucalyptus camaldulensis var. obtusa* leaves. *Phytochemistry.* 2000, 54, 861–865.

Soler, F.; Poujade, C.; Evers, M. Betulinic acid derivatives: A new class of specific inhibitors of human immunodeficiency virus type 1 entry, *J Med Chem.* 1996, 39, 1069–1083.

Solet, JM.; Bister, F.; Galons, H.; Spagnoli, R. *Glycosylation of Thiocolchicine by a Cell Suspension Culture of Centella asiatica Phytochemistry.* Pergamon Press, Oxford, New York, 1993, 33, 817–820.

Steele, JC.; Warhurst, DC.; Kirby, GC.; Simmonds, MS. In vitro and *in vivo* evaluation of betulinic acid as an antimalarial. *Phytother Res.* 1999, 13, 115–119.

Sultana, N.; Ata, A. Oleanolic acid and related derivatives as medicinally important compounds. *J Enzyme Inhib Med Chem.* 2008, 23, 739–756.

Tang, J.; Jones, SA.; Jeffery, JL.; Miranda, SR.; Galardi, CM.; Irlbeck, DM. et al. 2014. Synthesis and biological evaluation of macrocyclized betulin derivatives as a novel class of anti-HIV-1 maturation inhibitors. *Open Med Chem J.* 2014, 8, 23–27.

Tezuka, Y.; Stampoulis, P.; Banskota, AH.; Awale, S.; Tran, KQ.; Saiki, I.; Kadota, S. Constituents of the Vietnamese medicinal plant *Orthosiphon stamineus*. *Chem Pharm Bull*. 2000, 48, 1711–1719.

Thurnher, D.; Turhani, D.; Pelzmann, M.; Wannemacher, B.; Knerer, B.; Formanek, M.; Wacheck V.; Selzer E. Betulinic acid: A new cytotoxic compound against malignant head and neck cancer cells. *Head Neck*. 2003, 25, 732–740.

Tiwari, R.; Puthli, A.; Mishra, KP. Betulinic acid-induced cytotoxicity in human breast tumor cell lines MCF-7 and T47D and its modification by tocopherol. *Cancer Invest*. 2014, 32, 402–408.

Withers, ST.; Keasling, JD.; Biosynthesis and engineering of isoprenoid small molecules. *Appl Microbiol Biotechnol*. 2007, 73, 980–990.

Yasukawa, K.; Takido, M.; Matsumoto, T. Sterol and triterpene derivatives from plants inhibit the effects of a tumor promoter, and sitosterol and betulinic acid inhibit tumor formation in mouse skin two-stage carcinogenesis *Oncology*. 1991, 48, 72–76.

Yi, JH.; Zhang, GL.; Li, BG.; Chen, YZ. Two glycosides from the stem bark of *Tetracentron sinense*. *Phytochemistry*. 2000, 53, 1001–1003.

Yogeeswari, P.; Sriram, D. Betulinic acid and its derivatives: A review on their biological properties. *Curr. Med. Chem*. 2005, 12, 657–666.

Zuco, V.; Supino, R.; Righetti, SC.; Cleris, L.; Marchesi, E.; Gambacorti-Passerini, C.; Formelli, F. Selective cytotoxicity of betulinic acid on tumor cell lines, but not on normal cells. *Cancer Lett*. 2002, 175(1), 17–25.

CHAPTER 9

Optimizing the Performance of Wastewater Treatment Plants and Effluent Quality Using Evolutionary Algorithms

Abimbola Motunrayo Enitan, Josiah Adeyemo, Gulshan Singh, and Folasade Adeyemo

CONTENTS

ABSTRACT

Optimization of wastewater treatment processes is not an easy task due to the complexity and variations in the organic contents of different wastewaters. Like many other real world problems, treatment of wastewater requires making different decisions and setting the optimum operational parameters that are conflicting in nature. Evolutionary algorithms are of interest in optimizing the wastewater treatment plant (WWTP) to generate Pareto-optimal solutions by spontaneously minimizing and/or maximizing different desirable and undesirable properties. Different types of optimization tools have been developed and applied in predicting and simulating wastewater treatment processes. This study reviews different types of evolutionary algorithms as a computer-based and biologically inspired optimization algorithm for solving continuous, noncontinuous, non-differentiable, and multimodal optimization problems based on natural selection principles. Different studies on real-world applications of different evolutionary algorithms for finding global optimum solutions that could be used for making decisions during plant design and operation of WWTPs have been reviewed.

INTRODUCTION

With the competing demand on water resources and water reuse, discharge of wastewater into the aquatic environment has become an important issue (Kovoor et al., 2012). Much attention has been placed on the impact of wastewater on water bodies worldwide due to the accumulation of organic and inorganic suspended matters (Phiri et al., 2005; Islam et al., 2006; Baig et al., 2010; Ipeaiyeda and Onianwa 2012). In the last few decades, wastewater treatment plants (WWTPs) that are based on complex biological treatment configurations have been developed for efficient removal of nutrients and organic matter in wastewater. The most widely used biological WWTP is the activated sludge system in treating both industrial and domestic wastewater. The new modification to the original configuration for activated sludge system is the combination of anaerobic, anoxic, and aerobic reactors, with internal recycling between the different unit processes (Rivas et al., 2008).

The WWTP depends on the interaction of different operational and suitable microbial communities that are complex to integrate. With the increasing regulations and difficulties in operating a satisfactory WWTP due to several varied operational conditions and concerns on effluent quality, operational costs, investment costs, safety, emission of greenhouse gases (GHG), suitable plant design, and other factors has resulted in a search for predicting tools to improve plant performance without wasting resources and time (Mjalli et al., 2007). Satisfying several objectives during the treatment of wastewater has increased the complexity of WWTP problems. The selection of the most appropriate plant conditions has become a very difficult task, even for experienced designers. Therefore, a more reliable model that can serve as a predicting tool for controlling the performance of WWTP operation is very essential.

Owing to these facts, WWTP mathematical modeling and simulation have become increasingly popular as a tool to evaluate and predict organic matter and nutrient removal from wastewater since the mid-1990s (Gernaey et al., 2004; Morales-Mora et al., 2015; Van Loosdrecht et al., 2015). Some models have been developed for the selection of appropriate design parameters, with special attention to parameters that can significantly affect the final design such as the fraction of nonbiodegradable solids in the influent, for the maximum nitrification rate in WWTP (Jones et al., 2014). Development of these models could be divided into two, namely deterministic (white-box) and data-driven (black-box) models (Gernaey et al., 2004). The most widely used deterministic models that have been widely applied to activated sludge system are the activated sludge models: ASM1, ASM2, and ASM3, developed by Henze (Henze et al., 1987, 1995, 1999; Gujer et al., 1999; Gernaey et al., 2004).

Among the parameters that ASM1 considered are the nitrogen and organic carbon compound removal with simultaneous consumption of nitrate and oxygen as electron acceptors. The model further describes sludge production and chemical oxygen demand (COD) as the concentration of organic matter. ASM2 extends the capacity of ASM1 by considering biological phosphate removal via precipitation (Henze et al., 1995). Furthermore, little modification was done to ASM2 and named ASM2d. The ASM2d model gives a better description of nitrate and phosphate by adding the denitrifying activity of PAOs (Henze et al., 1999). Lastly, ASM3 was developed for the biological removal of nitrogen (Gujer et al., 1999). This model corrects some of the deficiencies that were discovered by ASM1 users. Some of the drawbacks of ASM1 that are covered by ASM3 are discussed in detail in Gujer et al. (1999) and Gernaey et al. (2004). Other models for different unit processes on WWTPs have been reported in the literature (Batstone et al., 2002). Most of these existing activated sludge models that have been reported are a set of differential equations that describe the chemical and biological reactions within the activated sludge system and are based on assumptions (Mjalli et al., 2007). In addition to different modification to activated sludge models, comprehensive knowledge of the WWTP, and several simulations with the sequential procedure that are very tedious have shown that mechanistic approaches are not sufficient for predicting plant performance.

Recently, several studies on statistical analysis of uncertainties in model-based WWTP design and control based on the combination of deterministic mathematical and Monte Carlo trials models have been proposed for WWTP (Flores-Alsina et al., 2012). Some of the shortcomings of using WWTP statistical model prediction include the limited information that are available to the plant designer concerning the load variations expected in the influent of the proposed plants, the biochemical activity of the new processes, definition of all constraints due to the design parameters (the effluent requirements, the volume of units physically available, and the maximum cost), and repeated selection of the most appropriate combination of design and operational parameters until the constraints of the problem are fulfilled (Rivas et al., 2008).

However, the increasing report of uncertainty in engineering design, especially in the modeling of environmental systems, due to the complexity of the treatment processes, the integration of different parameters, linear and nonlinear equations with single and multi-objective functions under different constraints reflected the need for an alternative optimization and control strategies (Jolma and Norton, 2005; Wagener and Kollat, 2007; Rivas et al., 2008). A new optimization and control strategy with immense benefits to manage wastewater treatment for better effluent quality that will reduce pollution to the environment is essential for effective reactor operation.

OPTIMIZATION OF WWTP PROBLEMS

Optimization can be defined as the "art of finding one or more feasible solutions corresponding to extreme values of one or more objectives problems, while satisfying specific constraints" (Babu et al., 2005). Optimization problems are divided into two, namely single- and multi-objective optimization (Fister et al., 2013). The single-objective optimization problem involves one objective function to which heuristic-based and gradient-based search techniques are applied in order to solve the single-objective optimization problem. This involves finding the minimum and maximum of a single-variable function. The single-objective optimization method is used for finding an optimum of a first- and second-order derivative of a function. It may also involve finding the true optimum in the presence of constraints to obtain solutions to real-world problems (Adeyemo and Otieno, 2009).

In contrast, multi-objective optimization problem (MOOP) is an optimization problem solving method that has more than one objective functions. It involves finding one or more optimum solutions to more than one objective optimization problems that are conflicting in nature (Deb, 2011). The aim of MOOP is to simultaneously optimize a set of conflicting objectives to obtain a group of alternative trade-off solutions called Pareto-optimal or non-inferior solutions which must be considered equivalent in the absence of specialized information concerning the relative importance of the objectives (Deb, 2011). With regard to all objectives, there is no best solution, rather they are equally good solutions. Meanwhile, most real-world search and optimization problems are multi-objectives in nature with all the objective functions being very important (Fister et al., 2013).

Over the last few decades, application of metaheuristics is used as an iterative generation process that combines different intelligent theories for exploring and exploiting the search space to find efficient near-optimal solutions to a particular problem (Zufferey, 2012). Metaheuristics as an optimization tool for different types of wastewater treatment process problems have recently been reported (e.g., design and operation of wastewater plants, effluent quality, operational cost, and model calibration). There are two types of metaheuristic techniques. These include ant colony optimization (ACO), differential evolution (DE), evolutionary strategies, particle swarm optimization (PSO), and genetic algorithms (GAs) among others as the population-based algorithms; and tabu search, simple evolutionary strategies, simulated annealing (SA), trajectory or local search methods among others as the single point-based methods (Maier et al., 2014).

EVOLUTIONARY ALGORITHMS

Recent optimization tool and control strategy with respect to external influences and different process disturbances for efficient operation of WWTPs is the power of computational intelligence of evolutionary algorithms (EAs). EAs are computer-based and biologically inspired optimization algorithms that are commonly used for solving non-differentiable, noncontinuous, and multimodal optimization problems based on Darwin's natural selection principle. EA is one of the stochastic search methods and class of direct search algorithms has proven to be an important tool in solving optimization problems. EAs as the most well-established class of metaheuristics search approach do not use the traditional deterministic search approaches of one at a time sampling pattern, they rather perform random sampling of the search space using evolutionary resembling operations (selection, reproduction, and mutation), which are applied to individuals in a population (Adeyemo and Otieno, 2009).

EAs are widely adopted for solving both constrained and unconstrained multi-objective problems to deliver global optimal solutions and to resolve shortcomings encountered in traditional methods (Sarker and Ray, 2009; Matijasevic et al., 2010; Enitan and Adeyemo, 2011). Some of the advantages of EAs for solving problems encountered in wastewater treatment technology include the following:

1. Optimization of real-world problems using EAs do not have much mathematical requirements, rather an evaluation of the objective function is performed.

2. The EAs have the capacity to link simulation models in solving problems with difficult mathematical properties and reduces the need for problem simplification that are very common to many of traditional optimization algorithms. EAs are able to deal with nonlinearities such as like discontinuities (e.g., convergent/iterative algorithms), nonlinear, and linear programming (Reed et al., 2013).

3. EAs find better and near globally optimal solutions to the actual problem, rather than globally optimal solutions to a simplified problem, especially when the simplified problem misses the key relevant properties (Maier et al., 2014).

4. The basic analogies for EA optimization strategies are conceptually easy to understand.

In the last few decades, different modern heuristics EA tools that help in the optimization of difficult and impossible problems have emerged. LP, nonlinear programming (NLP), stochastic dynamic programming (SPD), dynamic programming (DP), and heuristic programming such as artificial neural networks (ANNs), DE, GA, genetic programming, PSO, shuffled complex evolution SA, ACO, fuzzy logic (FL), and adaptive neuro-fuzzy inference system (ANFIS) are some of the promising optimization techniques that have been used in handling different real-world problems. The stages involved in using EAs for optimization of real-world problems are shown in Figure 9.1.

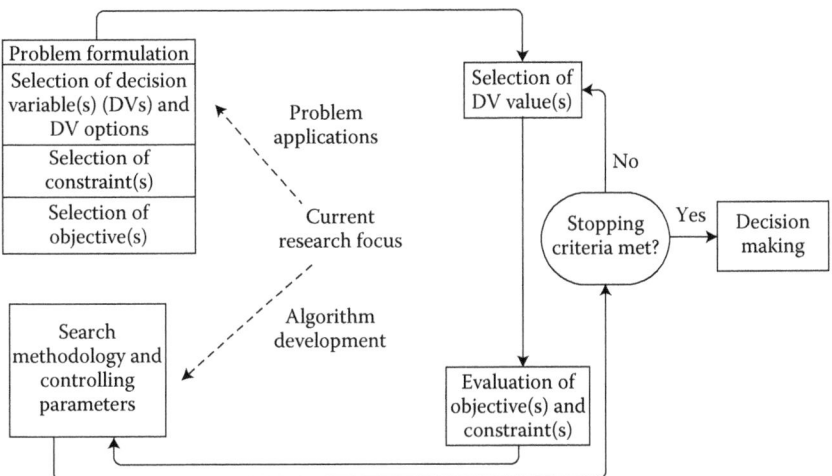

FIGURE 9.1 Steps involved in optimizing real-world problems using EA. Steps involved in EA process are represented with the square shapes while the oval shape represents a decision point. (From Maier, H. et al., 2014. *Environmental Modelling and Software*, 62, 271–299.)

Owing to the complexity and variations in the organic contents in both domestic and industrial wastewaters, EAs help in generating Pareto-optimal solutions to problems that may arise during the treatment processes. Over the past decade, GA, DE, ANNs, FL, and PSO algorithms have gained much popularity as problem solvers and as design tools in biological wastewater treatment technology due to their versatility and ability to optimize complex multi-objective problems (Enitan, 2014). They are the most commonly used machine learning algorithms for nonlinear modeling and optimization tools for wastewater treatment processes. They have been used to solve many conflicting wastewater problems. Recently, some of the applications are the reduction of organic matter to meet the discharge standard (Babu et al., 2005), forecasting the performance of effluent quality (Tashaouie et al., 2012), prediction of sludge volume index in municipal WWTP (Djeddou and Achour, 2015), and modeling of paper-making wastewater tretament plant using ANN and ANFIS; optimization of wastewater treatment design using GA (Iqbal and Guria, 2009). "Some of the benefits of these techniques is that they do not have much mathematical requirements about the optimization problem; all they need is an evaluation of the objective function with the presence of the constraints that makes the solution of the problem rather difficult" (Balku and Berber, 2006).

GENETIC ALGORITHM

GA is more popular and can be viewed as a general purpose search technique due to its earlier introduction than other algorithms. GA is based on biological evolution using Darwinian principles. The first report on GA was by Holland (1975) to mimic the natural selection of an individual as parent within the population and survival of the fittest. Chromosomes are used to represent candidate solutions called population and each chromosome is evaluated for fitness based on the fitness value that are scaled from the best to the worse. GA produces new solutions by mimicking the principle of natural selection through the application of genetic operators, selection, crossover, and mutation (Enitan and Adeyemo, 2011; Kachitvichyanukul, 2012). Figure 9.2 shows the application of three genetic operators to become the parents and uses crossover operators to produce new chromosomes (the offspring) through perturbation of solutions in the old population to generate new population for the next generation. Each solution in the population is evaluated for fitness.

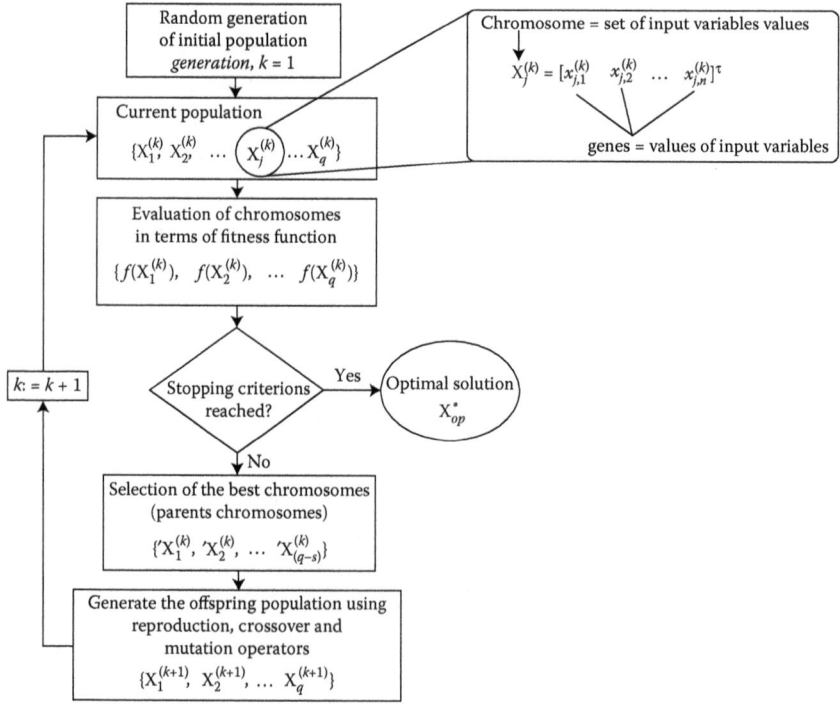

FIGURE 9.2 Principal steps for a typical GA.

However, there is a tendency for the new solutions to be very similar after generation without any diversity. Therefore, mutation helps in bringing in the diversity without stagnation of population by injecting new elements into the chromosomes. Several decisions must be made during GA optimization which include

1. Determination of population size.

2. Maximization of iteration number.

3. Selection method.

4. Probability assignment mechanism based on fitness.

5. Crossover method and crossover probability.

6. Mutation method and mutation probability that helps in maintaining population diversity.

An in-depth information on the principle of GA could be found in the literature (Gen and Cheng 1997; Enitan and Adeyemo 2011). Owing to different limitations of GA, different genetic versions of GA that use real numbers for coding and genetic operators to generate new solutions until a stopping criterion is satisfied have emerged (Mohebbi et al., 2008). There are different improved versions of the original GA that have been reported in the literature such as the elitist non-dominated sorting genetic algorithm (NSGA-II) (Deb et al., 2000), compressed-objective genetic algorithm (COGA-II) (Boonlong, 2013), and multi-objective uniform-diversity genetic algorithm (MUGA) (Nariman-Zadeh et al., 2010). The principal steps involved in GA are given in Figure 9.2.

ANN AND FL

"Artificial neural network (ANN) is a collection of interconnecting computational elements which are simulated like neurons in biological systems" (Adeyemo and Enitan, 2011). It emulates the "connectivity of biological neurons in the same manner as the human brain in solving complex real-world problems" (Huang et al., 2010; Rosales-Colunga et al., 2010; Enitan and Adeyemo, 2011). ANN model mimics the human brain by creating linkages between a number of input and output nodes with a number of hidden nodes. It generates mathematical models with a number of neurons called nodes with several groups of neurons or nodes called layers. The problem that needs to be solved determine the number of input and output nodes in ANN. In reality, ANNs are a more complex type of statistical (black-box) or regression model that mimics the neural behavior of real systems. Measured or pre-defined sets of input and output data are used in training ANN in an efficient way to develop a model that deals with the system's intrinsic nonlinearities (Goñi et al., 2008; Rosa et al., 2010; Enitan and Adeyemo, 2011).

ANNs can easily be used to train different types and structure of data without prior knowledge of the relationship between the data. In this manner, without a prediction equation or a mathematical model, ANNs could be used to predict or estimate system behaviors that are associated with the physical problem (Ramesh et al., 1996). They are widely used in pattern recognition and pattern classification, diagnosis and control, function approximation and optimization (Bose and Liang, 1996). The design of ANN layers and nodes have a great significant influence on its data-processing capability. Its major connection topologies that define the flow of data between the ANN input, hidden, and output nodes are divided into two main networks: feed-forward networks and feedback or recurrent networks (Loucks et al., 2005), which could be a multilayer neural network. For solving nonlinear regression problems, the most widely used neural network is the multilayer feed-forward neural network called the multilayer perceptron (MLP). The backpropagation (BP) algorithm is used in MLP during ANNs training and this involves the minimization of a performance function commonly called the mean square error (MSE). Furthermore, ANNs have the capacity to detect between independent and dependent variables with possible interaction between predicted variables. ANN model being a "black-box" model have a greater computational burden and they are prone to overfitting (Fernández-Navarro et al., 2010; Adeyemo and Enitan, 2011). The general procedure for preparing and training ANN was suggested by Smith (1993) in Loucks et al. (2005) as

1. To design a network.

2. To divide the data set into training, validation, and testing subsets.

3. To train the network using the training data set.

4. To periodically stop the training and measure the error on the validation data set.

5. To save the weights of the network.

6. To repeat Steps 2, 3, and 4 until the validation data set error starts increasing. This is the moment where the overfitting has started.

7. Then, go back to the weights that produced the lowest error on the validation data set, and use these weights for the trained ANN.

8. Test the trained ANN using the testing data set. If it shows good performance use it. If not, redesign the network and repeat the entire procedure from Step 3.

In contrast to ANN, FL modeling that uses both scientific and heuristic approaches with the ability to utilize expert knowledge and past data more convincingly than conventional methods were developed by Zadeh (1975). FL mimics human control logic by using the fuzzy set theory in problem optimization. "It uses an imprecise but very descriptive language as a human operator to deal with the input data" (Huang et al., 2010). An increasing interest to enhance the ability of fuzzy systems with learning capabilities by soft-computing methods such as genetic fuzzy systems with various applications in solving different problems is developing in the world (Liao et al., 2001; Mohebbi et al., 2008) which has led to the development of an ANFIS to treat the problems in which a source of vagueness data occurs.

PSO ALGORITHM

In the mid-1990s, an attempt to investigate the collective intelligence of bird flocking, herds of animals, and schools of fishes to adapt to their natural environment, search rich sources of foods and avoid predators by employing an information sharing approach was carried out by Kennedy and Eberhart (1995). This triggered the study on the simulation of motion of swarms of birds in the biological population (Peyvandi et al., 2011). In 1995, Kennedy and Eberhart published a paper on the PSO algorithm and other papers on the use of PSO to solve different difficult optimization problems have been published. PSO algorithms have simple conceptual framework and visualization of search process based on bird flocking using the basic flowchart shown in Figure 9.3.

In the PSO algorithm, a particle represents a solution and swarm of particles represents a set of randomly general solutions in the population of the design space over a number of iterations. Position and velocity to move are the two main properties of a particle. Each particle uses velocity to move from one position to the new position, then update the best

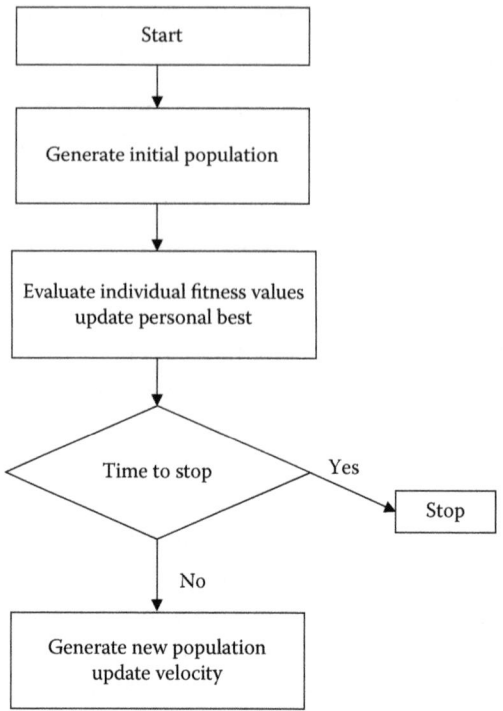

FIGURE 9.3   Basic flowchart of the PSO algorithm.

position of each particle, and the swarm are recorded. However, velocity adjustment for each particle is made based on the particle experience. The process is repeated by adjusting the velocity of each particle based on the particle experience until a stopping criterion is met (Kachitvichyanukul, 2012). Similar to GA is PSO algorithm that uses population-based search methods of looking for the optimal solution in the search space by updating generations. Like GA, the process of generating initial swarm of particles called initialization is the first process in PSO. The concept of solution representation, where each particle is initialized with a velocity and random position as the determining factors is very important in the PSO algorithm, then evaluated for fitness value; however, PSO does not require sorting of fitness values of solutions. Best fitness value of the whole swarm, global best, velocity, and personal best positions of the particles are updated to create a new swarm, if a stopping criterion is not met (Kachitvichyanukul, 2012).

The PSO uses update of velocity and update of position as the two key operations. Velocity update is performed based on three components: The personal best position or cognitive of self-learning term (experience of an individual particle), the global best positions or experience of the whole swarm, and the old velocity (momentum or inertia term) with an associated weight constant to each term. The number of required weight constants for basic PSO algorithm is three (Kachitvichyanukul, 2012). Both PSO and GA employ population-based search approaches to find a solution to a particular objective function using different computational methods and strategies. When large population size is involved, PSO algorithm shows a significant computational advantage of not sorting the fitness values of solutions during the optimization process as compared with GA. Simple arithmetic of real numbers is used by the PSO algorithm to update the velocity and position of each particle and swarm of particles. Owing to the advantages of PSO, different software in various forms are available to implement PSO. The PSO algorithm software ranged from user-friendly modifiable source code to a black box. Nguyen et al. (2010) designed and implemented C++ ET-Lib by which the user could define objective functions and representative particles. This application has been successfully implemented (Kasemset and Kachitvichyanukul, 2012; Sooksaksun et al., 2012).

OPTIMIZATION OF TREATMENT PLANT PERFORMANCE USING EAS

Evaluation and simulation of performance of WWTP for treating both domestic and industrial wastewater using ANN have been reported (Gontarski et al., 2000; Çinar, 2005; Zidan et al., 2015). Study on the use of ANN as black-box model to obtain information of a real WWTP was reported by Mjalli et al. (2007). The study has shown that "ANNs are capable of capturing the plant operation characteristics with a good degree of accuracy" (Mjalli et al., 2007). The ANN model prediction was accurate for predicting biological oxygen demand (BOD), COD, and total suspended solids (TSS) in the final effluent of the WWTP. The actual data from the treatment plant and ANN predicted data were compared and the developed graphical user interface was adopted as a tool for assessing the performance of WWTP. This served as an assessment tool for both decision makers and plant operators for operating WWTP. It could further increase the performance of the treatment plant such that the treated effluent will be controlled and meet discharge standards. Moreover, it will minimize the operational costs and help in quick stability of environmental balance of WWTPs. ANN was used to predict fecal coliform

(FC) bacteria in an aquatic environment, instead of direct measurement of FC concentrations (Brion and Lingireddy, 2003). The authors used ANN to identify nonpoint sources of microbial contamination removals using common process variables. Chandramouli et al. (2007) also used ANN for backfilling missing microbial concentrations in a riverine database. Neural network model was used to predict the concentration of *Giardia* and *Cryptosporidium* in the Delaware River by Neelakantan et al. (2002). However, due to some deficiency of ANN, ANN coupled with some other optimization tools have been adopted (Pai et al., 2011; Zaki et al., 2015). To avoid direct measurement of FC bacteria concentration in a full-scale municipal activated-sludge WWTP, a grammar-based genetic programming called evolutionary process model induction system was developed to automatically discover mathematical multivariate dynamic inference models (Hong and Paik, 2012). The developed sequential models were used to predict the removal of FC bacteria from full-scale municipal activated sludge WWTP using common process variables. Using the mathematical model in predicting the concentration of FC bacteria is very important in water quality management decisions as well as in reducing public health risk of microbial pollution.

Chen and Lo (2010) predicted the quality of the effluent produced from a domestic WWTP of CASP using neural network and gray model (GM). Three types of GM were used to predict the effluent BOD, COD, and suspended solids (SS) from the domestic WWTP using conventional activated sludge process. The use of insufficient data in predicting plant performance in terms of variation in domestic effluent composition using GM and backpropagation neural network (BPNN) were tested in their study. The predicted results were compared for the two predicting tools. The simulation results as reported by the authors showed higher fitness value for BPNN in predicting BOD (34.77%) when large quantities of data were used for constructing the model. On the contrary, only a small amount of data was required by GM to predict the effluent quality and the prediction results were analogous, even lower than SS (30.11%) and COD (16.21%) predicted by BPNN. The authors concluded that GM could predict better than BPNN, when the data available from the WWTP are insufficient (Chen and Lo, 2010).

Pai (2008) predicted the COD, SS, and pH in the effluent produced from a conventional activated process of an industrial WWTP using ANN and GM. Although, ANN was able to predict the effluent quality in terms of COD, SS, and pH, prediction of GM was much better than

ANN prediction for effluent quality. In another study conducted by the same authors, fuzzy inference system (FIS) was used to improve the performance of neural network in predicting effluent quality from hospital WWTP (Pai et al., 2009). Three types of ANFIS and ANN were used to predict effluent COD and SS coming from the WWTP treating hospital wastewater to the environment. The structure of ANFIS's consists of both FL and ANN that include linguistic expression of membership functions and if–then rules. This is to overcome the limitations of traditional neural network and increase the prediction performance. The authors reported high and better performance of ANFIS in predicting effluent quality as compared with ANN performance with the minimum mean absolute percentage errors of 12.75% and 11.99% COD_{eff} and SS_{eff} with a maximum correlation coefficient of 0.92 and 0.75, respectively. The minimum root mean square errors (RMSE) of 4.42 and 0.41 as well as the MSE of 19.58 and 0.17 for COD_{eff} and SS_{eff}, respectively were reported by the authors. These values show the efficiency of ANFIS in predicting the quality of the effluent produced from the WWTP (Pai et al., 2009).

In another study, ANFIS was used to develop models that could predict the removal of SS and COD from a full-scale WWTP treating paper mill wastewater with good control performances and forecast of better effluent quality (Wan et al., 2011). Recently, Yel and Yalpir (2011) predicted effluent quality of a municipal WWTP using the FIS method. Other studies that used ANFIS approach to select input parameters and predicting better effluent quality have been reported in the literature (Erdirencelebi and Yalpir, 2011; Mullai et al., 2012; Huang et al., 2013).

The performance evaluation and composition of effluent quality of a submerged membrane bioreactor (SMBR) treating both industrial and municipal wastewater were simulated using a radial basis function artificial neural network (RBFANN) (Mirbagheri et al., 2015). There was positive correlation between the experimental and predicted values of effluent TP, NH_4^+-N, COD, and BOD with high coefficient of determination ($R^2 > 0.98$) and RMSE values ($\leq 7\%$) for trained and tested models.

Ma et al. (2015) reported the application of ANN coupled with GA to model COD removal in anoxic/oxic processes. Sequential modeling approach using ANN to predict effluent quality of a full-scale coking WWTP was carried out by Shia et al. (2015) and four independent models developed using ANN were optimized by GA. The sequential approach was used to predict effluent COD, cyanide, volatile phenol, and NH_4^+-N using the estimated values and other parameters. The coefficient of determination showed high efficiency of ANN models in predicting effluent quality. The authors further mentioned the importance of the proposed models as tools for predicting coking WWTP performance. Sweetapple et al. (2014) investigated the potential for multi-objective EA of NSGA-II for the reduction of GHG emissions from WWTP in a cost-effective manner. Three objective functions that were formulated and optimized in their study include minimization of effluent pollutant concentration, operational cost, and GHG emission. NSGA-II was used as the optimization tool. Sets of Pareto values for optimal control and operational parameters that could be used to improve the activated sludge system were obtained.

Direct search algorithm developed by Himmelblau (1972) was used to design and obtain optimum synthesis of organic matter in a complex activated sludge processes by minimizing the global penalty function in combining total cost and effluent requirements (Ayesa et al., 1998). In another study, real-coded GA was used by Revollar et al. (2005) to optimize an activated sludge unit and Murphy et al. (2009) likewise optimized commercial wastewater bioreactor to control ammonia removal by minimizing the compressed air requirement. Multi-objective optimization of an operating domestic WWTP using elitist NSGA-II was performed by Iqbal and Guria (2009) for efficient plant monitoring based on performance parameters. Objective functions that were considered for optimization using NSGA-II were maximization of influent flow rate and minimization of effluent quality from the reactor as well as the economic criteria of minimizing the plant operational cost. Important decision variables and constraints were also set by the authors for optimization. Unique solution and set of non-dominated solutions were obtained for the MOOPs formulated and these results could be used by plant operators for successful operation of WWTP (Iqbal and Guria, 2009).

Furthermore, the study has shown that the PSO algorithm is a good training algorithm that could be used as a predicting tool for real world problems (Zhou et al., 2006; Huang et al., 2009) predicted the effluent quality of WWTP based on an improved least square support vector machine for regression (LS-SVR). The input–output data were generated and the forecast model was built for effluent parameters, BOD, COD, SNH (ammonium nitrogen), TN (total nitrogen), and TSS using Benchmark Simulation Model No.1 (BSM1). To obtain a more accurate model, the authors further used PSO to optimize LS-SVR parameters for better prediction. The authors concluded that the improved LS-SVR model approach could predict WWTP effluent parameters better with a good degree of accuracy.

In another study by Balku and Berber (2006), EA was used to tackle the constrained optimization problem of an activated sludge system during start-up in a continuous operational mode. The objective function was to find the optimum aeration schedule for minimum energy consumption. Three constraint methods of using Deb's method (tournament selection), penalizing infeasibles, and rejection of infeasibles were used to handle the constraints without using any penalty parameters. The authors reported that the efficiency of EA in determining the optimum aeration profile for carbon removal, denitrification, and nitrification during wastewater treatment using activated sludge plant can be achieved. Their results have shown that all the constraints methods used for treatment optimization with energy savings of about 12.90, 17.44, and 18.98% for penalizing, rejection and Deb's methods used performed better. However, the shortcoming of the study was its high computational times due to the integration of the large set of differential equations at every stage of the algorithm, which causes the optimization not to reach global optima. High computational times cause deviations from the discharge standards and the DO concentration that must be controlled. However, using Deb's method an optimum aeration profile of 18.98% for energy saving was achieved using EA in comparison to a constant arbitrary aeration profile and 5.81% in comparison to an optimized constant profile (Balku and Berber, 2006).

CONCLUSIONS

Modeling of WWTP is difficult to accomplish due to the high nonlinearity of the plant and the non-uniformity and variability of the crude supply as well as the nature of the biological treatment. To overcome the shortcomings of conventional, one-at-a-time optimization methods, optimization tools have been developed. This paper reviews different types of approaches and the importance of optimization as a tool in the biological treatment of wastewater. The significance of these tools for better and efficient plant performance during wastewater treatment has been mentioned. Different types of EAs such as the metaheuristics approach that combines different intelligent theories for exploring and exploiting the search space to find efficient near-optimal solutions to a particular problem were described. The principles of ANN and FL as black-box models as well as GA and PSO algorithms among the EAs are further discussed in this chapter. The EAs were discussed in relation to their abilities and applications in predicting and searching for Pareto set of solutions to complex and uncertain wastewater problems. Their applications in predicting the design, efficiency, effluent quality, and bacteria concentration from the WWTPs are further discussed.

REFERENCES

Adeyemo, J. and Enitan, A. 2011. Optimization of fermentation processes using evolutionary algorithms-A review. *Scientific Research and Essays*, 6, 1464–1472.

Adeyemo, J. and Otieno, F. 2009. Application of multi-objective differential evolution algorithm (MDEA) to irrigation planning. In World Environmental and Water Resources Congress 2009@ sGreat Rivers ASCE, 4689–4698.

Ayesa, E., Goya, B., Larrea, A., Larrea, L., and Rivas, A. 1998. Selection of operational strategies in activated sludge processes based on optimization algorithms. *Water Science and Technology*, 37, 327–334.

Babu, B. V., Chakole, P. G., and Mubeen, J. H. S. 2005. Multiobjective differential evolution (MODE) for optimization of adiabatic styrene reactor. *Chemical Engineering Science*, 60, 4822–4837.

Baig, S., Mahmood, Q., Nawab, B., Hussain, A., and Nafees, M. 2010. Assessment of seasonal variation in surface water quality of Chitral river, North West Frontier Province, (NWFP), Pakistan. *World Applied Science Journal*, 9, 674–680.

Balku, S. and Berber, R. 2006. Dynamics of an activated sludge process with nitrification and denitrification: Start-up simulation and optimization using evolutionary algorithm. *Computers and Chemical Engineering*, 30, 490–499.

Batstone, D., Keller, J., Angelidaki, I., Kalyuzhnyi, S., Pavlostathis, S., Rozzi, A., Sanders, W., Siegrist, H., and Vavilin, V. 2002. The IWA Anaerobic Digestion Model No.1 (ADM1). *Water Science Technology*, 45, 65–73.

Boonlong, K. 2013. Multiobjective optimization of a vehicle vibration model using the improved compressed-objective genetic algorithm with convergence detection. *Advances in Mechanical Engineering*, 2013, 14 pp, Article ID 131495.

Bose, N. K. and Liang, P. 1996. Neural network fundamentals with graphs. *Algorithms and Applications*. Series in electrical and computer engineering, McGraw-Hill, New York, NY, pp. 447–462.

Brion, G. M. and Lingireddy, S. 2003. Artificial neural network modelling: A summary of successful applications relative to microbial water quality. *Health-Related Water Microbiology*, 47, 235–240.

Chandramouli, V., Brion, G., Neelakantan, T., and Lingireddy, S. 2007. Backfilling missing microbial concentrations in a riverine database using artificial neural networks. *Water Research*, 41, 217–227.

Chen, H.-M. and Lo, S.-L. 2010. Prediction of the effluent from a domestic wastewater treatment plant of CASP using gray model and neural network. *Environmental Monitoring and Assessment*, 162, 265–275.

Çinar, Ö. 2005. New tool for evaluation of performance of wastewater treatment plant: Artificial neural network. *Process Biochemistry*, 40, 2980–2984.

Deb, K. 2011. Multi-objective optimization using evolutionary algorithms: An introduction. In: *Multi-objective Evolutionary Optimisation for Product Design and Manufacturing*, Springer, London, pp. 3–34.

Deb, K., Agrawal, S., Pratap, A., and Meyarivan, T. 2000. A fast elitist non-dominated sorting genetic algorithm for multi-objective optimization: NSGA-II. *Lecture Notes in Computer Science*, 1917, 849–858.

Djeddou, M. and Achour, B. 2015. The use of a neural network techniques for the prediction of sludge volume index in municipal wastewater treatment plant. *LARHYSS Journal*, 2(24), 351–370.

Enitan, A. M. 2014. *Microbial Community Analysis of a UASB Reactor and Application of an Evolutionary Algorithm to Enhance Wastewater Treatment and Biogas Production.* Doctoral dissertation, Durban University of Technology, South Africa. https://ir.dut.ac.za/bitstream/handle/10321/1276/ENITAN_2015.pdf?sequence=1&isAllowed=y.

Enitan, A. M. and Adeyemo, J. 2011. Food processing optimization using evolutionary algorithms. *African Journal of Biotechnology*, 10, 16120–16127.

Erdirencelebi, D. and Yalpir, S. 2011. Adaptive network fuzzy inference system modeling for the input selection and prediction of anaerobic digestion effluent quality. *Applied Mathematical Modelling*, 35, 3821–3832.

Fernández-Navarro, F., Valero, A., Hervás-Martínez, C., Gutiérrez, P. A., García-Gimeno, R. M., and Zurera-Cosano, G. 2010. Development of a multi-classification neural network model to determine the microbial growth/no growth interface. *International Journal of Food Microbiology*, 141, 203–212.

Fister, I., Yang, X. S., and Brest, J. 2013. Modified firefly algorithm using quaternion representation. *Expert Systems with Applications*, 40, 7220–7230.

Flores-Alsina, X., Corominas, L., Neumann, M. B., and Vanrolleghem, P. A. 2012. Assessing the use of activated sludge process design guidelines in wastewater treatment plant projects: A methodology based on global sensitivity analysis. *Environmental Modelling and Software*, 38, 50–58.

Gen, M. and Cheng, R. 1997. *Genetic Algorithms and Engineering Design.* Wiley, New York, NY.

Gernaey, K. V., Van Loosdrecht, M. C., Henze, M., Lind, M., and Jørgensen, S. B. 2004. Activated sludge wastewater treatment plant modelling and simulation: State of the art. *Environmental Modelling and Software*, 19, 763–783.

Goñi, S. M., Oddone, S., Segura, J. A., Mascheroni, R. H., and Salvadori, V. O. 2008. Prediction of foods freezing and thawing times: Artificial neural networks and genetic algorithm approach. *Journal of Food Engineering*, 84, 164–178.

Gontarski, C. A., Rodrigues, P. R., Mori, M., and Prenem, L. F. 2000. Simulation of an industrial wastewater treatment plant using artificial neural networks. *Computer and Chemical Engineering*, 24, 1719–1723.

Gujer, W., Henze, M., Mino, T., and Loosdrecht, M. 1999. Activated sludge model no. 3. *Water Science Technology*, 39, 183–193.

Henze, M., Grady, C. P. L. Jr, Gujer, W., Marais, G. V. R., and Matsuo, T. 1987. Activated Sludge Model No. 1. IAWQ Scientific and Technical Report No. 1. London, UK.

Henze, M., Gujer, W., Mino, T., Matsuo, T., Wentzel, M. C. M., and Marais, G. V. R. 1995. Activated Sludge Model No. 2. IWA Scientific and Technical Report No. 3. London, UK.

Henze, M., Gujer, W., Mino, T., Matsuo, T., Wentzel, M. C., Marais, G. V. R., and Loosdrecht, M. C. M. V. 1999. Activated sludge model no. 2d, ASM2D. *Water Science and Technology*, 39, 165–182.

Himmelblau, D. M. 1972. *Applied Nonlinear Programming.* McGraw-Hill, New York, NY.

Hong, Y.-S. T. and Paik, B.-C. 2012. Inference model derivation with a pattern analysis for predicting the risk of microbial pollution in a sewer system. *Stochastic Environmental Research and Risk Assessment*, 26, 695–707.

Holland, J. H. 1975. *Adaptation in Natural and Artificial Systems.* University of Michigan Press, Ann Arbor, MI.

Huang, M., Wan, J., Hu, K., Ma, Y., and Wang, Y. 2013. Enhancing dissolved oxygen control using an on-line hybrid fuzzy-neural soft-sensing model-based control system in an anaerobic/anoxic/oxic process. *Journal of Industrial Microbiology and Biotechnology*, 40, 1393–1401.

Huang, Y., Lan, Y., Thomson, S. J., Fang, A., Hoffmann, W. C., and Lacey, R. E. 2010. Development of soft computing and applications in agricultural and biological engineering. *Computers and Electronics in Agriculture*, 71, 107–127.

Huang, Z., Luo, J., Li, X., and Zhou, Y. 2009. Prediction of effluent parameters of wastewater treatment plant based on improved least square support vector machine with PSO. *Proceedings of the 2009 First IEEE International Conference on Information Science and Engineering*, December 26–28, 2009, Nanjing, China. IEEE Computer Society.

Ipeaiyeda, A. R. and Onianwa, P. C. 2012. Impact of brewery effluent on water quality of the Olosun river in Ibadan, Nigeria. *Chemistry and Ecology*, 25, 189–204.

Islam, M., Khan, H., Das, A., Akhtar, M., Oki, Y., and Adochi, T. 2006. Impacts of industrial effluents on plant growth and soil properties. *Soil & Environment*, 25, 113–118.

Iqbal, J. and Guria, C. 2009. Optimization of an operating domestic wastewater treatment plant using elitist non-dominated sorting genetic algorithm. *Chemical Engineering Research and Design*, 87, 1481–1496.

Jolma, A. and Norton, J. 2005. Methods of uncertainty treatment in environmental models. *Environmental Modelling and Software*, 20, 979–980.

Jones, N., Grace, M., and Clifford, E. 2014. Developing an AQUASIM biofilm model to simulate a novel batch biofilm passive aeration technology. *International Environmental Modelling and Software Society (iEMSs), 7th International Congress on Environmental Modelling and Software*. http://www.iemss.org/society/index.php/iemss-2014-proceedings

Kachitvichyanukul, V. 2012. Comparison of three evolutionary algorithms. *Industrial Engineering and Management Systems*, 11, 215–223.

Kasemset, C. and Kachitvichyanukul, V. 2012. A PSO-based procedure for a bi-level multi-objective TOC-based job-shop scheduling problem. *International Journal of Operational Research*, 14, 50–69.

Kennedy, J. and Eberhart, R. C. Particle swarm optimization. Proceedings of IEEE International Conference on Neural Networks (ICNN), 1995, Perth, Australia. 1942–1948.

Kovoor, P. P., Idris, M. R., Hassan, M. H., and Yahya, T. F. T. 2012. A study conducted on the impact of effluent waste from machining process on the environment by water analysis. *International Journal of Energy and Environmental Engineering*, 3, 21.

Liao, C. T., Tzeng, W. J., and Wang, F. S. 2001. Mixed-integer hybrid differential evolution for synthesis of chemical processes. *Journal of the Chinese Institute of Chemical Engineers*, 32, 491.

Loucks, D. P., Van Beek, E., Stedinger, J. R., Dijkman, J. P., and Villars, M. T. 2005. *Water Resources Systems Planning and Management: An Introduction to Methods, Models and Applications*. UNESCO, Paris.

Ma, Q., Qu, Y., Shen, W., Zhang, Z., Wang, J., Liu, Z., Li, D., Li, H., and Zhou, J. 2015. Bacterial community compositions of coking wastewater treatment plants in steel industry revealed by Illumina high-throughput sequencing. *Bioresource Technology*, 179, 436–443.

Maier, H., Kapelan, Z., Kasprzyk, J., Kollat, J., Matott, L., Cunha, M., Dandy, G., Gibbs, M., Keedwell, E., and Marchi, A. 2014. Evolutionary algorithms and other metaheuristics in water resources: Current status, research challenges and future directions. *Environmental Modelling and Software*, 62, 271–299.

Matijasevic, L., Dejanovic, I., and Spoja, D. A. 2010. Water network optimization using MATLAB—A case study. *Resources, Conservation and Recycling*, 54, 1362–1367.

Mirbagheri, S. A., Bagheri, M., Bagheri, Z., and Kamarkhani, A. M. 2015. Evaluation and prediction of membrane fouling in a submerged membrane bioreactor with simultaneous upward and downward aeration using artificial neural network-genetic algorithm. *Process Safety and Environmental Protection*, 96, 111–124.

Mjalli, F. S., Al-Asheh, S., and Alfadala, H. E. 2007. Use of artificial neural network black-box modeling for the prediction of wastewater treatment plants performance. *Journal of Environmental Management*, 83, 329–338.

Mohebbi, M., Barouei, J., Akbarzadeh-T, M. R., Rowhanimanesh, A. R., Habibi-Najafi, M. B., and Yavarmanesh, M. 2008. Modeling and optimization of viscosity in enzyme-modified cheese by fuzzy logic and genetic algorithm. *Computers and Electronics in Agriculture*, 62, 260–265.

Morales-Mora, M., Paredes, J., Montes Deoca, J., Mendoza-Escamilla, V., and Martínez-Delgadillo, S. 2015. Modeling and performance evaluation of a full scale petrochemical wastewater treatment process. *International Journal of Environmental Research*, 9, 77–84.

Mullai, P., Rene, E. R., Park, H. S., and Sabarathinam, P. 2012. Adaptive network based fuzzy interference system (ANFIS) modeling of an anaerobic wastewater treatment process. *Handbook of Research on Industrial Informatics and Manufacturing Intelligence: Innovations and Solutions,* 252–270.

Murphy, R., Young, B., and Kecman, V. 2009. Optimising operation of a biological wastewater treatment application. *ISA Transactions,* 48, 93–97.

Nariman-Zadeh, N., Salehpour, M., Jamali, A., and Haghgoo, E. 2010. Pareto optimization of a five-degree of freedom vehicle vibration model using a multi-objective uniform-diversity genetic algorithm (MUGA). *Engineering Applications of Artificial Intelligence,* 23, 543–551.

Neelakantan, T., Lingireddy, S., and Brion, G. M. 2002. Effectiveness of different artificial neural network training algorithms in predicting protozoa risks in surface waters. *Journal of Environmental Engineering,* 128, 533–542.

Nguyen, S. and Kachitvichyanukul, V. 2010. Movement strategies for multi-objective particle swarm optimization. International *Journal of Applied Metaheuristic Computing,* 1(3), 59–79.

Pai, T., Wan, T., Hsu, S., Chang, T., Tsai, Y., Lin, C., Su, H., and Yu, L. 2009. Using fuzzy inference system to improve neural network for predicting hospital wastewater treatment plant effluent. *Computers and Chemical Engineering,* 33, 1272–1278.

Pai, T.-Y. 2008. Gray and neural network prediction of effluent from the wastewater treatment plant of industrial park using influent quality. *Environmental Engineering Science,* 25, 757–766.

Pai, T. Y., Yang, P. Y., Wang, S. C., Lo, M. H., Chiang, C. F., Kuo, J. L., Chu, H. H. et al. 2011. Predicting effluent from the wastewater treatment plant of industrial park based on fuzzy network and influent quality. *Applied Mathematical Modelling,* 35, 3674–3684.

Peyvandi, M., Zafarani, M., and Nasr, E. 2011. Comparison of particle swarm optimization and the genetic algorithm in the improvement of power system stability by an SSSC-based controller. *Journal of Electrical Engineering and Technology,* 6, 182–191.

Phiri, O., Mumba, P., Moyo, B., and Kadewa, W. 2005. Assessment of the impact of industrial effluents on water quality of receiving rivers in urban areas of Malawi. *International Journal of Environmental Science and Technology,* 2, 237–244.

Ramesh, M. N., Kumar, M. A., and Rao, P. N. S. 1996. Application of artificial neural networks to investigate the drying of cooked rice. *Journal of Food Process Engineering,* 19, 321–329.

Reed, P. M., Hadka, D., Herman, J. D., Kasprzyk, J. R., and Kollat, J. B. 2013. Evolutionary multiobjective optimization in water resources: The past, present, and future. *Advances in Water Resources,* 51, 438–456.

Revollar, S., Lamanna, R., and Vega, P. 2005. Algorithmic synthesis and integrated design for activated sludge processes using genetic algorithms. *Computer Aided Chemical Engineering,* 20, 739–744.

Rivas, A., Irizar, I., and Ayesa, E. 2008. Model-based optimisation of wastewater treatment plants design. *Environmental Modelling and Software*, 23, 435–450.

Rosa, S. M., Soria, M. A., Vélez, C. G., and Galvagno, M. A. 2010. Improvement of a two-stage fermentation process for docosahexaenoic acid production by *Aurantiochytrium limacinum* SR21 applying statistical experimental designs and data analysis. *Bioresource Technology*, 101, 2367–2374.

Rosales-Colunga, L. M., García, R. G., and De León Rodríguez, A. 2010. Estimation of hydrogen production in genetically modified *E. coli* fermentations using an artificial neural network. *International Journal of Hydrogen Energy*, 35, 13186–13192.

Sarker, R. and Ray, T. 2009. An improved evolutionary algorithm for solving multi-objective crop planning models. *Computers and Electronics in Agriculture*, 68, 191–199.

Shia, S., Qub, Y., Mab, Q., Zhangb, X., Zhoub, J., and Ma, F. 2015. Performance and microbial community dynamics in bioaugmented aerated filter reactor treating with coking wastewater. *Bioresource Technology*, 190, 159–166.

Smith M. 1993. *Neural Networks for Statistical modelling*. Van Nostrand Reinhold, New York.

Sooksaksun, N., Kachitvichyanukul, V., and Gong, D.-C. 2012. A class-based storage warehouse design using a particle swarm optimisation algorithm. *International Journal of Operational Research*, 13, 219–237.

Sweetapple, C., FU, G., and Butler, D. 2014. Multi-objective optimisation of wastewater treatment plant control to reduce greenhouse gas emissions. *Water Research*, 55, 52–62.

Tashaouie, H. R., Gholikandi, G. B., and Hazrati, H. 2012. Artificial neural network modeling for predict performance of pressure filters in a water treatment plant. *Desalination and Water Treatment*, 39, 192–198.

Van Loosdrecht, M., Ekama, G., Wentzel, M., Hooijmans, C., Lopez-Vazquez, C., Meijer, S., and Brdjanovic, D. 2015. Introduction to modelling of activated sludge processes. *Applications of Activated Sludge Models*, 1.

Wagener, T. and Kollat, J. 2007. Numerical and visual evaluation of hydrological and environmental models using the Monte Carlo analysis toolbox. *Environmental Modelling and Software*, 22, 1021–1033.

Wan, J., Huang, M., Ma, Y., Guo, W., Wang, Y., Zhang, H., Li, W., and Sun, X. 2011. Prediction of effluent quality of a paper mill wastewater treatment using an adaptive network-based fuzzy inference system. *Applied Soft Computing*, 11, 3238–3246.

Yel, E. and Yalpir, S. 2011. Prediction of primary treatment effluent parameters by fuzzy inference system (FIS) approach. *Procedia Computer Science*, 3, 659–665.

Zadeh, L. A. 1975. The concept of a linguistic variable and its application to approximate reasoning. *Information Sciences*, 8, 199.

Zaki, M. R., Varshosaz, J., and Fathi, M. 2015. Preparation of agar nanospheres: Comparison of response surface and artificial neural network modeling by a genetic algorithm approach. *Carbohydrate Polymers*, 122, 314–320.

Zhou, J., Duan, Z., Li, Y., Deng, J., and Yu, D. 2006. PSO-based neural network optimization and its utilization in a boring machine. *Journal of Materials Processing Technology*, 178, 19–23.

Zidan, M., Sagheer, A., and Metwally, N. 2015. Autonomous Perceptron Neural Network Inspired from Quantum computing. *arXiv preprint arXiv:1510.00556*.

Zufferey, N. 2012. Metaheuristics: Some principles for an efficient design. *Computer Technology and Applications*, 3, 446–462.

CHAPTER 10

Production of Fructooligosaccharides as Ingredients of Probiotic Applications

Future Scope and Trends

Ruby Yadav, Puneet Kumar Singh, and Pratyoosh Shukla

CONTENTS

ABSTRACT

Fructooligosaccharides (FOSs) are naturally occurring functional food ingredients which constitute one of the most established groups of prebiotic oligosaccharides. These are oligomers of fructose made up of 1-kestose (GF_2), nystose (GF_3), and 1^F-fructofuranosyl nystose (GF_4), respectively, in which units of fructosyl residues are connected by β-2,1 glycosidic linkages. FOSs are desirable food ingredients as they are low-calorie artificial sweeteners which enhances food flavors. FOS occurs in many natural foods such as onion, rye, banana, agave, barley, garlic, tomato, asparagus, honey, wheat, etc. as reserve carbohydrates. Owing to their beneficial effects on proliferating intestinal bacteria and symbiotic association with probiotic bacteria, FOS can be widely used in the treatment and prevention of various infectious diseases. They can be synthesized by transfructosylation activity of fructosyltransferase (FTase) enzymes. The industrial production of FOS is gaining great attention nowadays. Microbial production is advantageous over chemical synthesis as it is more feasible, convenient, and economical. This chapter gives a summary of the occurrence, production, and applications of FOS. FTases enzymes involved in FOS synthesis, and their plant and microbial sources are also discussed.

INTRODUCTION

FOSs are natural compounds obtained from vegetables, fruits, and cereals. These are also known as alternative sweeteners as they are low-calorie molecules from sucrose. They consist of a number of desirable features such as bifidus-stimulating function, safe for diabetics, effective and safe for human consumption as they do not exhibit any genotoxic, carcinogenic, or toxicological effects. Naturally, FOS is made up of plant sugar chains formed by joining of glucose and fructose molecules. They can also be manufactured with fructosyltransferase (FTase) by transfructosylation activity. Commonly FOS is the name specifically used for oligomers of fructose, made up of GF_2, GF_3, and GF_4 that is, 1-kestose, nystose, 1^F-fructofuranosyl nystose, respectively, in which units of fructosyl residues are connected by β-2,1 glycosidic linkages (Yun 1996). FOSs are nondigestible by human digestive enzymes and pass through upper gastrointestinal tract in unmetabolized form. Moreover, they enter the intestine and are used by beneficial microbes for enhancing colon health and assist in the digestion process, meaning that they have prebiotic effect. FOSs are selectively used by lactobacilli and bifidobacteria which are members of the intestinal microflora. The intestinal flora contains fermentative bacteria and putrefactive bacteria. Fermentative or beneficial bacteria include bifido-bacteria which uses FOS as a substrate to promote bifidogenic flora. Their metabolism by fermentative bacteria produces acids (lactic acid and short-chain fatty acids), which leads to a drop in intestinal pH. This low pH acts as an ideal medium for bifidogenic flora development and decreases the development of pathogenic bacteria. As FOS represents a selective nutrient media for beneficial bacteria, they have potential to enhance the effectiveness of probiotic products (Hidaka et al. 1986). Probiotics are microbial food supplements used to facilitate advantageous effect on intestinal microbiological flora for good health (Goyal et al. 2013; Yadav and Shukla 2015). Considering all the products introduced in the market so far, FOSs are given special attention and credited for growth of sugar market by numerous factors. These factors are simple mass production and their incredible sweet taste similar to sucrose which is a common traditional sweetener. FOSs are commonly used for constipation, traveler's diarrhea, and reduction of high cholesterol levels in blood, as prebiotic, probiotic supplements, etc. (Losada and Olleros 2002). Due to the low production yield of FTases obtaining from plant sources and mass production restriction by the seasonal effects, industrial production of the enzyme is mainly dependent on microbial sources. The functional properties of microbial and plant FTases implicated in FOS synthesis, FOS production methods, and application as prebiotics and probiotics are described in this chapter.

Chemical Structure

Chemically, FOSs are inulin-type oligosaccharides consisting of β-D fructan chains in which units of fructosyl residues are attached by β-2, 1 glycosidic bonds that carries a D-glucosyl residue at the chain end. FOS structures synthesized in cell-free systems of enzymes are basically indistinguishable from those which are whole cell system produced. A research group of Meiji Seika Kaisha Co (Tokyo, Japan) is the former producer of FOS. They introduced the FOS chemical structure produced by using the FTase enzyme derived from *Aspergillus niger*. The structure obtained was identified using analysis techniques such as gas liquid chromatography (GLC), gas chromatography-mass spectrometry (GC-MS), and nuclear magnetic resonance (NMR) (Hayashi et al. 1989). FOSs are fructose oligomers which consist of 1-kestose (GF_2), nystose (GF_3), and 1^F-fructo-furanosyl nystose (GF_4). Their synthesis in plant cells starts by fructosyl moiety transfer among two molecules of sucrose. Their β-(2-1) glycosidic linkage resists hydrolysis by human digestive enzymes (Rumessen et al.

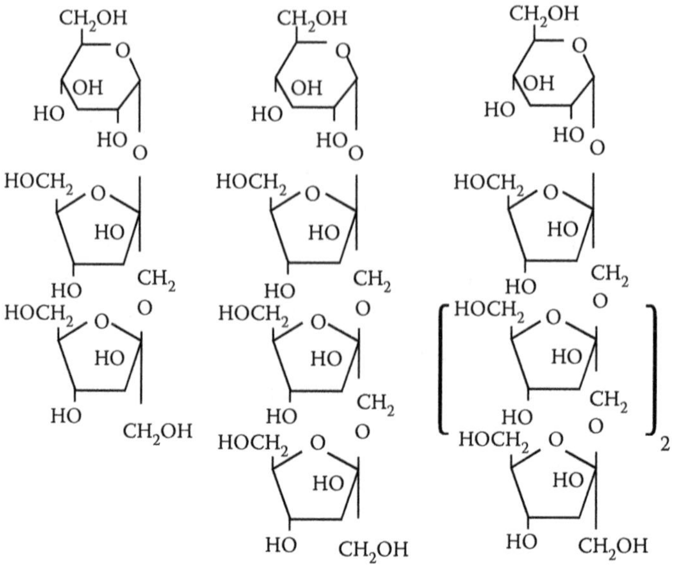

n 2 1-kestose

n 3 Nystose

n 4 Fructofuranosyl nystose

$1^F(1\text{-}\beta\text{-fructofuranosyl})_{n-1}\text{-sucrose}$

FIGURE 10.1 Chemical structure of FOS.

1990). However, recent reports suggest that they might be used as a nutrient medium by intestinal beneficial bacteria. Structural analysis of FOS is very important as their linkages and degree of polymerization differ from the enzyme source (Figure 10.1).

FOS SYNTHESIZING ENZYMES

There is an increased demand for FOS in functional food market, so discovery of novel FTase sources is always needed. FOS synthesizing enzyme sources could be divided into two main classes, plants and microbial sources. Table 10.1 shows list of microbial sources reported for FTase enzyme production. As the production yield from plant sources is low and limited by seasonal environment, industrial production of enzyme is mainly dependent on fungal or bacterial sources.

Bacterial and Fungal FTases

FOS synthesizing enzymes are extremely exceptional in bacterial strains. Enzyme transfructylase isolated from the *Bacillus macerans* EG-6 strain

TABLE 10.1 Microbial Sources of FTases

Microbial Sources of FOS Synthesizing Enzymes	References
Fungal Source	
Aspergillus niger	Ganaie et al. (2013)
Aspergillus japonicus	Mussatto and Teixeira (2010)
Aspergillus sydowii	Sangeetha et al. (2005)
Aspergillus foetidus	Sangeetha et al. (2005)
Aspergillus flavus	Ganaie et al. (2013)
Aspergillus aculeatus	Ghazi et al. (2007)
Aspergillus terreus	Ganaie et al. (2013)
Aureobasidium pullulans	Dominguez et al. (2012)
Claviceps purpurea	Sangeetha et al. (2005)
Fusarium oxysporum	Sangeetha et al. (2005)
Penicillium islandicum	Ganaie et al. (2013)
Penicillium frequentans	Sangeetha et al. (2005)
Penicillium citrinum	Hayashi et al. (2000)
Penicillium spinulosum	Sangeetha et al. (2005)
Phytophthora parasitica	Sangeetha et al. (2005)
Scopulariopsis brevicaulis	Sangeetha et al. (2005)
Bacterial Source	
Arthrobacter sp.	Yun et al. (1996)
Bacillus macerans	Park et al. (2001)
Zymomonas mobilis	Bekers et al. (2002)

by using sucrose source, selectively produces GF_5 and GF_6 FOSs. When 50% sucrose substrate was used, the yield reported was 33% (Park et al. 2001). *Zymomonas mobilis* produces levansucrase which produces FOS and levan. Levansucrases have been widely used as a biocatalyst in sugar industry which yields 24%–32% FOS. During 24 h of incubation, ethanol present in the sucrose syrup acts as a limiting factor for enzyme activity to 24%. Fructan syrup which is produced from sucrose has been used as a good prebiotic source (Beker et al. 2002). Strains of *Lactobacillus reuteri* 121 have been reported to produce 10 g/L FOS. The FTase enzyme isolated from this strain when allowed to incubate with sucrose produces FOS along with inulin. Numerous fungal strains specially *Aspergillus* sp. produces extracellular as well as intracellular fungal FTases. Strain *A. niger* AS 0023 has been reported to possess FTase activity and yields 54% FOS from 50% sucrose substrate. *Penicillium citrinum* produces more than 55 using 70% sucrose substrate (L'Hocine et al. 2000). *Aspergillus japonicas* mycelia immobilized in gluten was also reported to produce 61% FOS

TABLE 10.2   Plant Sources of FTases

Plant Sources of FOS Synthesizing Enzymes (Sangeetha et al. 2005)		
Agave Americana	Cichorium intybus	Lycoris radiata
Agave vera cruz	Crinum longifolium	Sugar beet leaves
Allium cepa	Helianthus tuberosus	Taraxacum officinale
Asparagus officinalis	Lactuca sativa	

(Chien et al. 2001). Various studies have reported *Aspergillus oryzae*, a filamentous fungus, as an innovative source of extracellular FTases enzyme production (Sangeetha et al. 2003).

Plant FTases

Many fructose oligomers obtained from sucrose occur as reserve carbohydrates in many higher plants. Barley, wheat, garlic, onion, tomato, sugar beet, chicory, asparagus, rye, and honey are common natural sources of FOS (Mussatto and Mancilha 2007). Among them, asparagus and onion are two most important sources. A group of researchers extensively studied extraction of FTase from *Asparagus officinalis* roots and isolated 11 FOS components and also synthesized it *in vitro*. Another study described oligosaccharides and fructans biosynthesis from *Agave vera cruz*. Agave-derived enzymes carry out the synthesis of inulotriose from inulobiose (Table 10.2).

PRODUCTION OF FOS

On industrial scale, FOS can be produced by inulin hydrolysis or transfructosylation of sucrose. A research group of Meiji Seika Kaisha Co (Tokyo, Japan) is the former producer of FOS since 1984. They introduced the first product in the market under the name Meioligo. Some other companies are also producing commercial FOS with a purity level above 95%. A summary of commercial FOS products and their industrial producers is given in Table 10.3.

TABLE 10.3   Commercially Available FOS Product

FOS Product	Industrial Producer	Country
Actilight® Profeed®	Beghin-Meiji Industries	Paris, France
Beneshine™ P-type	Victory Biology Engineering	Shanghai, China
Meioligo	Meiji Seika Kaisha	Tokyo, Japan
NutraFlora	GTC Nutrition	Golden, Colorado, USA
Oligo-Sugar	Cheil Foods and Chemicals	Seoul, Korea

It also involves enzymatic action with transfructosylating activity obtained from microbes. FOS production can be done either by using whole cell or immobilized cell. Fermentation methods such as solid-state fermentation (SSF) and submerged fermentation (SmF) for production are discussed in further sections.

Microbial Production of FOS

Many fungal species such as *Aspergillus niger, A. sydowii, A. foetidus, A. japonicas, A. oryzae, A. phoenicis, Aureobasidium pullulans, Fusarium oxysporum, Penicillium citrinum, P. rugulosum, P. frequentens, Scopulariopsis brevicaulis*, etc. have been reported to produce FTase enzyme using sucrose as substrate (Prata et al. 2010). Few studies have reported optimum pH (4.5–6.5) and temperature (50–60°C) conditions for fungal FTase enzyme activity (Maiorano et al. 2008). Additionally, bacteria such as *Zymomonas mobilis, Bacillus macerans, Lactobacillus reuteri, Arthrobacter* sp., and yeast species of *Candida* and *Kluyveromyces, Saccharomyces* are good sources of microbial production of FOS and there are several reports of perceptive inulinase modeling and docking toward substrate specific binding (Sguarezi et al. 2009; Singh and Shukla 2011). Mineral salts present in the fermentation media have been used by microbes which improves the FOS production. Studies have reported that in a medium containing 10% sucrose, 0.6 M NaCl concentration results in 3.5-fold increase in the production of FOS by *Z. mobilis* 113S strain. Sorbitol was also produced as a by-product of fermentation (Prapulla et al. 2000). Batch and continuous fermentation processes have been used for the industrial production of FOS as shown in Figure 10.2.

Fermentative Methods for FOS Production

There are two methods for FTase production by fermentation which are SmF and SSF. SSF offers low cost, high production, and holds tremendous advantage over SmF of enzymes in operation, productivity, less chances of contamination, and concentrated product. SSF is a very convenient process which requires less production costs and simpler downstream processing. Moreover the substrates used are cheap agro-industrial residues, which are converted into high-value commercial products. Many research groups have reported use of SmF for the production of FTases and have discussed in detail (Prapulla et al. 2000). However in case of using SSF as a method of choice, apple pomace substrate have been used widely (Hang et al. 1995). There are also reports of using cereal bran (wheat, rice, maize,

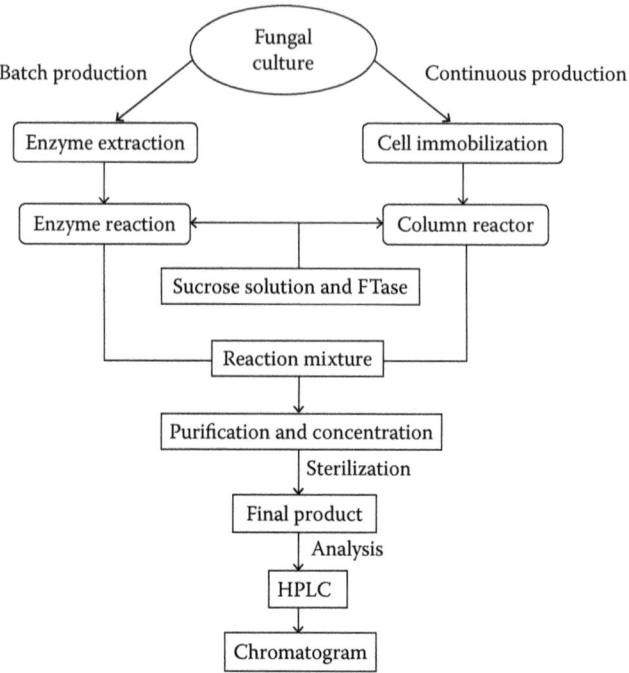

FIGURE 10.2 Industrial fermentation processes for FOS production.

etc.), corn products (corn flour), tea waste, by-products of coffee processing industries, bagasse of cassava and sugar as substrate for FTase enzyme production using *A. oryzae* strain under SSF conditions. As alternate sucrose sources sugarcane juice and jiggery were also used for the production of FOS using FTase enzyme that is procured from *A. oryzae* (Sangeetha et al. 2004).

Optimization of FOS Production

Using Plackett Burman design various parameters were selected which affects the yield of FOS production. The important parameters that influence FTase and FOS production are fermentation method and sucrose, potassium phosphate (KH_2PO_4) concentration in fermentation medium. Time and pH of reaction mixture were also found to have potential effect on FOS production. FOS yields were maximized by using response surface methodology (RSM) which is based on Shell design. Many experimental design models could be used to reduce the number of experiments under different conditions. Doehlert design is used to conclude the combined effect of the above parameters that is, pH and time, required for maximum

FOS production. RSM plays an important role in evaluating important parameters which influence FOS production. Optimized conditions for obtaining maximum yield were 90 h fermentation time, 0.9% KH_2PO_4, 10% sucrose, 18 h, and 5.15 pH of reaction mixture (Sangeetha et al. 2005).

FOS Analysis

Advanced analytical techniques such as chromatography particularly high performance liquid chromatography (HPLC) and thin-layer chromatography (TLC) have been used widely as a rapid and accurate method for FOS analysis. NMR, GC-MS, AOAC have also been used as a method of choice for the analysis. TLC involves quantitative FOS analysis by using isopropyl alcohol:ethyl acetate:water (2:2:1) solvent systems (Park et al. 2001). The colors produced by the products were observed by spraying plates with phenol sulfuric acid and then by heating the plates. As advancement in TLC, a routine method of using Diol HPTLC plates has been proposed for FOS analysis as a quick method for oligosaccharides detection in beet molasses and others. These plates use acetonitrile and acetone solvents. Developed Diol plates after heating for 15 min at 115.8°C showed a color change of spots, that is, from yellow to brown, corresponding to nonreducing FOS (Vaccari et al. 2000). A study reported FOS analysis by 13C NMR which is prepared from sucrose by *Penicillium citrinum* cells. This study reported the presence of nystose, neokestose, and 1-kestose in the products (Hayashi et al. 2000). GC-MS was also performed for the analysis by using Hitachi M-2000 AM instrument equipped with an OB 225 fused column of silicon at temperature 170–200°C with helium as the carrier gas. After clean up on silica gel column samples were methylated using methyl iodide and hydrolysis with 1 M H_2SO_4 for 1 h. Samples were reduced by addition of $NaBD_4$ and then alditol acetylated with acetic anhydride at 110°C for 3 h. HPLC is one of the most widespread methods to analyze FOS but lack of pure standards act as a major hinderance to determination. In support of this a new method has been reported named the AOAC method, based on sample treatment with inulinase enzymes. The products inulin and oligofructans were extracted from samples with the help of boiling water. Two aliquots were taken, one aliquot is left untreated as initial sample and the second aliquot of the extract is hydrolyzed by amyloglucosidase enzyme. Hydrolysate sample is kept as the second sample and the remaining sample is further hydrolyzed using inulinase enzyme (Fructo zyme SP 230). The quantity of glucose, sucrose, and fructose in these samples can be measured by various chromatographic methods such as HPLC, capillary gas chromatography, ion exchange chromatography, etc. The amount of inulin released during the process is calculated by subtracting the sugars present in first two aliquots from the third one (Sangeetha et al. 2005).

FOS AS PREBIOTIC

Prebiotics are nondigestible carbohydrates which beneficially affect the host by modulating the growth of beneficial intestinal bacteria. They act as food for probiotics. Since FOSs are nondigestible they pass through the upper gastrointestinal tract in unmetabolized form and undergoes fermentation in the large intestine. Their metabolism in the colon produces acids (lactic acid and short-chain fatty acids), which enhances the growth of bifidogenic flora and discourages the growth of putrefactive bacteria. FOS has been recognized as current leading valuable prebiotics all the way through *in-vitro* and *in-vivo* studies. Fermentation properties of prebiotics have been studied by monitoring gut bacterial population over a 24 h batch culture, using fluorescence in situ hybridization (FISH) technique. Also production of gas and short-chain fatty acid was calculated. As a result of this study, prebiotics increased bifidobacteria numbers and decreased clostridia number (Rycroft et al. 2001). In a similar study, *Bifidobacterium infantis* ATCC 15697 was grown on synthetic inulin-type fructans that is, oligofructose using glucose substrate and the difference in carbohydrate utilization patterns was also determined. Biomass production with fructose substrates was high with more metabolites as compared with sucrose and glucose substrates (Perrin et al. 2001). Two commercial strains of *Bifidobacterium spp.* (Bf-1 and Bf-6) were cultured with different FOS concentrations (0%–5.0% w/v) and their activity was determined. Growth and viability of strains were assessed following 4 weeks of refrigerated storage. Enhanced growth and viability was observed when *Bifidobacteria* were grown on an FOS medium. FOS consumption increases lactic acid bacteria (LAB) without changing the level of anaerobic bacteria. FOS diet could enhance microbial densities and the metabolic abilities of lactic acid bacteria which results in improved health, enhances host defense mechanisms, and accelerates recovery after gastrointestinal tract (GIT) disturbances (Kolida et al. 2002). FOSs at low doses show high bifidogenic activity. In addition, they act as a significant butyric acid which is a vital nutrient for the colonocytes. Synbiotics, which are a combination of probiotics and prebiotics, represent a promising and safe means against metabolic disorders.

CONCLUSION

The increasing consumer demand for safe and healthy food products s has gained significant attention within the food industry. FOS, the non-digestible carbohydrate, presents various health benefits to consumers. It results in their increasing demand as food ingredients in functional food market all over the world. Various studies on FOS have shown their use as efficient prebiotics. They exhibit scientifically proven health benefits such as in combating hypercholesterolemia, constipation, improving nutrient absorption, cholesterol reduction, fat and sugar replacement in food products, cancer control, etc. Synbiotics, which represent an association of prebiotic and probiotics, have been studied as efficient and safe tools in metabolic and intestinal disorders. Screening and identification of novel microbial FTase sources for FOS production are needed to be explored. Fungal FTases sources produce higher FOS yield than bacterial sources. The FTase purification from these sources leads to better understanding of properties and kinetics of the enzyme. Highly efficient, less expensive, and more reliable production methods using microfiltration, membrane reactor system, immobilized cells, or enzymes have led to advancement in FOS production.

REFERENCES

Beker, M., J. Laukevics, D. Upite, E. Kaminska, A. Vignats, U. Viesturs et al. 2002. Fructooligosaccharide and levan producing activity of *Zymomonas mobilis* and extracellular levan sucrase. *Process Biochemistry* 38:701–706.

Chien, C. S., W. C. Lee, and T. J. Lin. 2001. Immobilization of *Aspergillus japonicus* by entrapping cells in gluten for production of FOS. *Enzyme and Microbial Technology* 29:252–257.

Dominguez, A., C. Nobre, L. R. Rodrigues, A. M. Peres, D. Torres, I. Rocha et al. 2012. New improved method for fructooligosaccharides production by *Aureobasidium pullulans*. *Carbohydrate Polymers* 89:1174–1179.

Ganaie, M. A., U. S. Gupta, and N. Kango. 2013. Screening of biocatalysts for transformation of sucrose to fructooligosaccharides. *Journal of Molecular Catalysis B: Enzymatic* 97:12–17.

Ghazi, I., L. Fernández-Arrojo, H. Garcia-Arellano, M. Ferrer, A. Ballesteros, and F. J. Plou. 2007. Purification and kinetic characterization of a fructosyltransferase from *Aspergillus aculeatus*. *Journal of Biotechnology* 128:204–211.

Goyal, S., T. Raj, C. Banerjee, J. Imam, and P. Shukla. 2013. Isolation and ecological screening of indigenous probiotic microorganisms from curd and chili sauce samples. *International Journal of Probiotics and Prebiotics* 8:91.

Hang, Y. D., E. E. Woodams, and K. Y. Jang. 1995. Enzymatic conversion of sucrose to kestose by fungal extracellular fructosyltransferase. *Biotechnology Letters* 17:295–298.

Hayashi, S., K. Imada, Y. Kushima, and H. Ueno. 1989. Observation of the chemical structure of fructooligosaccharide produced by an enzyme from *Aureobasidium* sp. ATCC 20524. *Current Microbiology* 19:175–177.

Hayashi, S., T. Yoshiyama, N. Fuji, and S. Shinohara. 2000. Production of a novel syrup containing neo fructooligosaccharides by the cells of *Penicillium citrinum*. *Biotechnology Letters* 22:1465–1469.

Hidaka, H., T. Eida, T. Tarizawa, T. Tokunaya, and Y. Tashiro. 1986. Effects of fructo-oligosaccharides on intestinal flora and human health. *Bifidobacteria Microflora* 5:37–50.

Kolida, S., K. Tuohy, and G. R. Gibson. 2002. Prebiotic effects of inulin and oligo-fructose. *British Journal of Nutrition* 87:193–197.

L'Hocine, L., Z. Wang, B. Jiang, and S. Xu. 2000. Purification and partial characterization of fructosyl transferase and invertase from *Aspergillus niger* AS0023. *Journal of Biotechnology* 81:73–84.

Losada, M. A., and T. Olleros. 2002. Towards a healthier diet for the colon: The influence of fructooligosaccharides and lactobacilli on intestinal health. *Nutrition Research* 22:71–84.

Maiorano, A. E., R. M. Piccoli, E. S. Silva, and M. F. A. Rodrigues. 2008. Microbial production of fructosyltransferases for synthesis of pre-biotics. *Biotechnology Letters* 30:1867–1877.

Mussatto, S. I., and I. M. Mancilha. 2007. Non-digestible oligosaccharides: A review. *Carbohydrate polymers* 68: 587–597.

Mussatto, S. I., and J. A. Teixeira. 2010. Increase in the fructooligosaccharides yield and productivity by solid-state fermentation with *Aspergillus japonicus* using agro-industrial residues as support and nutrient source. *Biochemical Engineering Journal* 53:15–157.

Park, J., T. Oh, and J. W. Yun. 2001. Purification and characterization of a novel transfructosylating enzyme from *Bacillus macerans* EG-6. *Process Biochemistry* 37:471–476.

Perrin, S., M. Warchol, J. P. Grill, and F. Schneider. 2001. Fermentations of fructooligosaccharides and their components by *Bifidobacterium infantis* ATCC 15697 on batch culture in semi-synthetic medium. *Journal of Applied Microbiology* 90:859–865.

Prapulla, S. G., V. Subhaprada, and N. G. Karanth. 2000. Microbial production of oligosaccharides: A Review. In *Advances in Applied Microbiology*, ed. A. L. Laskin, J. W. Bennet, and G. Gadd, 299–337. New York: Academic Press.

Prata, M. B., S. I. Mussatto, L. R. Rodrigues, and J. A. Teixeira. 2010. Fructooligosaccharide production by *Penicillium expansum*. *Biotechnology Letters* 32:837–840.

Rumessen, J. J., S. Bode, O. Hamberg, and E. G. Hoyer. 1990. Fructans of Jerusalem artichokes: Intestinal transport, absorption, fermentation and influence on blood glucose, insulin and C-peptide responses in healthy subjects. *American Journal of Clinical Nutrition* 52:675–681.

Rycroft, C. E., M. R. Jones, G. R. Gibson, and R. A. Rastall. 2001. A comparative *in vitro* evaluation of the fermentation properties of prebiotic oligosaccharides. *Journal of Applied Microbiology* 91:878–887.

Sangeetha, P. T., M. N. Ramesh, and S. G. Prapulla. 2003. Microbial production of fructooligosacharides. *Biotechnology and Environmental Sciences* 53:313–318.

Sangeetha, P. T., M. N. Ramesh, and S. G. Prapulla. 2004. Production of fructosyl transferase by *Aspergillus oryzae* CFR202 in solid-state fermentation using agricultural by-products. *Applied Microbiology and Biotechnology* 65:530–537.

Sangeetha, P. T., M. N. Ramesh, and S. G. Prapulla. 2005. Maximization of Fructooligosaccharide production by two stage continuous process and its scale up. *Journal of Food engineering* 68:57–64.

Sguarezi, C., C. Longo, G. Ceni, G. Boni, M. F. Silva, M. Luccio, M. A. Mazutti, F. Maugeri, M. I. Rodrigues, and H. Treichel. 2009. Inulinase production by agro-industrial residues: Optimization of pretreatment of substrates and production medium. *Food Bioprocess Technology* 2:409–414.

Singh, P. K., and P. Shukla. 2011. Molecular modeling and docking of microbial inulinases towards perceptive enzyme-substrate interactions. *Indian Journal of Microbiology* 52:373–380.

Vaccari, G., G. Lodi, E. Tamburini, T. Bernardi, and S. Tosi. 2000. Detection of oligosaccharides in sugar products using planar chromatography. *Food Chemistry* 74:99–110.

Yadav, R., and P. Shukla. 2015. An overview of advanced technologies for selection of probiotics and their expediency: A review. *Critical Review in Food and Nutrition* 52: 373–380. DOI: 10.1080/10408398.2015.1108957.

Yun, J. W. 1996. Fructooligosaccharides—Occurrence, preparation and application. *Enzyme and Microbial Technology* 19:107–117.

CHAPTER 11

Avenues in Ophthalmic Optical Coherence Tomography in Medical Biotechnology

Prospects and Future Trends

Raju Poddar, Vinod Aggarwal, Varun Gogia, Mayank Bansal, Shika Gupta, Rohan Chawla, and Pradeep Venkatesh

CONTENTS

ABSTRACT

In this chapter we describe different technological background of optical coherence tomography (OCT) technique. Then next generation high speed, dense three dimensional ophthalmic imaging with spectral and swept source OCT was incorporated. Details application of OCT technique in medical biotechnology for imaging of anterior and posterior segment was discussed. State-of-art intraoperative OCT was also included. Finally, we introduce, *in-vivo* microvascular imaging of the human chorio-retinal complex and corneo-scleral junction (limbus) using phase variance OCT based angiography.

INTRODUCTION

Imaging techniques are of tremendous importance in biology and medicine. In recent years, interdisciplinary studies, particularly in the realm of optical physics, have contributed to rapid and significant improvements in the field of medical diagnostics. Some of the optical imaging techniques that have been influenced by advances in technology include confocal microscopy, fluorescent imaging, scanning and multifocal laser microscopy, endoscopy, and optical tomography. Optical imaging techniques have superior resolution compared with other imaging techniques. They are also noninvasive, easy to maintain, and cost effective relative to other imaging techniques. Optical coherence tomography (OCT) is one such imaging technique; it has the ability to generate the internal microstructure images of the biological samples, that is, it enables *in-vivo* imaging of various tissues.

The word tomography is a composition of two Greek words "tomo" and "graphō." *Tomo* means "slice or section" and *graphō* means "to write or record," and taken together it means "recording cross sections." OCT means recording/imaging cross-sectional structures of the samples with the help of low-coherence light. An OCT works based on an optical principle termed low-coherence interferometry. Imaging with low-coherence interferometer was first performed by the David Huang et al. group in 1991 at the Massachusetts Institute of Technology (MIT), Cambridge, MA, USA.[1-4]

OCT enables image capture of tissues at a fairly reasonable depth and has high resolution. The resolution of OCT is far superior to that of ultrasonography (USG) and marginally less than confocal microscopy. Figure 11.1 depicts the comparison between various imaging modalities with regard to depth of tissue penetration and resolution.

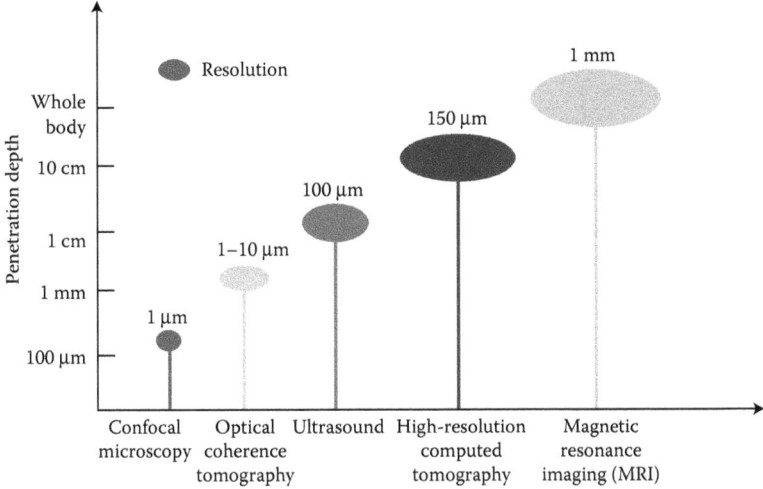

FIGURE 11.1 Picture depicts the resolution and imaging depths of various imaging techniques, the circle size represents the resolution, and lines length represents the penetration depth.

Before going into a detailed description of OCT it would be useful to understand the terms, axial resolution and lateral resolution. The resolution of any system could be defined as up to how small it can distinguish two points in an image obtained by the system. Like most imaging systems, OCT also has two kinds of resolutions, axial resolution and lateral resolution. In OCT, these resolutions are independent unlike confocal microscopy.[5] Axial resolution is the ability to distinguish point separation along the axis of propagation (Figure 11.2) and in OCT, it is determined by the coherence function. The latter is in turn directly proportional to the square of central wavelength of light source, and inversely proportional to the bandwidth of the laser source used in the OCT. Mathematically, it could be explained by the following equation:

$$\Delta Z = \frac{2\ln 2}{\pi} \frac{\lambda_0^2}{\Delta\lambda} \approx 0.44 \frac{\lambda_0^2}{\Delta\lambda}$$

where λ_0 is the central wavelength in the given band and $\Delta\lambda$ is the wavelength bandwidth. If the chosen central wavelength of the laser is lower or the band width of the source is higher axial resolution increases. In an OCT literature the most reported light sources are super luminescent diode (SLD) lasers, femtosecond lasers, and photonic crystal/fiber-based

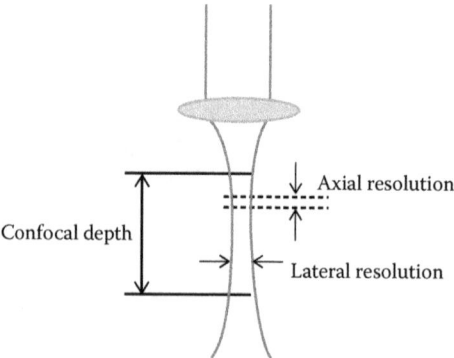

FIGURE 11.2   The schematic diagram of TD-OCT comprising of a broad band light source, beam splitter, reference arm, and sample arm. Reference arm comprises lenses and reference mirror. Sample arm comprises an XY-axis scanner, focusing lens, and tissue sample to be imaged. The output components include a photodetector, data acquisition, and processing system.

supercontinuum lasers.[5] Resolution perpendicular to the propagation direction of beam is called lateral resolution. The confocal parameter of focusing lens at the end of sample arm would decide the lateral resolution of the OCT. If the focusing lenses focal length is f and D is the diameter of the lens, then mathematically lateral resolution would be expressed as

$$\Delta x = \frac{1.22\lambda_0 f}{D}.$$

Here λ_0 is the central wave length of the light source. Ultimately, the diffraction and aberrations limit the lateral resolution. Confocal depth is the one where the resolution of the imaging system is constant. In OCT, resolution is enhanced by a design in which the reflected light coming from the sample tissue passes through two filters, coherence gating and confocal gating.

OCT has been broadly categorized into two types: time domain (TD-OCT) and Fourier domain (FD-OCT) based on their operating and detection mechanism. In TD-OCT, the light emitted from the low-coherence (broadband) light source passes into an interferometer. This light is broken into two arms by the beam splitter. One part of the split beam travels into the reference arm, along which lies a reference mirror that is capable of moving back and forth. The second part of the split beam, called sample arm, is directed toward the tissue being studied. Interference of light reflected from both reference and sample would only occur when the effective path difference is less than or equal to coherence length. To scan the depth of a given sample, the position of the reference is accurately determined and this corresponds to the depth information from where the signal is reflected from within the sample. One cycle movement of reference mirror would give one A-line, next the XY-axis scanner steers the beam onto an adjacent position in the same plane and in the reference arm, the reference mirror repeats the back and forth movement. The whole process repeats until scanning along one plane is complete. The system is so designed that the signals reflected from the reference mirror and sample arm would interfere at the beam splitter. The optical signals produced by this interference are detected by a photodetector. The photodetector then converts the optical signal into an electrical signal which is then digitized and processed with the data acquisition processing module. Ultimately a B-scan image of the biological tissue is generated. A schematic diagram of TD-OCT is as shown in Figure 11.3.

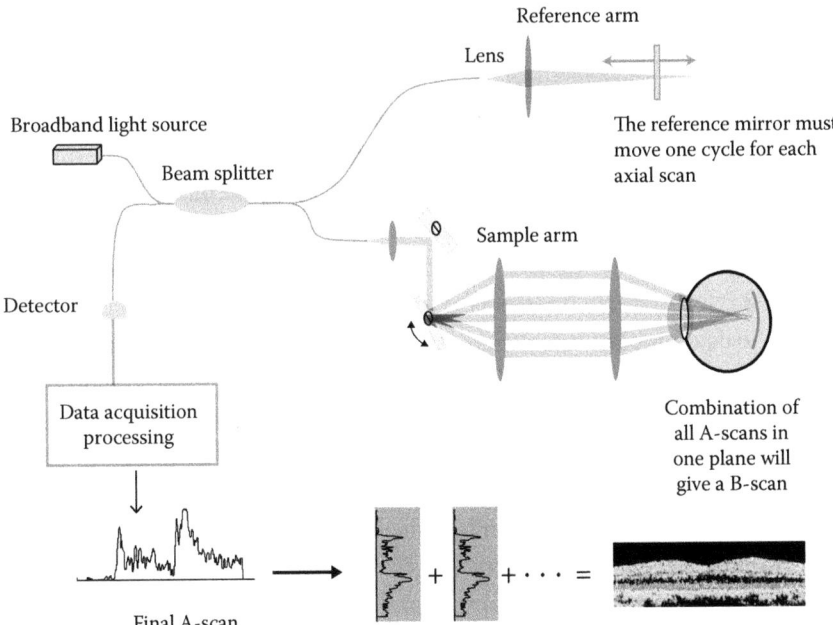

FIGURE 11.3   Schematic diagram of SD-OCT, which consists of a broadband light source, a beam splitter, focusing lens, fixed reference mirror, XY-axis scanner, focusing lens, a spectrometer module which comprises a collimator, grating, and a CCD/CMOS detector.

This technique was first used for measuring the thickness of layers in a sample by Fercher et al. in 1990; they used only one A-line for this purpose.[3] In 1991, Fujimoto's group at MIT expanded the use of TD-OCT and for the first time generated tissue images.[1] They generated images by using multiple A-line intensity profiles reflected from the sample. The back and forth movement of the reference mirror was found to impede the speed and accuracy of image capture during *in-vivo* imaging of the eye. To improve the imaging speed, signal-to-noise ratio (SNR) and resolution researchers introduced FD-OCT.[6-9]

In the FD-OCT, distance information of the sample is encoded in terms of frequencies, which modifies the spectral shape.[5] FD-OCT itself is categorized into two types, spectral domain (SD-OCT) and swept source (SS-OCT), based on the method of signal detection. In SD-OCT, the wavelength spectrum is divided into individual wavelengths with the help of grating and spatial detection is carried out with the help of spectrometers and charge-coupled device (CCD)/complementary metal oxide semiconductor (CMOS) camera.[6] In SS-OCT on the other hand, no spectrometers are used. Instead, two photodetectors enable identification of the signal which is then encoded in time.[9,10]

In SD-OCT, a broad spectrum of light from the source travels to the beam splitter wherein it splits into two parts with one part of light entering into the reference arm. Unlike in TD-OCT, here the reference arm is fixed and this provides stability to the system and improves the speed of image capture. The second part of the split beam travels into the sample arm, which carries an XY-axis scanner. The reflected light from the sample and the fixed reference arm interferes at the beam splitter; this interfered signal would be decomposed into different wavelengths by the spectrometer grating, and identified by the CCD/CMOS camera. The detected signal would be processed and at the end of processing, the inverse Fourier would give the reflection profile of the sample.[10,11] A schematic diagram of SD-OCT is shown in Figure 11.4. Though SD-OCT has been successfully implemented for *in vivo* imaging, limitations do exist. High-speed detectors are not available for all spectral wavelengths and interference fringe washout is also a problem while integrating signal with CCD/CMOS detectors. The more recently introduced SS-OCT overcomes some of these limitations.

In SS-OCT, the swept-source laser fed into the fiber-based Michelson interferometer, travels into the optical circulator and emerges as two equally intense beams. The coupler splits the light intensity into 50:50 ratio; hence 50% of the input light goes into the reference arm. The reference

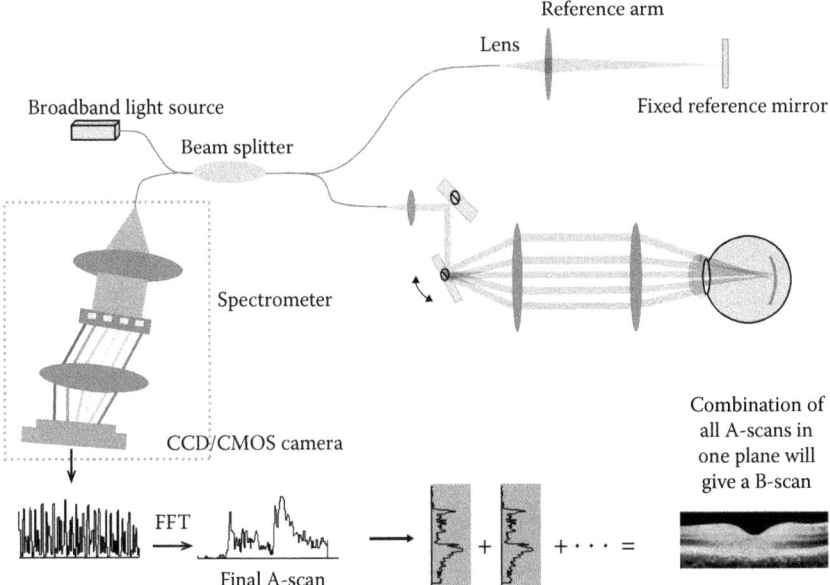

FIGURE 11.4 Schematic diagram of SS-OCT setup. The setup comprises a broadband swept source tunable laser, circulator, beam splitter, a fixed reference arm, and sample arm. The reference arm comprises a lens and a fixed mirror. The sample arm consists of an XY-axis galvo scanner and a focusing lens system. Detection components include a balanced photodetector, data acquisition system, and galvo driver.

arm comprises a collimator, lens, and mirror. The rest 50% amount of an input light travels into the sample arm, which also carries a collimator. The latter generates a collimated beam which incidents on the XY-axis galvo scanner. The light from the XY-axis scanner is directed onto the sample to be imaged by a focusing lens. The reflected light from the sample and the reference arm produces an interference signal at the 50:50 coupler. Finally, the OCT signal is detected by the balanced photodetector and processed further by the data acquisition card. A schematic diagram of SS-OCT is shown in Figure 11.5.

The method of signal processing in OCT and image generation is shown in Figure 11.6. The process is described in brief herein. OCT signal has static noise which degrades the quality of the images, and hence needs to be removed. This is performed by averaging and subtraction method. Spectral reshaping is then done by multiplying the static noise free signal with the Gaussian window, which is useful for suppressing noise and generating

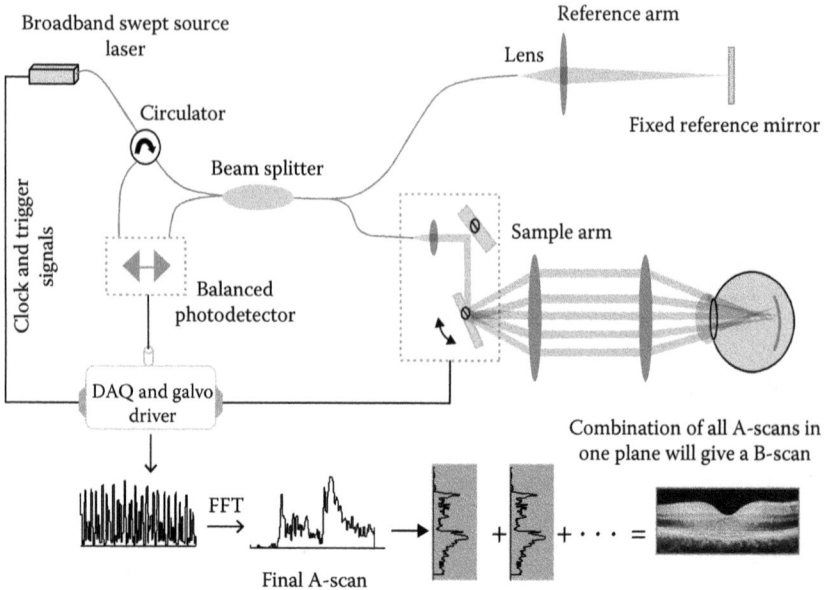

FIGURE 11.5 The above figure depicts the lateral resolution, an axial resolution, and the confocal depth parameters.

optimal point spread function (PSF). The SS-OCT signal is acquired with respect to time and this signal needs to be rescaled into the wave number space in order to remove the nonlinearities and to get the evenly spaced SS-OCT signal in the wavelength domain. In the SS-OCT system light travels through the fiber and sample, hence it disperses. This dispersion needs to be compensated for getting good resolution. The numerical dispersion compensation is done by multiplying $I(k) = \exp(-j(a_1(k - k_0)^2 + a_2 (k - k_0)^3))$, here k is the wave number; a_1, a_2 are arbitrary coefficients, which can be varied for optimizing the dispersion compensation; k_0 is the central wavelength. Zero padding is a standard step before applying Fourier transformation, which is performed by adding zeros and interpolate data to convert into multiples of 1024 points. This step helps to preserve the number of data points after removal of complex conjugate part from the Fourier-transformed SS-OCT signal. The Fourier transformation on K-space SS-OCT signal converts a signal into the Z-space signal, which gives the depth reflection profile or distance information of the sample.

Since its invention the application of OCT has exponentially increased. Applications can be categorized into three types, namely, medical applications, biological science applications, and material science applications. In

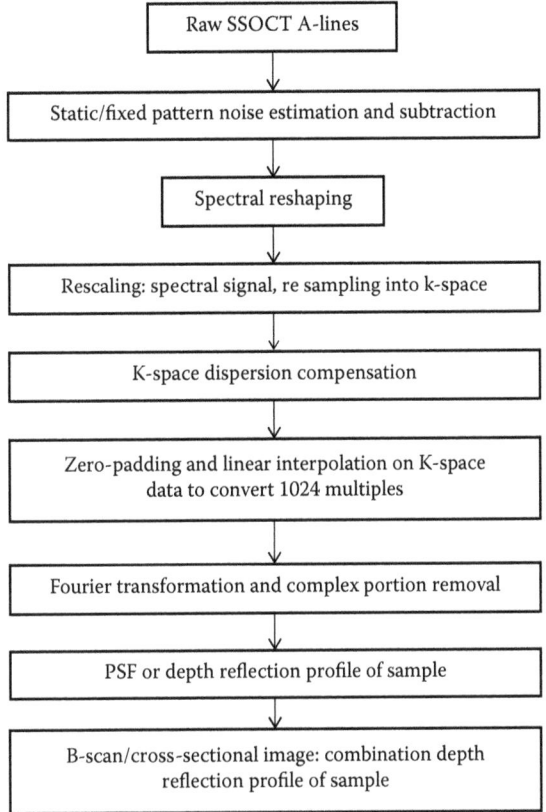

FIGURE 11.6   A flow chart diagram for SS-OCT signal processing algorithm.

the next section, we describe the most recent prospects and trends in OCT in the field of ophthalmology. Included in this section is anterior segment OCT (AS-OCT), posterior segment imaging with SD and SS-OCT, intraoperative OCT, OCT angiography (OCTA), and Doppler OCT.

PROSPECTS AND TRENDS IN OPHTHALMIC OCT

Anterior Segment Imaging

There are various modalities for imaging the anterior segment which include slit lamp biomicroscopy, ultrasound biomicroscopy (UBM), Scheimpflug imaging, and AS-OCT. However, at present, no anterior segment imaging technique provides all information regarding morphology and microstructure of tissue. Imaging with each device is associated with certain merits and demerits as presented in Figure 11.1. Using AS-OCT, anterior segment dimensions can be accurately measured owing to the large (15 × 15 mm) transverse angle and depth of scanning.

The uses of AS-OCT span from the evaluation of chamber angle configurations, measurement of anterior chamber dimensions, follow-up of corneal refractive surgery, after phakic intraocular lens (pIOL) implantation, corneal collagen crosslinking (CXL), intrastromal corneal ring segment (ICRS) implantation, glaucoma filtering surgery to the evaluation of certain anterior segment pathologies such as keratitis, uveitis, and iris tumors. The currently available AS-OCT devices include Visante OCT (TD AS-OCT), slit-lamp-adapted OCT, FD-OCT, and SS-OCT (Table 11.1).

Latest innovations in AS-OCT include attempts at creating ultrahigh resolution OCT systems. Ultrahigh resolution OCT devices include Envisu (Bioptigen) and the SOCT Copernicus HR (Optopol Technologies). A broadband supercontinuum light source with a mean bandwidth of about 375 nm (range 625–1000 nm) is used for scanning in these technologies. In corneal *in vivo* imaging, these can achieve an axial resolution of 1.1 μm. Due to the high magnification, the field of view is limited. This disadvantage can be overcome using ultrahigh resolution OCT systems. As a result, a precise visualization of different corneal layers, viz, epithelium, basement membrane, Bowman's layer, stroma, Descemet's membrane, and endothelium can be done. Corneal lesions, such as corneal edema, Bowman's layer irregularities, breaks, or scars (keratoconus), stromal opacities, deposits, fibrosis, and Descemet's membrane guttae can be identified. The full-field OCT (FF-OCT) incorporates ultrahigh resolution OCT imaging with visualization of cellular structures with 1 μm resolution in all directions. To distinguish the corneal epithelium and delineate the position of Bowman's membrane, a 1060 nm imaging system is useful as it has an axial resolution of 4.5 μm. Figure 11.7 is a collage of AS-OCT images highlighting its role in the imaging of various anterior segment.

Posterior Segment Imaging
SD-OCT-Based Imaging
Spectral domain OCT scan of retina has the following advantages—higher resolution, reduction of motion artifacts, ability to obtain three-dimensional (3D) images, image overlay, volumetric analysis, integration into multimodality imaging systems, and improved follow-up reproducibility of image capture and analysis.

However, the sensitivity is not constant throughout the image. As the tissue of interest moves further from the point of maximum sensitivity

TABLE 11.1 Comparison between Different Imaging Modality of OCT for Anterior Segment

	Central Wavelength (nm)	Axial Resolution (μm)	Lateral Resolution (μm)	A-Scan Rate per second	Lateral Range (mm)	Depth Range (in tissue, r. i.~ 1.4) (mm)	Three Dimensional Imaging Capability
Time-Domain OCT (TDOCT)							
Visante (Carl Zeiss Meditec)	1310	18	60	2048	16 × 6	7.0	No
Fourier-Domain—Spectrometer-Based OCT (SDOCT)							
RT Vue (Optovue)	830	5	15	26,000	10 × 10	2.0	Yes
Spectralis (Heidelberg)	870	4	15	40,000	16 × 6	2.0	Yes
Fourier-Domain—Swept Source OCT (SSOCT)							
Casia (Tomey)	1310	10	30	30,000	16 × 16	7.0	Yes
Poddar et al. [25,26]	1050	4.7	13–30	100,000	20 × 20	2.5–8.5	Yes

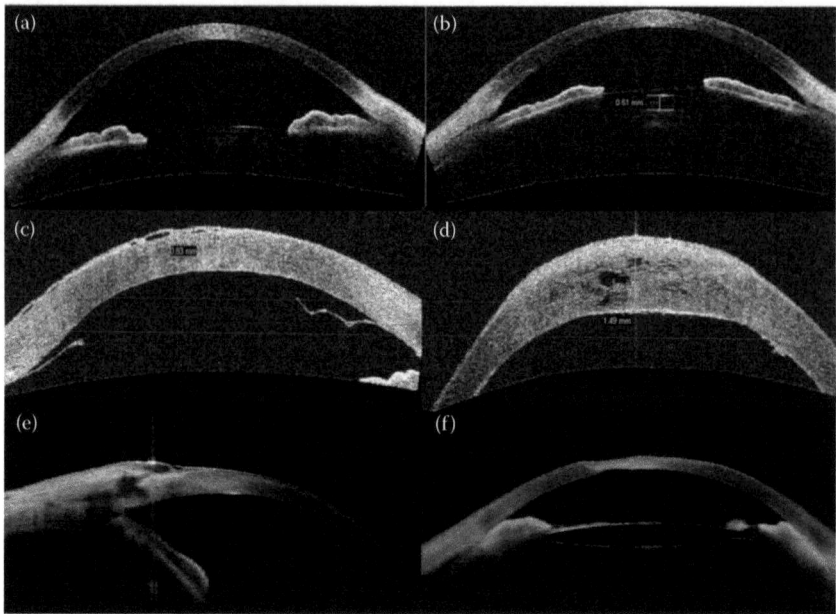

FIGURE 11.7 (a) Scan showing AS-OCT of a normal anterior segment (Visante OCT). (b) Scan represents a well-positioned implantable Collamer lens within the ciliary sulcus with measurements showing ICL vault. (c) Scan shows thickened cornea (bullous keratopathy with epithelial bullae and a Descemet's membrane detachment). (d) Scan shows a thickened cornea with intracorneal vacuoles as seen in acute hydrops in a case of advanced keratoconus. (e) Scan represents a filtering bleb with multiple intra-bleb cystic spaces indicating anatomical health of the bleb. (f) Scan shows a case of pseudophakia with malignant glaucoma with shallow anterior chamber and closed angles.

(called the "zero-delay line"), the quality of the image can decrease, and more noise can appear. One can choose to acquire images depending on the pathology such that the vitreous or choroidal side is closer to the point of maximal sensitivity. Thus, recently, certain modifications in SD-OCT have enabled a more detailed visualization of the choroid through incorporation of enhanced depth imaging (EDI) and en-face imaging.[12-14] Various advances in SD-OCT imaging which is creating an impact in the diagnosis and management of various retinal and choroidal disorders as well as becoming the part of future vitreoretinal surgery has been described below.

EDI OCT is the scanning protocol which allows a more detailed visualization of the choroid. By placing the SD-OCT device closer to the eye and inverting the retinal image, the choroid is placed closer to the zero-delay line resulting in higher depth of field and resolution of choroidal details. This allows us to measure choroidal thickness (CT) *in vivo.*

SD-OCT uses an 800 nm source for imaging and achieves an axial resolution of 5 μm. At this axial resolution there is greater visibility of all layers of the retina compared with images obtained using the earlier TD-OCT. The images obtained however still lack tissue-specific contrast. Moreover, the ability to capture details of structures underneath the retinal pigment epithelium (RPE) is far from satisfactory. The Heidelberg Spectralis has an axial resolution of 7 μm at 40 kHz and was unable to image deeper layers (Figure 11.8). However, their new technology EDI comes with deeper imaging capability by focusing in deeper. Using a 1050 nm imaging source, scattering and absorption by the RPE is reduced and this allows better imaging of the choroidal morphology.

SS-OCT-Based Imaging

SS-OCT is the latest addition to the fast advancements occurring in the technology of OCT. It is the modified Fourier-domain technology that

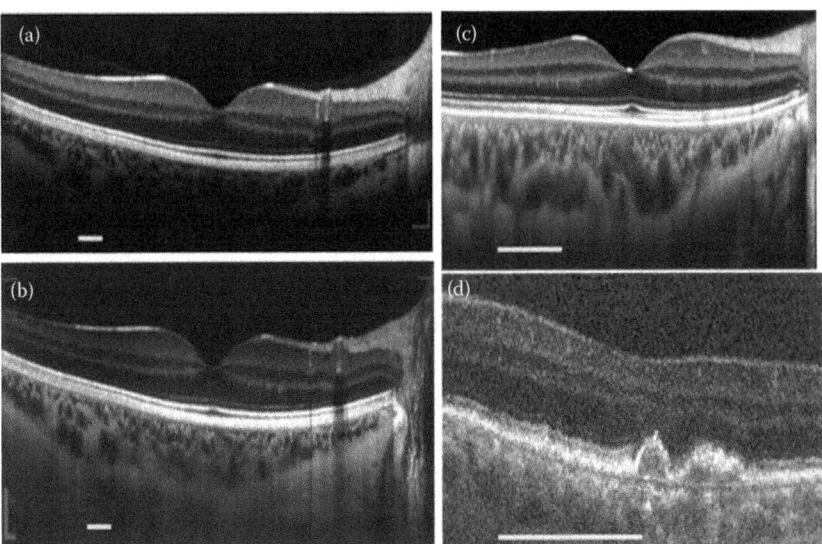

FIGURE 11.8 (a) OCT B-scan of human posterior eye segment with Heidelberg Spectralis™, (b) OCT B-scan at same position with Heidelberg Spectralis EDI™, (c) Image captured with high-resolution 1060 nm SS-OCT, (d) SS-OCT B-scan of an AMD patient at central 3° fovea. Scale bar: 1 mm.

offers higher imaging depth and tissue detection capability. DRI-OCT (Topcon) was the first commercially available SS-OCT for ophthalmic imaging of the retina.

With the advent of SD-OCT we were able to acquire high-resolution images of retina and vitreoretinal interface. The focus of the retina specialists then shifted to choroidal imaging as choroidal vessels supply the outer layers of the retina. Thus CT may be altered in a host of clinical situations. Imaging of the choroid using OCT was first described by Spaide et al., when he reported that inverting the OCT image brings the choroid close to the zero-delay line.[15] This was based on the already known fact that the resolution of OCT was maximum near the zero-delay line. The major drawbacks of scattering of 840 nm wavelengths by the RPE however prevented good visualization of deeper choroidal structures.

This limitation has been overcome by SS-OCT, which utilizes a longer wavelength of 1050 nm. The recording of interference of backscattered light from the retina was also made better by using photodiode detectors instead of CCD cameras used in the SD-OCT. These features in combination with double the scan speed of that available in current SD-OCT lead to high-resolution (1 micron as compared with 3–5 microns of SD-OCT), 12 mm wide-field B-scan imaging of vitreous, retina, and choroid simultaneously. The higher raster scan density and deeper penetration also lead to better quality en-face reconstructions. Thus SS-OCT can provide us with excellent information on choroid, which was previously limited to indocyanine green angiography (ICGA). Another advantage of using 1050 nm wavelength in SS-OCT is the minimal effect of media opacities such as cataract on the quality of scans.

The SS-OCT since its existence has led to the revelation of many important ocular findings. One of its major contributions is toward the exact determination of CT. The study by Hirata et al., one of the earliest clinical studies on SS-OCT, measured the macular CT in normal adults to be 191.5 ± 74.2 microns.[16] Copete et al. have shown that SD-OCT was able to reproducibly measure the CT in 74.4% of eyes-only versus 100% of eyes with SS-OCT.[17] Other important areas of interest that can be studied with SS-OCT is vitreoretinal interface, posterior precortical vitreous pocket (PPVP), and suprachoroidal space.[18–21] The latter could be of great benefit in evaluating the effects of drugs delivered to this space in future.

Intraoperative OCT

OCT has added a new dimension to the imaging ability of the retina. It provides *in vivo* microscopic morphology of not only individual retinal layers but also of the sub-components of individual layers such as the inner and outer segment junction integrity. The vitreous and abnormal vitreoretinal interface pathologies can also be clearly visualized. Posterior segment surgeons are fascinated by the idea of all this information being available to them real time during the surgery.

To achieve this, initially handheld OCT devices were designed which could capture at least the preoperative and immediate postoperative images of the retina.[22] This would enable the surgeon to assess the surgical result as it is being performed and accordingly modify the surgical plan. Further, the concept of integrating the OCT within the microscope itself evolved. The major challenges to attain this were OCT hardware incorporation into the microscope, incorporation of a separate real-time focusing lens for OCT independent of surgeon focus, software changes in OCT to analyze and display continuous real-time OCT data, integrating the display into the microscope eyepiece adjacent to the actual view of the surgeon, and providing the surgeon with the ability to easily obtain OCT scan from the desired area in the field of view.

Few systems that have been able to address these issues and are commercially available are Lumera 700 with integrated Rescan 700 (Carl Zeiss Meditec, Inc.) and HS Hi-R Neo 900 with OPMedT iOCT (Haag-Streit Surgical GmbH). At our center we have the Rescan 700 (Carl Zeiss) system. It is extremely helpful in corneal surgery in identifying the separate corneal layers and in precise assessment of depth of corneal incisions intraoperatively. Interface changes can also be visualized in real time and appropriate surgical decisions can be taken. In the posterior segment it has been helpful in certain cases in identifying pre-foveal vitreous remnants (Figure 11.9). It is also useful to confirm macular hole closure on the table itself (Figure 11.10). A major limitation of intraoperative OCT is that the present surgical instruments interfere with OCT imaging and cause shadowing of the underlying structures. There have been some studies underway to design instruments using materials which do not interfere with OCT imaging.[23] Another limitation is the speed of image acquisition. Surgeons may find speed of image acquisition with a spectral domain OCT a little slow than what they would like for a real-time OCT. Swept-source OCT has faster image acquisition

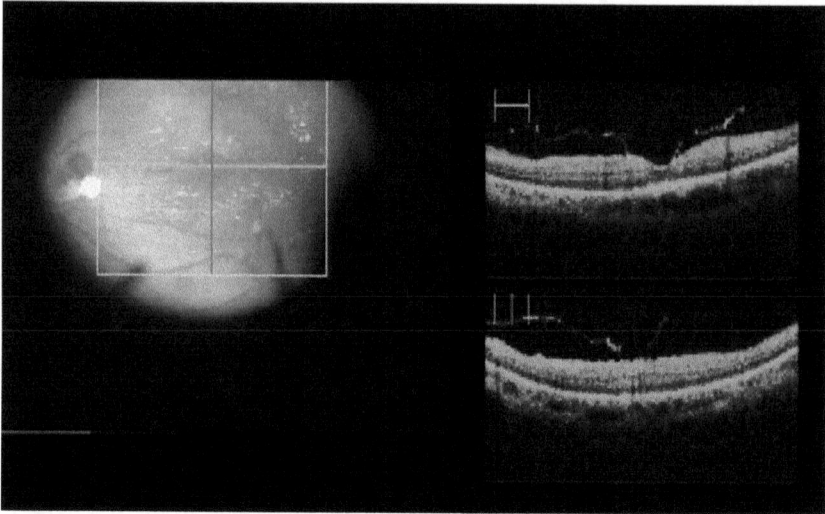

FIGURE 11.9   Identification of pre-foveal vitreous remnants with intraoperative OCT.

and incorporation of an SS-OCT into the microscope may help solve this issue.

Integration of OCT into the surgical microscope is indeed a major step in enhancing the surgeon's ability to view the microscopic details of the operative field. However, at present we lack surgical instrumentation and therapeutic concepts which may enable us to deal with each of these layers

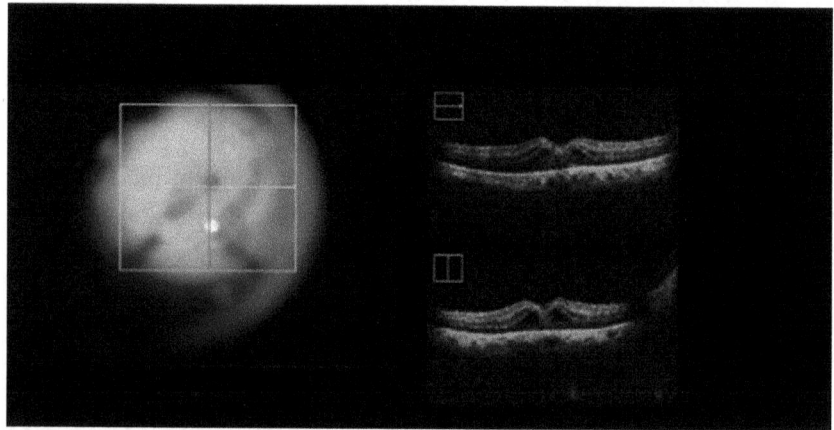

FIGURE 11.10   Confirmation of macular hole closure with intraoperative OCT.

separately in treating retinal disorders. OCT-integrated microscopes may pave the way for the development of intraretinal, retinal layer specific surgical concepts.

OCT Angiography

OCTA is an imaging modality that generates 3D images of blood flow within the eye by using motion contrast. It is based on rapid OCT scanning of the eye and compares repeat scans acquired at the same position in the retina to look for changes in the scan. Compared with stationary areas of the retina, the movement of erythrocytes within a vessel generates a different signal. The split-spectrum amplitude-decorrelation angiography (SSADA) algorithm improves the signal-to-noise ratio by splitting the source spectrum into four parts and averaging the resultant four signals. Mapping these areas of blood flow by point-to-point (or A scan–to–A scan), comparison of two or more OCT volumetric cubes provides incredibly detailed maps of the vasculature of the retina rapidly, in a noninvasive manner, using OCT scanning alone and without the use of any kind of exogenous dye.

The advantages of OCTA over traditional angiography are that it is completely noninvasive, does not involve the risks of exogenous dye injection, unlike traditional angiography, which uses a fluorescent dye and provides limited two-dimensional information. Even though one can identify the superficial retinal capillary plexus using fluorescein angiography, this angiographic technique poorly visualizes the intermediate and deep plexuses that are a critical focus of retinal vascular disease. Therefore, the use of OCTA could greatly enrich our understanding of the ischemic processes affecting different layers of the retinal vasculature, such as exudative age-related macular degeneration (AMD), diabetic retinopathy, and vascular occlusions.

In addition, the angiography image comes cross-registered with structural OCT B scans (because the angiography image is generated by comparison of the corresponding OCT B scans in a volumetric scan). This process allows for precise registration and therefore correlation of the vasculature to the structural scans. OCTA can also generate data on vascular flow. This feature can be used to assess tissue perfusion in the absence of obvious morphological changes. For example, a flow index of the optic nerve head has been studied to ascertain disk perfusion.

The OCTA image is typically displayed as an en-face map of the vasculature, which offers the advantage of allowing for visualization of the vasculature over the entire region of the scan in one image. This depiction corresponds to what ophthalmologists are familiar with seeing on retinal examinations and on fluorescein angiography. The en-face vascular image can include all the vessels seen throughout the retina or can be used to isolate the vessels in the inner retinal layers, the middle retina, and the outer retina. Deeper segmentation allows for the visualization of the choriocapillaris, although there may be loss of resolution at this level because of reduced penetration of the SD-OCT signal beyond the RPE. SS-OCT-based systems are likely to overcome this handicap (Figure 11.11). The superficial, intermediate, and deeper capillary network lie within the nerve fiber–ganglion cell layer, inner plexiform–inner nuclear layer, and outer plexiform–outer nuclear layer, respectively.

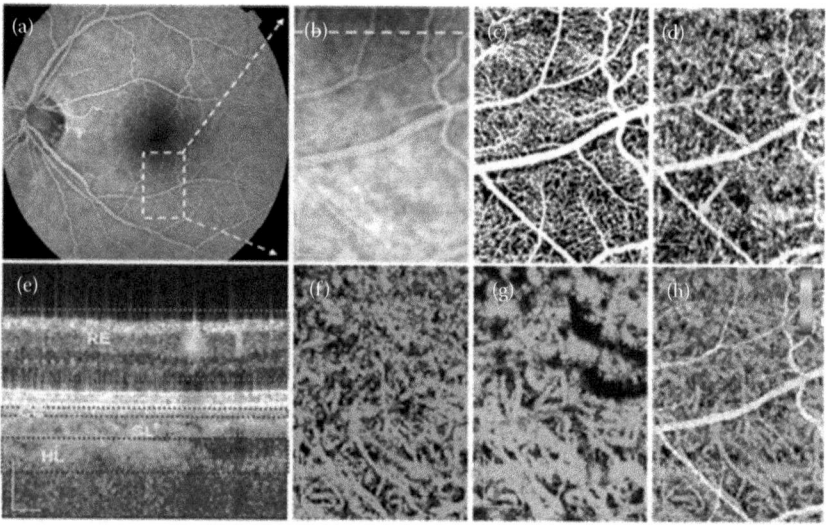

FIGURE 11.11   SSOCA volumetric scan over 1.5×1.5 mm^2 at 6° nasal retina. (a) FA image of retina, (b) zoomed FA image within white rectangle, (c) corresponding segmented depths for en-face projection images for retinal layers (RE), (d) 6 µm below Bruch's membrane for choriocapillaris (CC), (e) B-scan image showing flow signal in different retinal and choroidal layers overlaid over intensity image (position mentioned by dotted lines from "b"), (f) 26 µm to 34 µm below Bruch's membrane for Sattler's layer (SL), (g) 61 µm to 90 µm below Bruch's membrane for Haller's layer (HL), and (h) total en-face projection of retina and choroid. Arrows in "d" show the shadow artifact from big retinal vessels projected on choriocapillaris layer. Scale bars: 300 µm.

A few limitations of OCTA include its inability to visualize vessels that have no flow or that have flow slower than the detection threshold of the OCTA. In addition, OCTA is unable to identify leakage points.

Doppler OCT

Doppler shift allows detection of blood-flow velocity by calculating the relative angle between the reflected OCT beam and blood vessel.[24] Unlike conventional ICG and fluorescein angiography, Doppler OCT has depth resolution and so the morphology and location of vascular abnormalities, both within the retina and choroid, are more precisely evident. Using this technology, blood-flow measurement can be assessed from the transection of all branch retinal arteries and veins by eight circular scans, acquired in ~2 s, each composed of 3,000 axial scans and then summing all blood flow measured in the veins.[24] The background axial inner retinal tissue boundary motion is compared with that of the vessel wall to attain the net Doppler shift, produced by blood flow. Doppler OCT may impact management in vascular disorders such as proliferative diabetic retinopathy, ischemic optic neuropathy, and glaucoma.

There are several methods of increasing the contrast of vessels using the Doppler effect, such as Doppler standard deviation imaging, optical coherence angiography (OCA) phase-variance OCT, and joint spectral and TD-OCT. However, Poddar et al.[25,26] developed a 3D high-resolution phase-variance optical coherence angiography (pv-OCA) system powered by 1060 nm band SS-OCT (Figures 11.12 through 11.15). It is

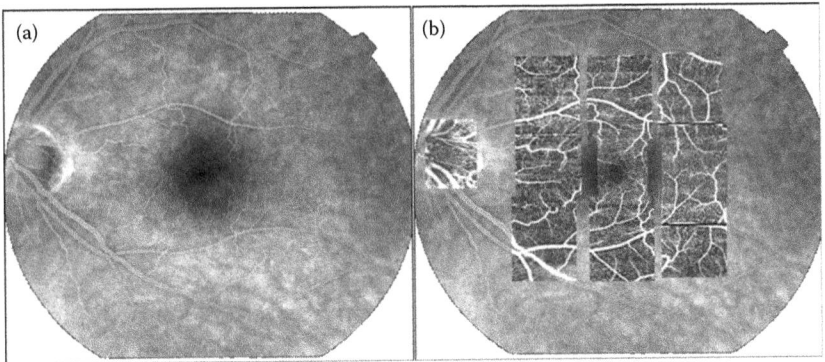

FIGURE 11.12   Large field of view stitched pvOCT imaging overlaid on a fundus FA. (a) Fluorescein angiography. (b) Projection view of retinal contributions to pvOCT imaging with the 10 volumes.

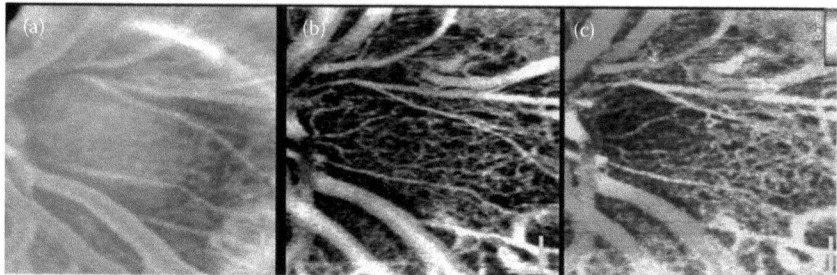

FIGURE 11.13 Retinal capillary network at ONH (optic nerve head) area (1.5 × 1.5 mm²). (a) Fluorescein angiography, (b) total projection view of pvOCT in gray scale, (c) total projection view of pvOCT in gray scale with higher intensity threshold. Top layer denoting a superficial layer, then capillaries as an intermediate vascular bed, and finally microcapillaries appears as a deeper vascular layer. Scale bar: 250 μm.

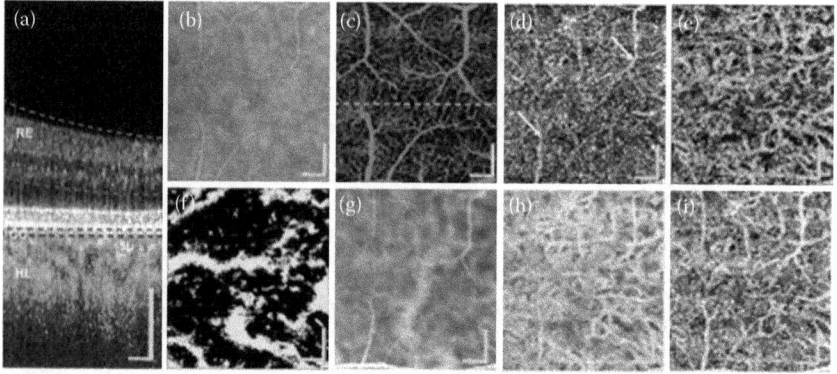

FIGURE 11.14 pv-SSOCT volumetric scan over 1.5 × 1.5 mm² at 6° temporal retina. (a) Composite B-scan image showing flow signal in different retinal and choroidal layers overlaid over intensity image, and (b) corresponding FA image. Segmented depths for en-face projection images are (c) retinal layers, (d) 6 μm below Bruch's membrane for choriocapillaris, (e) 26 μm to 34 μm below Bruch's membrane for Sattler's layer, (f) 61 μm to 90 μm below Bruch's membrane for Haller's layer, (g) corresponding ICGA image, (h) total *en-face* projection of choroidal layer in gray scale, and (i) total *en-face* projection of retina and choroidal layer in gray scale. Arrows in (d) show the shadow artifact from large retinal vessels projected onto the choriocapillaris layer. Scale bars: 300 μm.

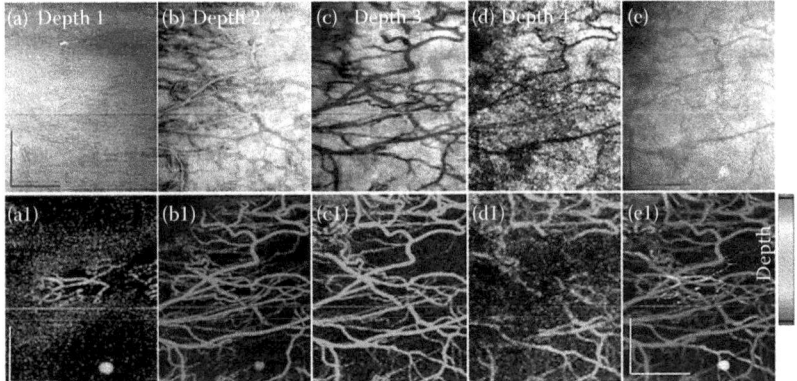

FIGURE 11.15   Virtual C-scans (projections) from intensity (b, c, d, e) and phase-variance (b1, c1, d1, e1) data set showing vascular networks in conjunctiva (1), episclera (2, 3) and (4) sclera, respectively. (e) En-face projection (1.5 × 1.5 mm²) of structural images (a), (e1) Projection image of phase-variance OCT scleral layers, Scale bar: 1 mm.

based on high penetration 1 μm range SS-OCT and has several advantages over SD-OCT, including significant reduction of motion artifacts, increased imaging speed and range of depth, and k-linear sampling and compactness.

REFERENCES

1. D. Huang, E. A. Swanson, C. P. Lin, J. S. Schuman, W. G. Stinson, W. Chang, M. R. Hee, T. Flotte, K. Gregory, and C. A. Puliafito, Optical coherence tomography, *Science* 1991, 254(5035), 1178–81.
2. M. Wojtkowski, High-speed optical coherence tomography: Basics and applications, *Appl. Opt.* 2010, 49(16), D30–61.
3. M. E. Brezinski, G. J. Tearney, B. Bouma, S. A. Boppart, C. Pitris, J. F. Southern, and J. G. Fujimoto, Optical biopsy with optical coherence tomography, *Ann. NY Acad. Sci.* 1998, 838, 68–74.
4. J. A. Izatt, and M. A. Choma, *Theory of Optical Coherence Tomography, Biological and Medical Physics, Biomedical Engineering* (Springer, Berlin and Heidelberg, 2008), 47–72.
5. W. Drexler, and J. G. Fujimoto (ed.) *Optical Coherence Tomography Technology and Applications* (Springer, Berlin and Heidelberg, 2008).
6. R. Leitgeb, C. Hitzenberger, and A. Fercher, Performance of fourier domain vs. time domain optical coherence tomography, *Opt. Express* 2003, 11(8), 889–94.
7. M. Wojtkowski, R. Leitgeb, A. Kowalczyk, T. Bajraszewski, and A. F. Fercher, In vivo human retinal imaging by Fourier domain optical coherence tomography, *J. Biomed. Opt.* 2002, 7(3), 457–63.

8. H. Lim, M. Mujat, C. Kerbage, E. C. Lee, Y. Chen, T. C. Chen, and J. F. de Boer, High-speed imaging of human retina *in vivo* with swept-source optical coherence tomography, *Opt. Express* 2006, 14(26), 12902–8.

9. S. Yun, G. Tearney, J. de Boer, N. Iftimia, and B. Bouma, High-speed optical frequency-domain imaging, *Opt. Express* 2003, 11(22), 2953–63.

10. R. Poddar, D. E. Cortés, J. S. Werner, M. J. Mannis, and R. J. Zawadzki, Three-dimensional anterior segment imaging in patients with type 1 Boston Keratoprosthesis with switchable full depth range swept source optical coherence tomography, *J. Biomed. Opt.* 2013, 18(8), 86002.

11. B. Potsaid et al., Ultrahigh speed 1050 nm swept source/Fourier domain OCT retinal and anterior segment imaging at 100,000 to 400,000 axial scans per second, *Opt. Express* 2010, 18(19), 20029–48.

12. R. Margolis, and R. F. Spaide, A pilot study of enhanced depth imaging optical coherence tomography of the choroid in normal eyes, *Am. J. Ophthalmol.* 2009, 147(5), 811–5.

13. R. F. Spaide, Age-related choroidal atrophy, *Am. J. Ophthalmol.* 2009, 147(5), 801–10. PMID: 19232561.

14. V. Manjunath, J. Goren, J. G. Fujimoto, and J. S. Duker, Analysis of choroidal thickness in age-related macular degeneration using spectral-domain optical coherence tomography, *Am. J. Ophthalmol.* 2011, 152(4), 663–8.

15. R. F. Spaide, H. Koizumi, and M. C. Pozzoni, Enhanced depth imaging spectral-domain optical coherence tomography, *Am. J. Ophthalmol.* 2008, 146(4), 496–500.

16. M. Hirata, A. Tsujikawa, A. Matsumoto, M. Hangai, S. Ooto, K. Yamashiro, M. Akiba, and N. Yoshimura, Macular choroidal thickness and volume in normal subjects measured by swept-source optical coherence tomography, *Invest. Ophthalmol. Vis. Sci.* 2011, 52(8), 4971–8.

17. S. Copete, I. Flores-Moreno, J. A. Montero, J. S. Duker, and J. M. Ruiz-Moreno, Direct comparison of spectral-domain and swept-source OCT in the measurement of choroidal thickness in normal eyes, *Br. J. Ophthalmol.* 2014, 98(3), 334–8.

18. H. Itakura, S. Kishi, D. Li, and H. Akiyama, Observation of posterior precortical vitreous pocket using swept-source optical coherence tomography. *Invest. Ophthalmol. Vis. Sci.* 2013, 54(5), 3102–7.

19. D. Li, S. Kishi, H. Itakura, F. Ikeda, and H. Akiyama, Posterior precortical vitreous pockets and connecting channels in children on swept-source optical coherence tomography. *Invest. Ophthalmol. Vis. Sci.* 2014, 55(13), 2412–6.

20. P. E. Stanga et al. In vivo imaging of cortical vitreous using 1050-nm swept-source deep range imaging optical coherence tomography, *Am. J. Ophthalmol.* 2014, 157(2), 397–404.

21. Z. Michalewska, J. Michalewski, Z. Nawrocka, K. Dulczewska-Cichecka, and J. Nawrocki, Suprachoroidal layer and suprachoroidal space delineating the outer margin of the choroid in swept-sourceoptical coherence tomography, *Retina.* 2015, 35(2), 244–9.

22. P. Dayani, R. Maldonado, S. Farsiu, and C. Toth, Intraoperative use of hand-held spectral domain optical coherence tomography imaging in macular surgery, *Retina*. 2009, 29(10), 1457–68.

23. J. P. Ehlers et al. Integrative advances for OCT-guided ophthalmic surgery and intraoperative OCT: Microscope integration, surgical instrumentation, and heads-up display surgeon feedback. *PLoS ONE* 2014, 9(8), e105224. doi: 10.1371/journal.pone.0105224.

24. Y. Wang, A. Lu, J. Gil-Flamer, O. Tan, J. A. Izatt, and D. Huang, Measurement of total blood flow in the normal human retina using Doppler Fourier-domain optical coherence tomography, *Br. J. Ophthalmol.* 2009, 93(5), 634–7.

25. R. Poddar, D. Y. Kim, J. S. Werner, and R. J. Zawadzki, In vivo imaging of human blood circulation in the chorioretinal complex with new phase stabilized 1 μm swept-source phase-variance optical coherence tomography (pv-SSOCT), *J. Biomed. Opt.* 2014, 19(12), 26010.

26. R. Poddar, D. E. Cortés, J. S. Werner, M. J. Mannis, and R. J. Zawadzki, In-vivo volumetric depth-resolved vasculature imaging of human limbus and sclera with 1 μm swept source phase-variance optical coherence angiography, *J. Opt.* 2015, 17(6), 065301.

Index

C

Milton Keynes UK
Ingram Content Group UK Ltd.
UKHW021819071024
449327UK00021B/1352